國際經貿專業英語
理論、實務與方法

黃載曦、姚星 編著

財經錢線

序

　　《國際經貿專業英語：理論、實務與方法》，讓學生專業素養和英語能力齊頭並進，解決學生在雙語教學和全英文教學中面臨的「對專業術語不認識、對專業理論不瞭解、對英語應用不到位」的「三不」問題，提高學生對雙語課程和全英文課程的學習興趣和能力。

　　與其他類似的教材相比較，《國際經貿專業英語：理論、實務與方法》的特色體現在專業性、前沿性和方法論三方面。「專業性」主要表現在該教材對國際經貿核心專業，具體包括經濟學、管理學、國際貿易、國際商務、國際金融、國際行銷、國際投資和國際商法等進行了較為全面的介紹，內容既包括對相關專業理論進行框架式梳理，也包括對主要的實務操作做重點突出的介紹，通過理論與實務兼顧，既能滿足高水準財經專業人才專業理論素養培養的需要，又能同時兼顧對其實際操作能力的培養。「前沿性」主要體現在內容選取了最新的國際經濟訊息和美國等國家財經領域中最新的研究成果，緊跟專業發展前沿，進一步注重材料的實效性和適用性，力圖更適合財經類專業英語教學的需求和變化趨勢。「方法論」主要表現在本教材不僅僅向學生傳播知識和理論，更希望能夠對財經專業英語的學習方法和技巧做一個總結和梳理，主要目的是幫助學生理解原文，提高其學習專業英語聽力、閱讀、寫作以及翻譯的技巧。

編寫小組成員都是在學院長期從事雙語教學、全英文教學的老師，他們不僅獲得了經濟學和外語相關專業的博士和碩士學位，而且有出國留學、訪學的經歷，既具備深厚的經濟學專業的基礎，又具備熟練運用英語的能力，同時還有豐富的教學經驗和較強的科研能力。本書編寫具體分工如下：第一章，黃載曦、熊英；第二章，周茂；第三章，姚星；第四章，徐永安；第五章，蔣海曦；第六章，李雨濃；第七章，張贇；第八章，徐聰、羅小豔。附錄：高小娟、姜鑒濤。全書統稿由黃載曦、姚星完成。感謝王玲、張寧寧、劉薇、高慧、徐冬陽、王勝蘭、李佳虹對本書提供資料收集和實務內容指導。

　　感謝國際商學院對本書撰寫的支持。

　　由於編者水準有限，書中的錯誤和疏漏之處在所難免，敬請廣大讀者提出寶貴意見，以便我們進一步修改和完善。

黃載曦

目　錄

Chapter 1	**Economics**	（1）
Chapter 2	**Management**	（34）
Chapter 3	**International Trade**	（67）
Chapter 4	**International Business**	（105）
Chapter 5	**International Finance**	（156）
Chapter 6	**International Marketing**	（186）
Chapter 7	**International Investment**	（222）
Chapter 8	**International Business Law**	（265）
附錄 1	國際經貿專業英語的特點	（298）
附錄 2	國際經貿英語的學習方法	（305）
附錄 3	國際經貿專業英語詞彙英漢對照表	（325）
附錄 4	國際經貿專業重點期刊和報紙目錄	（346）
附錄 5	國際經貿專業重要網站	（349）

Chapter 1
Economics

Chapter checklist

When you have completed your study of this chapter, you will be able to
1. Define economics and explain the kinds of questions that economists try to answer.
2. Explain the core ideas that define the economic way of thinking.
3. Understand the main theories of microeconomics and macroeconomics.

In this chapter, we will concentrate on the basic concepts and theories of economics, pay attention to analyzing the economic problems in everyday life and learn about some ways of economic thinking. There are three sections in this chapter. They are introduction to economics, microeconomics, macroeconomics. The more important and difficult parts include microeconomics and macroeconomics.

Section 1 An Introduction to Economics

Lesson 1 What Does Economics Mean?

Economics may appear to be the study of complicated tables and charts, statistics and numbers, but, more specifically, it is the study of what constitutes rational human behavior in the endeavor to fulfill needs and wants. Economics is the social science that seeks to describe the factors which determine the production, distribution and consumption of goods and services. The word 「economy」 comes from the Greek word oikonomos meaning 「one who manages a household」. This makes some sense because in the economy we are faced with many decisions (just as a household is). The fundamental economic problem comes from the fact that resources are scarce. All economic questions and problems arise because human wants exceed the resources available to satisfy them. Our inability to satisfy all our wants is called scarcity. Faced with scarcity, we must make choices. We must choose among the available alternatives. Economics is the social science that studies the choices that individuals, businesses, governments, and entire societies make as they cope with scarcity and the incentives that influence and reconcile those choices. In a word, economics is the study of how society allocates its scarce resources.

As an individual, for example, you face the problem of having only limited resources with which to fulfill your wants and needs. As a result, you must make certain choices with your money. You'll probably spend part of your money on rent, electricity and food. Then you might use the rest to go to the movies and/or buy a new pair of jeans. Economists are interested in the choices you make and inquire into why, for instance, you might choose to spend your money on a new DVD player instead of replacing your old TV. They would want to know whether you would still buy a carton of cigarettes if prices increased by $2 per pack. The underlying essence of economics is trying to understand how both individuals and nations behave in response to certain material constraints.

When we talk about economics, we often want to think as an economist. But how to do it, there are some good ideas as follows:

Trade-off is a key word in discussing economic ways of thinking. 「There is no such thing as a free lunch.」 This famous saying tells us that making decisions requires trading one goal for another. That is, to get one thing that we like, we usually have to give up another thing that we like. Consider a girl who must decide how to allocate her most valuable resource—her time. She can spend all of her time studying economics; she can spend all of her time studying psychology; or she can divide her time between the two fields. For every hour she studies one subject, she gives up an hour she could have used to study the other.

Costs in economics are quite different from what we think are as usual. The cost of something is what you give up to get it. Making decisions requires individuals to consider the benefits and costs of some action. Consider the decision to go to college. The benefit is intellectual enrichment and a lifetime of better job opportunities. What is the cost? We might add up the money we spend on tuition, books, room and board. But this total does not truly represent what we give up to spend a year in college. We cannot count room and board (at least all of the cost) because the student would have to pay for food and shelter even if he was not in school. We would want to count the value of the student's time because he could be working for pay instead of attending classes and studying. For most students, the wages given up to attend school are the largest single cost of their education. So we must take opportunity cost into account. The opportunity cost of something is the best thing you must give up to get it.

Economists generally assume that people are rational. Rational people systematically and purposefully do the best they can to achieve their objectives, given the opportunities they have. For example, consumers want to purchase the goods and services that allow them the greatest level of satisfaction given their incomes and the prices they face. And firm managers want to produce the level of output that maximizes the profits the firms earn. Many decisions in life involve incremental decisions: Should I remain in school this semester? Should I take another course this semester? Should I study another hour for tomorrow's ex-am? Economists use the term marginal changes to describe small incremental adjustments to an existing plan of action. Keep in mind that margin, means edge, so marginal changes are adjustments around the edges of what you are doing. Rational people often make decisions by comparing marginal benefits and marginal costs. There is an example: Suppose that flying a 200-seat plane across the country costs the airline $100,000, which means that the average cost of each seat is $500. Suppose that the plane is minutes from departure and a passenger is willing to pay $300 for a seat. Should the airline sell the seat for $300? Of course it should. If the plane has empty seats, the cost of adding one more passenger is miniscule. Although the average cost of flying a passenger is $500, the marginal cost is merely the cost of the bag of peanuts and can of soda that the extra passenger will consume. As long as the standby passenger pays more than the marginal cost, selling the ticket is profitable.

Besides the traditional concern in production, distribution, and consumption in an economy, economic analysis may be applied throughout society, as inbusiness, finance, health care, and government. Economic analyses may also be applied to such diverse subjects as crime, education, the family, law, politics, religion, social institutions, war, science, and the environment. Education, for example, requires time, effort, and expenses, plus the foregone income and experience, yet these losses can be weighted against future benefits education may bring to the agent or the economy. At the turn of the 21st century, the expanding domain of economics in the social sciences has been described as economic imperialism.

New Words, Phrases and Expressions
economic agents   n. 經濟代理人
reconcile   v. 調和，和解，妥協
trade-off   n. 權衡取捨
distribution   n. 分配
rational people   n. 理性人

Notes

1. Economics is the social science that seeks to describe the factors which determine the production, distribution and consumption of goods and services.

經濟學是一門社會學科，致力於研究那些決定商品和服務的生產、分配與消費的因素。

2. All economic questions and problems arise because human wants exceed the resources available to satisfy them.

人類的慾望超出了可以滿足這些慾望的各種資源，這是所有經濟問題產生的根源。

3. Microeconomics examines the behavior of basic elements in the economy, including

individual agents and markets, their interactions, and the outcomes of interactions. Individual agents may include, for example, households, firms, buyers, and sellers.

個體經濟學考察經濟主體的各種行為，包括個體行為人和市場，他們之間的相互影響以及相互影響的後果。個體行為人包括家庭、廠商、買家和賣家。

4. Macroeconomics analyzes the entire economy (meaning aggregated production, consumption, savings, and investment) and issues affecting it, including unemployment of resources (labor, capital, and land), inflation, economic growth, and the public policies that address these issues (monetary, fiscal, and other policies).

總體經濟學分析整體經濟（總產量、總消費、儲蓄和投資）以及影響整體經濟的問題，後者包括各種資源（勞動力、資本和土地）的失業（未被利用）問題、通貨膨脹、經濟增長和針對這些問題提出的公共政策（貨幣政策、財政政策和其他政策）。

5. The cost of something is what you give up to get it. Making decisions requires individuals to consider the benefits and costs of some action.

做某件事情的成本是你為了得到它而放棄的東西。個人在做決策的時候需要考慮這一行為的成本和收益。

6. Economic analyses may also be applied to such diverse subjects as crime, education, the family, law, politics, religion, social institutions, war, science, and the environment. Education, for example, requires time, effort, and expenses, plus the foregone income and experience, yet these losses can be weighted against future benefits education may bring to the agent or the economy.

經濟分析方法還可以被運用到各種領域，比如犯罪、教育、家庭、法律、政治、宗教、社會制度、戰爭、科學和環境等。舉個例子，教育需要花費時間、個人努力以及花錢加上預先放棄的收入和經歷。然而，這些損失可以和教育將來帶給個體或整個國家經濟的收益相權衡比較。

Lesson 2　Ways to Learn Economics

How to learn economics? Maybe this is a question troubling many students. Before we start to learn economics, we must try to grasp the important role of assumption, economic model, positive and normative analysis in learning economics.

Assumptions

Assumptions are very important in learning economics. Assumptions can simplify the complex world and make it easier to understand. To study the effects of international trade, for example, we may assume that the world consists of only two countries and that each country produces only two goods. Of course, the real world consists of dozens of countries, each of which produces thousands of different types of goods. But by assuming two countries and two goods, we can focus our thinking on the essence of the problem. Once we understand international trade in an imaginary world with two countries and two goods, we are in a better position to understand international trade in the more complex world in which we live.

Economists use different assumptions to answer different questions. For example, when we want to find out the most important relationship between price and quantity demanded, we find the law of demand, that is, other things being equal, the quantity demanded of a good falls when the price of the good rises. 「Other things being equal」is a usual expression of assumptions, which means that other factors affecting quantity demanded, such as income, price of related goods, expectation, preference and so on so forth are not taken into account and price of the goods is the only factor analyzed in the model. Ceteris paribus, a Latin phrase, is often used to stand for 「other things being equal」, which means all variables other than the ones being studied are assumed to be constant.

Economic models

Economists use models to learn about the world; they are most often composed of diagrams and equations. Economic models omit many details to allow us to see what is truly important. These models are stylized and they omit many details. Yet despite this lack of realism—indeed, because of this lack of realism—studying these models is useful for learning how the economic world works.

As we use models to examine various economic issues, we will see that all the models are built with assumptions. Economists assume away many of the details of the economy that are irrelevant for studying the question at hand. All models simplify reality to improve our understanding of it.

Positive and Normative Analysis

Suppose that two people are talking about rent control. Here are two statements you might hear:

Tom: Rent control causes a shortage of housing.

Mary: Government should stop rent control.

Ignoring whether you agree with these statements, notice that Tom and Mary differ in what they are trying to do. Tom is speaking like a scientist: he is making a claim about how the world works. Mary is speaking like a policy adviser: she is making a claim about how she would like to change the world.

In general, statements about the world are of two types. One type, such as Tom's, is positive. Positive statements are descriptive. They make a claim about how the world is. A second type of statement, such as Mary's, is normative. Normative statements are prescriptive. They make a claim about how the world ought to be.

A key difference between positive and normative statements is how we judge their validity. We can, in principle, confirm or refute positive statements by examining evidence. An economist might evaluate Tom's statement by analyzing data on changes in rent control and changes in housing rent over time. By contrast, evaluating normative statements involves values as well as facts. Mary's statement cannot be judged using data alone. Deciding what is good or bad policy is not merely a matter of science. It also involves our views on ethics, religion, and political philosophy. Positive and normative statements are fundamentally different, but they are often closely intertwined in a person's set of beliefs. As we study economics, keep in mind the distinction between positive and normative statements because it will help we stay focused on the task at hand.

New Words, Phrases and Expressions
assumptions   n. 假設前提
other things being equal   其他條件不變
ceteris paribus   其他條件不變
validity   n. 有效性
Notes

1. Once we understand international trade in an imaginary world with two countries and two goods, we are in a better position to understand international trade in the more complex world in which we live.

一旦我們瞭解了只有兩個國家、兩種商品這樣虛擬世界的國際貿易，我們就能夠更好地理解我們生活中更加複雜的真實世界裡的國際貿易。

2. Yet despite this lack of realism——indeed, because of this lack of realism——studying these models is useful for learning how the economic world works.

雖然這是和現實格格不入的，但恰恰是因為和現實格格不入，我們才能夠通過研究這些模型來幫助我們瞭解經濟世界是如何運作的。

3. In general, statements about the world are of two types. One type, such as Tom's, is positive. Positive statements are descriptive. They make a claim about how the world is. A second type of statement, such as Mary's, is normative. Normative statements are prescriptive. They make a claim about how the world ought to be.

一般來說，我們可以用兩種方式來評價這個世界。一種方式，就如湯姆那樣，是實證分析。實證分析是描述性的，說明世界是怎麼樣的。第二種方式，就像瑪麗那樣，是對世界的規範分析。規範分析是關於世界應該是怎麼樣的一種規範性描述。

4. Deciding what is good or bad policy is not merely a matter of science. It also involves our views on ethics, religion, and political philosophy.

對一項政策好壞的評價並不僅僅涉及科學問題，它還涉及我們的道德觀、宗教觀和政治哲學觀等。

Section 2 Microeconomics

Lesson 1 A brief history of Economics and Theoretical Framework of Microeconomics

As early as the 18th century, economists were studying the decision-making processes of consumers, which maybe is the beginning of microeconomics. For some 200 years beginning in the mid-1700s, the dominant economic theory was Adam's Smith's (1723–1790) 「invisible hand」of self-interest. In his famous works, *Wealth of Nations*, Smith put forwards the theory that if the government does not tamper with the economy, a nation's resources will be most efficiently used, free-market problems will correct themselves and a country's welfare and best interests will be served.

Smith's views on the economy prevailed through two centuries, but in the late 19th and early 20th century, the ideas of Alfred Marshall (1842–1924) had a major impact on economic thought. In Marshall's book, *Principles of Economics*, he proposed a new idea of the study of specific, individual markets and firms as a means of understanding the dynamics of economics. Marshall also formulated the concepts of consumer utility, price elasticity of demand and the demand curve, all of which will be discussed in the following chapter.

At the time of Marshall's death, John Maynard Keynes (1883–1946), who would become the most influential economist of the 20th century starting in the 1930s, was already at work on his revolutionary ideas about government management of the economy. Keynes' contributions to economic theory have guided the thinking and policy-making of central bankers and government economists for decades; both globally and in the U. S. Keynes advocated government intervention into free markets and into the general economy when market crises warranted, an unprecedented idea when proposed during the Great Depression. Thus was born the modern science of macroeconomics—the big picture view of the economy—evolving in large part from what came to be called Keynesian economic theory. These are among the tools of microeconomics, and their principles, along with others, are still employed today by economists who specialize in this area.

According to a brief history of economics we put forward above, we can easily know that the field of economics is broken down into two distinct areas of study: microeconomics and macroeconomics. Microeconomics examines the behavior of basic elements in the economy, including individual agents and markets, their interactions, and the outcomes of interactions. Individual agents may include, for example, households, firms, buyers, and sellers. Macroeconomics analyzes the entire economy (meaning aggregated production, consumption, savings, and investment) and issues affecting it, including unemployment of resources (labor, capital, and land), inflation, economic growth, and the public policies that address these issues (monetary, fiscal, and other policies). The ultimate goal of economics is to improve the living conditions of people in their everyday life.

Microeconomics looks at the smaller picture and focuses more on basic theories of supply and demand and how individual businesses decide how much of something to produce and how much to charge for it. The platform on which microeconomic thought is built lies at the very heart of economic thinking—namely, how decision makers choose between scarce resources that have alternative uses.

Microeconomics includes some theories such as theories of supply and demand, consumer and producer's theories, market theory, factors theory and so on. Just as microeconomics is to learn how scarce resources are allocated, markets involved in producers and consumers make decisions. It is concerned with the interaction between individual buyers and sellers and the factors that influence the choices made by buyers and sellers. In particular, microeconomics focuses on patterns of supply and demand and the determination of price and output in individual markets. Starting from the discussion of supply, demand and the equilibrium price, microeconomics studies consumers' behavior to know how consumers make decisions about what to buy that best suits their wants, or in a more formal way, the consumer's theory is aimed to know how a consumer makes choices to maximize utility under budget constraint. The demand curve is decided by the way of consumers' behavior. The producers' behaviors to pursuit maximization of profit are learned in the cost theory, produce theory and market theory, and in this way the supply curve is decided. In a market, demand and supply are the opposing forces, when the quantity demanded equals the quantity supplied, that is, when buyers' and sellers' plans are in balance, market equilibrium occurs. The result of market equilibrium shows how scarce resources are allocated by the way of market mechanism, in which all people are looking for their own benefit.

A theoretical framework of microeconomics is following (Figure 1.1):

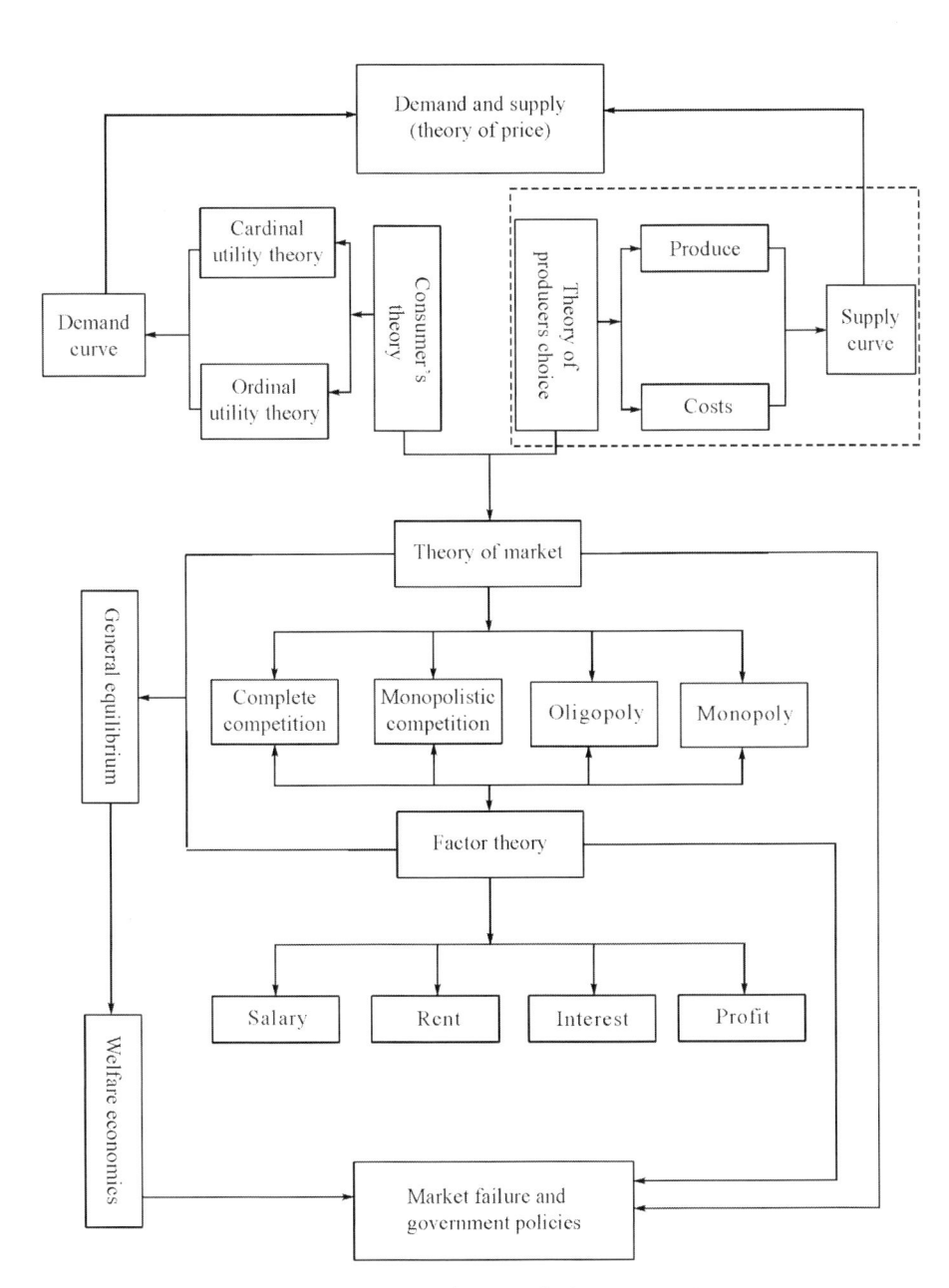

Figure 1.1　Theoretical framework of microeconomics

New Words, Phrases and Expressions
decision-making processes   n. 決策過程
self-interest   n. 自利
tamper with   v. 亂動，篡改
distinct   adj. 明顯的，不同的，獨特的
be concerned with   adj. 有關的
diminishing   adj. 遞減的

Notes

1. In his famous work, *Wealth of Nations*, Smith put forwards the theory that if the government does not tamper with the economy, a nation's resources will be most efficiently used, free-market problems will correct themselves and a country's welfare and best interests will be served.

亞當‧斯密在他的名著《國富論》中提出了這樣的理論：如果政府不對經濟橫加干預，那麼一個國家就會最有效地利用該國資源，自行解決自由市場中的各種問題，以及實現該國的福利和利益的最大化。

2. In Marshall's book, *Principles of Economics*, he proposed a new idea of the study of specific, individual markets and firms as a means of understanding the dynamics of economics.

馬歇爾在《經濟學原理》一書中，提出了一個新的觀點，即透過研究一個一個特殊的獨立的市場和廠商來理解經濟學產生的動力。

3. Keynes' contributions to economic theory have guided the thinking and policy-making of central bankers and government economists for decades; both globally and in the U. S. Keynes advocated government intervention into free markets and into the general economy when market crises warranted, an unprecedented idea when proposed during the Great Depression.

凱恩斯對經濟學理論的貢獻導致無論是美國還是全球的中央銀行家和政府經濟學家們為制定政策而冥思苦想。凱恩斯倡導，當出現市場危機的時候，政府應該對自由市場和實體經濟進行干預。這一前所未有的觀點是在大蕭條期間提出來的。

4. Microeconomics examines the behavior of basic elements in the economy, including individual agents and markets, their interactions, and the outcomes of interactions. Individual agents may include, for example, households, firms, buyers, and sellers.

微觀經濟學考察經濟中基本要素的行為，包括個體、市場以及個體之間的相互作用和相互作用的後果。這些個體包括家庭、企業、買者和賣者。

5. Macroeconomics analyzes the entire economy (meaning aggregated production, consumption, savings, and investment) and issues affecting it, including unemployment, inflation, economic growth, and the public policies that address these issues (monetary, fiscal, and other policies).

總體經濟學分析整體經濟（總產出、總消費、總儲蓄和總投資）以及影響整體

經濟的各種問題，包括失業、通貨膨脹、經濟增長以及為了解決這些問題所採取的公共政策（貨幣政策、財政政策和其他）。

Lesson 2 The Price Mechanism

Much of the study of microeconomics is devoted to analysis of how prices are determined in markets. A market is any system through which producers and consumers come together. In early subsistence economies, markets were usually physical locations where people would come together to trade. In more complex economic systems, markets do not depend on humans actually meeting one another, so many markets today arise when producers and consumers come together less directly, such as by post and on the internet. Producers and consumers generate forces that we call supply and demand respectively, and it is their interaction within the market that creates the price mechanism. This mechanism was once famously described as the 「invisible hand」 that guides the actions of producers and consumers. Markets are essential to produce the goods and services required for everyday life. Even if an individual can produce all the food needed to survive, that person will still need clothes, shelter and other necessities. Therefore, from very early times, communities learned that they would benefit from exchange. The crudest form of exchange was barter, but the evolution of money as a medium of exchange and unit of account accelerated the development of the process. But how would people know what they could charge, or what they should pay, for goods and services? Before any formal thought was given to this, traders soon discovered that if they fixed their prices too low they would soon run out of inventory, while if they set their prices too high they would not sell what they had produced. In physical markets there would often be perfect knowledge, as traders would be able to check the prices of those who had similar goods and services to trade, simply by walking around the stalls. Once markets became more remote, less perfect knowledge of prices was inevitable and the process became less certain. Much of price mechanism are concerned with demand, supply and equilibrium market as we will show following.

Demand is created by the needs of consumers, and the nature of demand owes much to the underpinning worth that consumers perceive the good or service to have. We all need necessities, such as basic foodstuffs, but other products may be highly sought after by some and regarded as worthless by others. The level of demand for a good or service is determined by several factors, including: the price of the good or service, prices of other goods and services, especially substitutes and complements, income, tastes, preferences and expectations. In orthodox economic analysis, these determinants are analyzed by testing the quantity demanded against one of these variables, holding all others to be constant (or ceteris paribus). The most common way of analyzing demand is to consider the relationship between quantity demanded and price. Assuming that people behave rationally, and that other determinants of demand are constant, the quantity demanded has an inverse relationship with price. Therefore, if price increases, the quantity demanded falls, and vice versa. This happens with most types of goods, with some bizarre exceptions. If we then relax the assumption that other variables (such as income and tax rates, etc.) are constant, what happens then? An increase in income will often cause the demand for a good or service to increase, and this will shift the whole curve away from the origin. Likewise, a reduction in the price of a substitute good will move the demand curve towards the origin as the good in question will then be less attractive to the consumer. While these generalizations are useful, it is important to remember that economic behavior is based on human decisions, and so we can never predict fully how people will act. For example, some very basic foodstuffs will become less popular as incomes increase and when consumers find that they no longer have to subsist on basic diets.

Supply refers to the quantity of goods and services offered to the market by producers. Just as we can map the relationship between quantity demanded and price, we can also consider the relationship between quantity supplied and price. Generally, suppliers will be prepared to produce more goods and services the higher the price they can obtain. Therefore, the supply curve—when holding other influences constant—will slope upwards from left to right. There is a direct relationship between price and quantity supplied. The determinants of supply are: price, prices of other goods and services, relative revenues and costs of making the good or service, the objectives of producers and their future expectations, technology. Generally, a firm will maximize profit when its marginal revenue (the revenue arising from selling one extra unit of production) equals its marginal cost (the cost of producing that one extra unit of production). However, a firm may continue to produce as long as the marginal revenue exceeds its average variable costs, as in doing so it will be making a contribution towards covering its fixed costs. Following the same rationale as applied earlier, a movement along the supply curve will be brought about by a change in price, but a movement of the whole curve will be caused by a determinant other than price.

Assuming all determinants of supply and demand are to be constant except price, a firm will produce where the supply curve intersects the demand curve. By definition, this is the point at which the quantity supplied equals the quantity demanded. Price is determined at the intersection of the supply and demand curves. If the price is set above the equilibrium price, this will result in the quantity supplied exceeding the quantity demanded. Therefore, in order to clear its inventory, the company will need to reduce its price. Conversely, if the price is set below the equilibrium price, this will result in an excess demand situation, and the only way to eliminate this is to increase the price.

In any economic system, scarce resources have to be allocated among competing uses. Market economics harness the forces of supply and demand to serve that end. Supply and demand together determine the prices of the economy's many different goods and services, prices in turn are the signals that guide the allocation of resources. For example, consider the allocation of beachfront land. Because the amount of this land is limited, not everyone can enjoy the luxury of living by the beach. Who gets this resource? The answer is whoever is willing and able to pay the price. The price of beachfront land adjusts until the quantity of land demanded exactly balances the quantity supplied. Thus, in market economies, prices are the mechanism for rationing scarce resources. Similarly, prices determine who produces each good and how much is produced. For instance, consider farming. Because we need food to survive, it is crucial that some people work on farms. What determines who is a farmer and who is not? In a free society, there is no government planning agency making this decision and ensuring an adequate supply of food. Instead, the allocation of workers to farms is based on the job decisions of millions of workers. This decentralized system works well because these decisions of millions of workers. The prices of food and the wages of farm workers adjust to ensure that enough people choose to be farmers. If a person had never seen a market economy in action, the whole idea might seem preposterous, Economies are enormous groups of people engaged in a multitude of interdependent activities. What prevents decentralized decision making from degenerating into chaos? What coordinates the actions of the millions of people with their varying abilities and desires? What ensures that what needs to be done is in fact done? The answer, in a word, is prices. If an invisible hand guides market economies, as Adam Smith famously suggested, then the price system is the baton that the invisible hand uses to conduct the economic orchestra.

New Words, Phrases and Expressions
subsistence   n. 實體，生存，存在
barter   n. 以物易物，貨物交換
inventory   n. 存貨
bizarre   adj. 奇異的；怪誕的
baton   n. 指揮棒，接力棒
orchestra   n. 管弦樂隊

Notes

1. Producers and consumers generate forces that we call supply and demand respectively, and it is their interaction within the market that creates the price mechanism. This mechanism was once famously described as the 「invisible hand」 that guides the actions of producers and consumers.

我們分別稱生產者和消費者的力量為供給與需求，正是市場中的供給與需求的相互作用才能夠產生價格機制。史上對這一機制的最著名的說法是「看不見的手」，這只「手」引導著生產者和消費者的行為。

2. Demand is created by the needs of consumers, and the nature of demand owes much to the underpinning worth that consumers perceive the good or service to have. We all need necessities, such as basic foodstuffs, but other products may be highly sought after by

some and regarded as worthless by others.

消費者們的需要是產生需求的根本，而需求的本質絕大多數源於消費者對持有的商品或者服務本身價值的考量。每個人都需要必需品，比如基本的食品，但其他商品卻可能被一些人趨之若鶩，而被另外的人視如草芥。

3. In orthodox economic analysis, these determinants are analyzed by testing the quantity demanded against one of these variables, holding all others to be constant (or ceteris paribus).

在複雜的經濟學分析中，一般這樣對（需求）的決定因素進行分析：假定其他因素不變（其他條件相同），分析需求量和其中一個決定因素的變量的關係。

4. An increase in income will often cause the demand for a good or service to increase, and this will shift the whole curve away from the origin.

收入的增加常常會導致對商品或者服務的需求增加，這將使得整個需求曲線偏離原來的位置。

Lesson 3 Theories of Consumers and Producers

We have discussed demand curve and supply curve and there is another question: where do supply and demand curves come from? In fact, behind the two curves, there are consumer theory and producer theory, which study the choices-making of consumers and producers. Microeconomics had practical appeal to economists because it sought to understand the most basic machinery of an economic system: consumer decision-making and spending patterns, and the decision-making processes of individual businesses.

The study of consumer decision-making reveals how the price of products and services affects demand, how consumer satisfaction—although not precisely measurable—works in the decision-making process, and provides useful information to businesses selling products and services to these consumers. In microeconomics, the consumer's choices are really about the utility maximization problem, which is the problem a consumer face:「how should I spend my money in order to maximize my utility?」It is a type of optimal decision problem. As the resources are scarce, the consumers must face their budget constraint. In short, the consumers' choice are problems how consumers make decision about what to buy in the face of budget constraint, or, given resources are limited, consumers buy a bundle of goods that best suit their desires. So, as we knew, the theory of consumers' choices includes three parts which are utility, budget constraint and optimal choice.

Economics concept that although it is impossible to measure the utility derived from a good or service, it is usually possible to rank the alternatives in their order of preference to the consumer. Economists distinguish between cardinal utility and ordinal utility. When cardinal utility is used, the magnitude of utility differences is treated as an ethically or behaviorally significant quantity. On the other hand, ordinal utility captures only ranking and not strength of preferences. An important example of a cardinal utility is the probability of achieving some target.

As to cardinal utility since this choice is constrained by the price and the income of the consumer, the rational consumer will not spend money on an additional unit of good or service unless its marginal utility is at least equal to or greater than that of a unit of another good or service. Therefore, the price of a good or service is related to its marginal utility and the consumer will rank his or preferences accordingly. As to ordinal utility, indifference curves are used to describe a consumer's preferences. Indifference curve is a line that shows combinations of goods among which a consumer is indifferent. A series of indifference curves tell us that the higher the indifference curve, the better off is the consumer. So the consumer's goal can be restated as to allocate his or her budget in such a way as to get onto the highest attainable indifference curve. So the best affordable point is indifference curve and budget line tangent.

The producers' theory is concerned with the behavior of firms in hiring and combining productive inputs to supply commodities at appropriate prices. In general, a firm is thought very simply as a black box, where inputs go in and outputs come out. Two sets of issues are involved in this process: one is the technical constraints, which limit the range of feasible productive processes, while the other is the institutional context such as the characteristics of the market where commodities and inputs are purchased and sold.

Producers' problem is maximizing profits subject to technological constraints just as consumers have a simple goal which is to maximize their utility. Profits are defined as revenue minus cost. The key to maximizing profits is going to produce goods as efficiently as possible. To decide how to efficiently produce goods, firm production functions are turned and discussed. That's essentially the technology by which a firm takes inputs, or what we call factors of production, and turns them into outputs. This is production function. Once again, to make life easy, firms are assumed only use two kinds of inputs: labor (l) and capital (k). Labor is just hours of work in production and capital is as everything else that goes into production, such as the machines, the buildings, the land or everything. The output (q) is produced by the firm. So basically, a production function is that q is some function of l and k.

After knowing the production function, the important distinction made here is between variable versus fixed inputs. Variable inputs are inputs that are easily changed, like how many hours somebody works. Fixed inputs are things which are harder to change quickly, like the size of the building that the workers are building in. And this will lead to a critical distinction for production theory, which is the short run versus the long run. The long run is the period over which all inputs are variable and the short run is a period over which some inputs are fixed.

In a short run, labor is regarded as a variable input and capital as a fixed input. That is, you have a given plant, but you can adjust how many workers you use every day in that plant. And now the producer has to decide, given that plant exists, how many workers should he hire to produce the goods? And the key concept that's going to determine that is something we'll call the marginal product of labor, which is the change in total output resulting from the next unit of labor used—that is, delta q, delta l—is the marginal product of labor. The reason each worker does less is because they only have the same amount of stuff to work with. For example, a certain amount of capital to work with, each additional worker just can't do as much as the one before.

In the long run, all inputs are variable. So now a firm doesn't just choose how many workers to hire, or how many hours of labor to buy. It chooses both l and k, and has to trade them off. If k and l are traded off and decided to produce, then what's called isoquants will get. Isoquants are sets of inputs along which production is the same. So along a given isoquant, q is fixed. Each of those isoquants is a different level of q, but they show how you can vary k and l to get the same amount of q. The slope of the isoquant is called the marginal rate of technical substitution—the rate at which you can substitute one input for another in a production function—which defined as delta k, delta l for a given q bar. Now, as with marginal rate of substitution, the marginal rate of technical substitution will change along the isoquant. Once again, the principle of diminishing marginal product, just like the principle of diminishing marginal utility, implies that the marginal rate of technical substitution is going to be falling as you go down the isoquant. Why is this? It's because of this diminishing marginal productivity. That is, as you add more and more labor, given capital, each unit of labor can do less and less.

What will happen if all inputs are increased proportionally? The answer is about the concept of returns to scale. If when a firm increases its plant size and labor employed by the same percentage, its output increases by a larger percentage, the firm's average total cost decreases. The firm experiences economies of scale. The main source of economies of scale is greater specialization of both labor and capital. If when a firm increases its plant size and labor employed by the same percentage, its output increases by a smaller or just the same percentage, the former is called diseconomies of scale and the latter is called constant returns to scale. Diseconomies of scale arise from the difficulty of coordinating and controlling a large enterprise. And constant returns to scale occur when a firm is able to replicate its existing production facility including its management system. These are basic analysis of production and costs of a firm, which is the basis of the decision-make of each firm in different markets.

New Words, Phrases and Expressions
optimal decision   n. 最優選擇
maximization   n. 最大化
attainable   adj. 可得到的，可達到的

tangent   adj. 相切的
commodity   n. 商品，產品
a black box   n. 黑箱，黑匣子
critical   adj. 關鍵的，至關重要的
stuff   n. 材料，原料

Notes

1. Microeconomics had practical appeal to economists because it sought to understand the most basic machinery of an economic system: consumer decision-making and spending patterns, and the decision-making processes of individual businesses.

個體經濟學對經濟學家有著現實的吸引力，這是因為它試圖解釋經濟系統最基本的原理：消費者的消費決策和消費模式以及企業個體的決策過程。

2. The study of consumer decision-making reveals how the price of products and services affects demand, how consumer satisfaction—although not precisely measurable—works in the decision-making process, and provides useful information to businesses selling products and services to these consumers.

對消費者購買決策的研究揭示了產品及服務價格是如何影響需求的以及影響消費者的滿意程度的——雖然不能夠精確度量，卻在消費者的決策過程中起到了作用，同時也為企業向消費者銷售商品和服務提供了有用的信息。

3. In short, the consumers' choice are problems how consumers make decision about what to buy in the face of budget constraint, or, given resources are limited, consumers buy a bundle of goods that best suit their desires.

簡而言之，消費者的選擇是關於消費者在面臨預算約束的情況下，如何做出購買決策的問題，或者，假定資源是有限的，消費者如何購買一組最能滿足自己偏好的商品。

4. When cardinal utility is used, the magnitude of utility differences is treated as an ethically or behaviorally significant quantity. On the other hand, ordinal utility captures only ranking and not strength of preferences. An important example of a cardinal utility is the probability of achieving some target.

在使用基數效用時，效用差異的範圍被視為一種數量差異，該差異具有道德上或者行為上的意義。另一方面，序數效用感興趣的只是排名，而不是偏好的強度。基數效用的一個重要例子就是實現某一目標的概率的大小。

5. Two sets of issues are involved in this process: one is the technical constraints, which limit the range of feasible productive processes, while the other is the institutional context such as the characteristics of the market where commodities and inputs are purchased and sold.

在這個過程中涉及兩個問題：一個是技術約束，這限制了生產過程的可行範圍，而另一個是制度環境，如商品和投入品的購買和出售的地方——市場的特點。

6. So there is a rule called as diminishing marginal product. That is, from a given

level of labor, the next worker you add increases your total product by less than the previous one.

因此，邊際產品遞減規律出現了。也就是說，在一個既定的勞動水準下，增加的一個工人所生產的產品比前一個工人所生產的產品數量更少。

7. If when a firm increases its plant size and labor employed by the same percentage, its output increases by a larger percentage, the firm's average total cost decreases. The firm experiences economies of scale.

如果一個企業以相同的百分比增大企業規模和增加所雇傭的工人，其產量增加的百分比大於企業規模增大的百分比，該企業的平均總成本下降，該企業就實現了規模經濟。

Lesson 4　Theory of Markets

The theory of markets is a branch of microeconomics that examines how a producer makes decisions of price and quantity of goods in different kinds of markets to maximize the profit. In other words, there are four different kinds of markets where producers take actions and the theory wants to know what are characteristics of each market and how firms act actually. The four kinds of markets are perfect competition, monopolistic competition, oligopoly and monopoly.

A perfectly competitive market is one in which: there are many firms producing homogeneous goods or services, there are no barriers to entry to the market or exit from the market, and both producers and consumers have perfect knowledge of the market place. Under such conditions, the price and level of output will always tend towards equilibrium as any producer that sets a price above equilibrium will not sell anything at all, and any producer that sets a price below equilibrium will obtain 100% market share. The demand curve is perfectly elastic, which means that it will be horizontal. As these conditions imply, there are few if any examples of perfectly competitive markets in real life. However, some financial markets approximate to this extreme model, and there is no doubt that in some fields of commerce the development of the internet as a trading platform has made the markets for some products, if not perfectly competitive, then certainly less imperfect. Economists assume that there are a number of different buyers and sellers in the perfectly competitive marketplace. This means that we have competition in the market, which allows price to change in response to changes in supply and demand. Furthermore, for almost every product there are substitutes, so if one product becomes too expensive, a buyer can choose a cheaper substitute instead. In a market with many buyers and sellers, both the consumer and the supplier have equal ability to influence price.

A monopoly arises when there is only one producer in the market. In other words, the single business is the industry. Entry into such a market is restricted due to high costs or other impediments, which may be economic, social or political. It should be noted the laws of many countries define a monopoly in less extreme terms, usually referring to firms that have more than a specified share of a market. Unlike perfect competition, monopolies can and do arise in real life. This may be because the producer has a statutory right to be the only producer, or the producer may be a corporation owned by the government itself. A monopoly enjoys a privilege in that it can strike its own price in the market place, which can give rise to what economists call「super-normal profits」. For this reason, monopolies are usually subject to government control, or to regulation by non-governmental organizations.

An oligopoly arises when there are few producers that exert considerable influence in a market. This select group of firms has control over the price and, like a monopoly, an oligopoly has high barriers to entry. The products that the oligopolistic firms produce are often nearly identical and, therefore, the companies, which are competing for market share, are interdependent as a result of market forces. As there are few producers, they are likely to have a high level of knowledge about the actions of their competitors, and should be able to predict responses to changes in their strategies. The minimum number of firms in an oligopoly is two, and this particular form of oligopoly is called a duopoly. There are several examples of duopolies, including the two major cola producers and, for several product lines, Unilever and Procter & Gamble. However, markets dominated by perhaps up to six producers could be regarded as oligopolistic in nature. Where a few large producers dominate a market, the industry is said to be highly concentrated. Although it is difficult to make generalizations across all oligopolistic markets, it is frequently noted that their characteristics include complex use of product differentiation, significant barriers to entry and a high level of influence on prices in the market place.

Monopolistic competition arises in markets where there are many producers, but they will tend to use product differentiation to distinguish themselves from other producers in the market. Therefore, although their products may be very similar, their ability to differentiate means that they can act as monopolies in the short run, irrespective of the actions of their competitors. For monopolistic competition to exist, consumers must know of differences in products sold by firms. There tend to be fewer barriers to entry or exit than in oligopolistic markets.

New Words, Phrases and Expressions
homogeneous   adj. 同質的，同類的
barriers   n. 障礙，屏障
approximate to 接近，近似
platform   n. 平臺，站臺，綱領
refer to 指的是，涉及，參考
statutory   adj. 法定的，法令的，按照法律的
Notes
1. Under such conditions, the price and level of output will always tend towards equi-

librium as any producer that sets a price above equilibrium will not sell anything at all, and any producer that sets a price below equilibrium will obtain 100% market share.

在這樣的條件下，產品的價格和數量就會趨於均衡。因為如果任何一個生產廠商把產品價格定在均衡價格之上，那麼他一件商品也賣不出去；而他如果將商品價格定在均衡價格之下，那麼就將獲得市場100%的份額。

2. A monopoly enjoys a privilege in that it can strike its own price in the market place, which can give rise to what economists call「super-normal profits」. For this reason, monopolies are usually subject to government control, or to regulation by non-governmental organizations.

壟斷享有在市場上制定自己商品價格的特權，這可能會產生經濟學家們所稱的「超正常利潤」。正是因為這個原因，壟斷常常被政府控制，或者受到非政府組織的管制。

3. As there are few producers, they are likely to have a high level of knowledge about the actions of their competitors, and should be able to predict responses to changes in their strategies.

由於生產廠商為數不多，因此它們彼此很可能充分瞭解到競爭對手將採取什麼樣的行動，同時也能夠預測當它們的策略發生變化後，對手會做出什麼樣的反應。

4. Although it is difficult to make generalizations across all oligopolistic markets, it is frequently noted that their characteristics include complex use of product differentiation, significant barriers to entry and a high level of influence on prices in the market place.

雖然很難總結出所有寡頭市場的一般化規律，但人們經常指出，寡頭市場的特點包括產品差異性的複雜運用、市場進入的明顯障礙以及對市場價格的影響程度較高。

5. Therefore, although their products may be very similar, their ability to differentiate means that they can act as monopolies in the short run, irrespective of the actions of their competitors.

因此，雖然它們的產品非常相似，但它們有差異化能力，意味著無論競爭對手採取什麼樣的行動，在短期內它們都可以像壟斷廠商一樣決策。

Section 3　Macroeconomics

Lesson 1　Introduction to Macroeconomics：the History and Schools

　　Macroeconomics, on the other hand, looks at the big picture (hence「macro」). It focuses on the national economy as a whole and provides a basic knowledge of how things work in the business world. For example, people who study this branch of economics would be able to interpret the latest Gross Domestic Product figures or explain why a 6% rate of unemployment is not necessarily a bad thing. Thus, for an overall perspective of how the entire economy works, you need to have an understanding of economics at both the micro and macro levels.

In general, economics is the study of how agents (people, firms, nations) use scarce resources to satisfy unlimited wants. Macroeconomics is the branch of economics that concerns itself with market systems that operate on a large scale. Where microeconomics is more focused on the choices made by individual actors in the economy (individual consumers or firms, for instance), macroeconomics deals with the performance, structure and behavior of the entire economy. When investors talk about macroeconomics, discussions of policy decisions like raising or lowering interest rates or changing tax rates are discussed. Some of the key questions addressed by macroeconomics include: What causes unemployment? What causes inflation? What creates or stimulates economic growth? Macroeconomics attempts to measure how well an economy is performing, understand how it works, and how performance can improve. While the term 「macroeconomics」 is not all that old, many of the core concepts in macroeconomics have been the focus of study for much longer. Topics like unemployment, prices, growth and trade have concerned economists almost from the very beginning of the discipline, though their study has become much more focused and specialized through the 1990s and 2000s. Likewise, it is difficult to name any sort of founder of macroeconomic studies. John Maynard Keynes is often credited with the first theories of economics that described or modeled the behavior of the economy, elements of earlier work from the likes of Adam Smith and John Stuart Mill clearly addressed issues that would now be recognized as the domain of macroeconomics. Although microeconomic ideas like game theory are clearly quite significant today and the decision-making process of individual agents like firms is still an important field of study, macroeconomics has arguably become the dominant focus of economics—at least as it applies to the investment process and financial markets.

The field of macroeconomics is organized into many different schools of thought, with differing views on how the markets and their participants operate.

Classical economists hold that prices, wages and rates are flexible and markets always clear. As there is no unemployment, growth depends upon the supply of production factors. Other economists built on Smith's work to solidify classical economic theory.

Keynesian economics was largely founded on the basis of the works of John Maynard Keynes. Keynesians focus on aggregate demand as the principal factor inissues like unemployment and the business cycle. Keynesian economists believe that the business cycle can be managed by active government intervention through fiscal policy (spending more in recessions to stimulate demand) and monetary policy (stimulating demand with lower rates). Keynesian economists also believe that there are certain rigidities in the system, particularly 「sticky」 wages and prices that prevent the proper clearing of supply and demand.

The Monetarist school is largely credited to the works of Milton Friedman. Monetarist economists believe that the role of government is to control inflation by controlling the money supply. Monetarists believe that markets are typically clear and that participants have rational expectations. Monetarists reject the Keynesian notion that governments can 「manage」 demand and that attempts to do so are destabilizing and likely to lead to inflation.

The New Keynesian school attempts to add microeconomic foundations to traditional Keynesian economic theories. While New Keynesians do accept that households and firms operate on the basis of rational expectations, they still maintain that there are a variety of market failures, including sticky prices and wages. Because of this 「stickiness」, the government can improve macroeconomic conditions through fiscal and monetary policy.

Neoclassical economics assumes that people have rational expectations and strive to maximize their utility. This school presumes that people act independently on the basis of all the information they can attain. The idea of marginalism and maximizing marginal utility is attributed to the neoclassical school, as well as the notion that economic agents act on the basis of rational expectations. Since neoclassical economists believe the market is always in equilibrium, macroeconomics focuses on the growth of supply factors and the influence of money supply on price levels.

The New Classical school is built largely on the Neoclassical school. The New Classical school emphasizes the importance of microeconomics and models based on that behavior. New Classical economists assume that all agents try to maximize their utility and have rational expectations. They also believe that the market clears at all times. New Classical economists believe that unemployment is largely voluntary and that discretionary fiscal policy is destabilizing, while inflation can be controlled with monetary policy.

The Austrian school is an older school of economics that is seeing some resurgence in popularity. Austrian school economists believe that human behavior is too idiosyncratic to model accurately with mathematics and that minimal government intervention is best. The Austrian school has contributed useful theories and explanations on the business cycle, implications of capital intensity, and the importance of time and opportunity costs in determining consumption and value.

New Words, Phrases and Expressions
perspective   n. 觀點，看法
dominant   adj. 占主導的，顯性的
solidify   v. 鞏固，固化
intervention   n. 干預，干涉
rigidity   n. 剛性，固化
marginalism   n. 邊際主義
resurgence   n. 中興，復興，復甦
idiosyncratic   adj. 獨特的，特殊的

discretionary　adj. 任意的；自由決定的；酌情行事的；便宜行事的

Notes

1. For example, people who study this branch of economics would be able to interpret the latest Gross Domestic Product figures or explain why a 6% rate of unemployment is not necessarily a bad thing.

例如，研究經濟學這一分支的人將能夠解釋最新的國內生產總值數字或者能夠解釋為什麼一個國家的失業率達到6%不一定是一件壞事。

2. Where microeconomics is more focused on the choices made by individual actors in the economy (individual consumers or firms, for instance), macroeconomics deals with the performance, structure and behavior of the entire economy. When investors talk about macroeconomics, discussions of policy decisions like raising or lowering interest rates or changing tax rates are discussed.

個體經濟學更側重於研究個體行為者（比如個人消費者或企業）的經濟選擇，總體經濟學則側重於研究整個經濟的績效、結構和行為。當投資者談論總體經濟學時，他們討論的都是諸如提高或降低利率、改變稅率等這樣的政策決策。

3. John Maynard Keynes is often credited with the first theories of economics that described or modeled the behavior of the economy, elements of earlier work from the likes of Adam Smith and John Stuart Mill clearly addressed issues that would now be recognized as the domain of macroeconomics.

約翰・梅納德・凱恩斯經常被譽為第一個對整個經濟行為進行描述或模型化的經濟學家。這方面最早的工作來自亞當・斯密和約翰・斯圖亞特・穆勒之流所討論的問題，這些問題現在被公認為屬於總體經濟學領域。

4. Keynesian economists believe that the business cycle can be managed by active government intervention through fiscal policy (spending more in recessions to stimulate demand) and monetary policy (stimulating demand with lower rates).

凱恩斯主義學派的經濟學家們認為，商業週期可以通過政府的積極干預來予以「管理」，而政府干預則通過財政政策（在經濟衰退時，政府通過增加支出來刺激需求）和貨幣政策（在經濟衰退時，政府通過降低利率來刺激需求）來實施。

5. The Austrian school has contributed useful theories and explanations on the business cycle, implications of capital intensity, and the importance of time and opportunity costs in determining consumption and value.

奧地利學派對商業週期、資本強度的影響以及時間和機會成本在確定消費和價值上的重要性等方面做出了有益的理論解釋。

Lesson 2　Unemployment and Inflation

Labor is a driving force in every economy—wages paid for labor fuel consumer spending, and the output of labor is essential for companies. Likewise, unemployed workers represent wasted potential production within an economy. Consequently, unemployment is a significant concern within macroeconomics.

「Official」 unemployment refers to the number of civilian workers who are actively looking for work and not currently receiving wages. Given that official unemployment statistics specifically exclude those who would like to work but have become discouraged and ceased looking for employment, the true unemployment rate is always higher than the official rate. Within the unemployment number are several sub-types of unemployment such as frictional unemployment, cyclical unemployment and structural unemployment.

Frictional unemployment results from imperfect information and the difficulties in matching qualified workers with jobs. A college graduate who is actively looking for work is one example. Frictional unemployment is almost impossible to avoid, as neither job-seekers nor employers can have perfect information or act instantaneously, and it is generally not seen as problematic to an economy. Cyclical unemployment refers to unemployment that is a product of the business cycle. During recessions, for instance, there is often inadequate demand for labor and wages are typically slow to fall to a point where the demand and supply of labor are back in balance. Structural unemployment refers to unemployment that occurs when workers are not qualified for the jobs that are available. Workers in this case are often out of work for much longer periods of time and often require retraining. Structural unemployment can be a serious problem within an economy, particularly in cases where entire sectors (manufacturing, for instance) become obsolete.

While high unemployment is undesirable, full employment (meaning zero unemployment) is neither practical nor desirable. When economists talk about full employment, frictional unemployment and some small percentage of structural unemployment are excluded. Economists do not generally believe it is practical or desirable to have 100% employment in an economy. In particular, the Phillips curve highlights why this is so. Generally there is a relationship between inflation and unemployment—the lower the rate of unemployment, the higher the rate of inflation. While a variety of factors can alter the curve (including productivity gains), the essential take-away is that neither a zero-unemployment or zero-inflation scenario is viable on a long-term basis.

There is also a trade-off between employment and efficiency. Businesses maximize their profits when they produce the largest number of goods possible at the lowest price possible. In some cases, though, labor is more expensive (less efficient) than capital equipment. Consequently, there is always a trade-off between the cost and productivity of labor and that of labor-substituting capital equipment and that effectively reduces the number of jobs available. Likewise, structural employment is a recurrent problem as technology progresses—workers find their skills no longer match the needs of the employers and must update their training as industries adopt new technologies.

Inflation is a key concept in macroeconomics, and a major concern for government policymakers, companies, workers and investors. Inflation refers to a broad increasein prices across many goods and services in an economy over a sustained period of time. Conversely, inflation can also be thought of as the erosion in value of an economy's currency (a unit of currency buys fewer goods and services than in prior periods).

In the United States, the Consumer Price Index (CPI) is among the most commonly-used measures of inflation. The CPI uses a so-called 「market basket」 of goods to measure the changes in prices experienced by average consumers in the economy. Economists and central bankers will often subdivide the CPI into so-called 「core inflation」, a measure that excludes the price of food and energy. The Producer Price Index (PPI) is a measure of inflation that tracks the prices that producers obtain for their goods. Though a long-followed economic statistic, the change in composition of some economies away from manufacturing and towards services is eroding the value of this statistic. The GDP deflator is another option for measuring prices and inflation. As the name suggests, the GDP deflator is a price measurement tool that is used to convert nominal GDP to real GDP. The GDP deflator is a broader measure than the CPI, as it includes goods and services bought by businesses and governments.

While there is little consensus on the 「right」 rate of inflation for an economy (or even if inflation is necessary at all), there is little disagreement in the differing impacts of expected and unexpected inflation. When inflation is expected, agents in the economy can plan for it and act accordingly—businesses raise prices, workers demand higher wages, lenders raise interest rates and so on. Unexpected inflation is considerably more problematic. When inflation is higher than expected, it tends to hurt workers, recipients of fixed incomes, and savers. In contrast, unexpected inflation often benefits companies (who can raise prices quickly without needing to raise wages in tandem) and borrowers (who can repay their debts with money that is now worth less than when they borrowed it).

Over the long term, unanticipated inflation can cause a number of problems for an economy. Businesses will invest less in long-term projects because of the uncertainty of returns, price information becomes distorted, and consumers will spend more time trying to protect themselves from inflation and less time engaging in productive activities. Periods of inflation also tend to redirect investment from businesses and toward hard assets, thus depriving companies of the capital they need to grow and expand.

What causes inflation is also a key argument in economic theory. Some economists believe that there are different types of inflation—cost-push and demand-pull inflation. Cost-push inflation is supposed to be a type of inflation caused by rising prices in goods or services with no suitable alternatives. An example of this inflation is the oil crisis of the 1970s. Cost-push inflation is largely a Keynesian argument, as monetarists do not believe that increased prices for goods and services lead to inflation absent an increase in the money supply. Demand-pull inflation is a rise in the price of goods and services created by aggregate demand in excess of aggregate supply, sometimes referred to as 「too much money chasing too few goods」. As with cost-push inflation, monetarists argue against the existence of demand-pull inflation absent changes in the money supply.

New Words, Phrases and Expressions
fuel   n. 燃料   v. 給……補充燃料
inadequate   adj. 不足的，不夠的
obsolete   adj. 過時的，廢棄的
scenario   n. 方案，場景
viable   adj. 可行的，可實現的
recurrent   adj. 復發的，可循環的，可週期發生的
recipient   adj. 容易接受的
tandem   n. 串聯，一個跟著一個

Notes

1.「Official」unemployment refers to the number of civilian workers who are actively looking for work and not currently receiving wages. Given that official unemployment statistics specifically exclude those who would like to work but have become discouraged and ceased looking for employment, the true unemployment rate is always higher than the official rate.

「官方」失業是指那些正在積極尋找工作，目前未能獲得工資收入的職工人數。鑒於官方的失業統計數據沒有包括那些想工作但已經喪失信心而停止繼續尋找工作的人，所以，真正的失業率總是高於官方失業率。

2. Frictional unemployment results from imperfect information and the difficulties in matching qualified workers with jobs. Frictional unemployment is almost impossible to avoid, as neither job-seekers nor employers can have perfect information or act instantaneously, and it is generally not seen as problematic to an economy.

摩擦性失業產生的原因是因為信息的不完全導致有技能的工人和相關的工作不能匹配。摩擦性失業幾乎是不可能避免的，因為無論是求職者還是雇主都不能獲得完全信息，也不能根據信息立即採取行動。人們通常不認為摩擦性失業是一個經濟問題。

3. During recessions, for instance, there is often inadequate demand for labor and wages are typically slow to fall to a point where the demand and supply of labor are back in balance.

例如，在經濟衰退期間，對勞動力的需求不足，工資逐漸降低到勞動力的需求和供給又回到均衡的那一點上面。

4. While high unemployment is undesirable, full employment (meaning zero unemployment) is neither practical nor desirable. When economists talk about full employment, frictional unemployment and some small percentage of structural unemployment are excluded.

雖然高失業率是人們不認可的，但充分就業（意味著失業率為零）既不現實，也不可取。當經濟學家談論充分就業時，並沒有包括摩擦性失業和輕微的結構性失業。

5. Likewise, structural employment is a recurrent problem as technology progresses—workers find their skills no longer match the needs of the employers and must update their training as industries adopt new technologies.

同樣，隨著技術的進步，結構性就業便成為一個經常性的問題。工人們發現，隨著各行各業採納了新技術，他們的技能不再符合雇主的需要，必須接受再培訓。

6. Conversely, inflation can also be thought of as the erosion in value of an economy's currency (a unit of currency buys fewer goods and services than in prior periods).

反過來說，通貨膨脹也可以被認為是對貨幣價值的侵蝕（一單位貨幣所購買的商品和服務數量少於通貨膨脹之前的數量）。

7. While there is little consensus on the「right」rate of inflation for an economy (or even if inflation is necessary at all), there is little disagreement in the differing impacts of expected and unexpected inflation.

雖然經濟學家對一個經濟體該維持什麼樣的通貨膨脹率才是「正確」的無法達成共識（或者說，即使他們認為通貨膨脹是必要的），但在可預期的通貨膨脹和不可預期的通貨膨脹所產生的不同影響這個問題上，卻很少有分歧。

8. Demand-pull inflation is a rise in the price of goods and services created by aggregate demand in excess of aggregate supply, sometimes referred to as「too much money chasing too few goods」.

需求拉動型通貨膨脹是由於總需求超過總供給而導致商品和服務的價格上漲，有時被稱為「太多的錢追逐太少的商品」。

Lesson 5　Economic Performance and Growth

Income is one of the most significant factors in measuring economic performance, and gross domestic product (GDP) is the most commonly used measure of a country's economic activity. In short, GDP reflects the value of all final goods and services legally produced in an economy in a given time period.

The distinction between final goods and intermediate goods is an important one. A tomato sold to a ketchup manufacturer would NOT be included in the GDP number, while a tomato sold in a store as produce would be included, as it represents the final use of that good. It is also worth noting that trade in illegal goods and services are also excluded from GDP figures. There are two ways of approaching GDP—the expenditure approach and the resource cost-income approach. The expenditure approach totals the amount spent on goods and services during a year, while the resource cost-income approach adds up the payments made to suppliers of resources and other inputs that go into goods and services. The expenditure approach is arguably more common, and it breaks GDP into four commonly watched components—personal consumption, gross profit domestic investment, government spending, and net exports to foreigners. Personal consumption has long been the largest component of GDP in the United States and is made of household spending on goods and services. Domestic investment refers to spending on fixed assets (capital expenditures) and additions made to inventories during the year. By comparison, under the resource cost-income approach, compensation paid to employees is the largest component of GDP, with depreciation, indirect taxes, interest, corporate profits, and the income of the self-employed following.

Going a step further, there are other types of analysis that can be applied to GDP. Real GDP is the broadest view of an economy's output, and the one most widely used by economists. Real GDP differs from nominal GDP in that it attempts to adjust for rising price levels to determine what amounts to changes in the volume of activity in an economy.

Economists observe not only the absolute level of GDP, but its growth over time. In fact, most discussions of GDP are undertaken in terms of growth. Going another step further, it is often useful to examine GDP in the context of its relationship to the number of participants in an economy. GDP per capita often correlates to the standard of living and the extent of economic development in a country. The U.S. and China, for instance, have similar GDPs in absolute terms, but the per capita GDP in each country is much different and reflects the much higher standard of living in the United States.

There are some drawbacks and limitations to the use of GDP. GDP excludes any unpaid activity as well as illegal activity, so it cannot account for the value of all economic activity in a nation. GDP also fails to account for reductions in quality of life and losses to natural disasters or crime. The destruction wrought by an earthquake, for instance, would not be accounted for in GDP accounting, but the rebuilding efforts would be counted. Likewise, even per-capita GDP does not necessarily reflect the wealth of the typical citizen, as a majority of low-earning citizens could be offset by a small group of very high earners.

There are other significant measures of income to consider. National income refers to the total income paid to owners of human capital (wages for labor) and physical capital, and includes both domestic and foreign income. National income can also be calculated as the sum of wages, interest, self-employment income, rent and business profits. Personal income, in contrast, is the income received by individuals; it excludes corporate profits and social security taxes, but adds back transfer payments (like Social Security), interest and dividend. Disposable income is personal income that is actually available for spending or saving; it is personal income net of personal income taxes.

While the large majority of economists basically agree with the use of metrics like GDP and GDP per capita as measures of growth, there is considerably less agreement in how to explain how economies grow over time. Some models hold that growth is exogenous—long-term growth is determined by factors external to the economy. Other models post that growth is endogenous—long-term growth is determined by factors within the system. More specifically, there are models like the Harrod-Domar model that examine the consequences of fixed capital and labor ratios and the propensities to save. This model highlights the problems of rigidities in the capital/labor ratio and savings rate. By comparison, the Solow model holds that growth in GDP is explained by population increases, technical progress and increased investment. As with many economic models, these are not so much predictive as explanatory; seeking to identify the impact of certain variables and conditions. While economic growth is clearly an important objective for most governments, most economies do not operate at their full potential. Often there is a gap between the amount of GDP actually produced and the potential GDP that the economy could produce with full employment and full resource utilization—this gap is called the output gap (or the GDP gap). (For examples of some fast growing economies)

New Words, Phrases and Expressions
ketchup   n. 番茄醬
expenditure   n. 花費，指出，費用
depreciation   n. 折舊，貨幣貶值
by comparison 相比之下，相形之下
wrought   v. 發生，運轉
metrics   n. 韻律學
utilization   n. 使用，利用

Notes

1. A tomato sold to a ketchup manufacturer would NOT be included in the GDP number, while a tomato sold in a store as produce would be included, as it represents the final use of that good.

將番茄賣給番茄醬製造商不算在國內生產總值之內，而在商店中將番茄作為商品出售則要計算在國內生產總值之中，因為這是最終產品。

2. The expenditure approach totals the amount spent on goods and services during a year, while the resource cost-income approach adds up the payments made to suppliers of resources and other inputs that go into goods and services.

支出法計算一年內在商品和勞務上花費的費用總額，而資源成本收入法則增加了支付給生產商品和提供勞務的廠商資源和投入的費用。

3. Real GDP differs from nominal GDP in that it attempts to adjust for rising price levels to determine what amounts to changes in the volume of activity in an economy.

實際國內生產總值與名義國內生產總值不同之處在於前者試圖調整物價上漲水準，以確定哪些因素決定了經濟活動總量的變化。

4. Going another step further, it is often useful to examine GDP in the context of its relationship to the number of participants in an economy.

進一步說，將國內生產總值和經濟體參與人人數聯繫起來研究兩者之間的關係是非常有用的。

5. The destruction wrought by an earthquake, for instance, would not be accounted for in GDP accounting, but the rebuilding efforts would be counted. Likewise, even per-capita GDP does not necessarily reflect the wealth of the typical citizen, as a majority of low-earning citizens could be offset by a small group of very high earners.

例如，地震造成的破壞不會計算在 GDP 之內，但重建費用卻會被計算在內。同樣地，即使人均 GDP 也不一定能反應出普通公民的財富水準，因為大多數低收入的公民可能會被少數高收入者的收入抵消。

6. Some models hold that growth is exogenous—long-term growth is determined by factors external to the economy. Other models post that growth is endogenous—long-term growth is determined by factors within the system.

一些模型認為，增長是外生的，長期的增長是由經濟外部因素決定的。其他模型則認為增長是內生的，即長期的增長是由系統內部因素決定的。

7. As with many economic models, these are not so much predictive as explanatory; seeking to identify the impact of certain variables and conditions.

許多經濟模型都是解釋性的，而不是預測性的，這些模型都在努力確定某些變量和條件是如何產生影響的。

Exercises

Ⅰ. Define the following concepts.
1. economics
2. microeconomics
3. GDP
4. Official unemployment

Ⅱ. Translate English into Chinese.

1. Economics may appear to be the study of complicated tables and charts, statistics and numbers, but, more specifically, it is the study of what constitutes rational human behavior in the endeavor to fulfill needs and wants.

2. There are some drawbacks and limitations to the use of GDP. GDP excludes any unpaid activity as well as illegal activity, so it cannot account for the value of all economic activity in a nation. GDP also fails to account for reductions in quality of life and losses to natural disasters or crime.

3. Smith put forwards the theory that if the government does not tamper with the economy, a nation's resources will be most efficiently used, free-market problems will correct themselves and a country's welfare and best interests will be served.

4. Some of the key questions addressed by macroeconomics include: What causes unemployment? What causes inflation? What creates or stimulates economic growth? Macroeconomics attempts to measure how well an economy is performing, understand how it

works, and how performance can improve.

5. Price is determined at the intersection of the supply and demand curves. If the price is set above the equilibrium price, this will result in the quantity supplied exceeding the quantity demanded. Therefore, in order to clear its inventory, the company will need to reduce its price.

Further Reading Comprehension

India's salvation

IN MAY America's Federal Reserve hinted that it would soon start to reduce its vast purchases of Treasury bonds. As global investors adjusted to a world without ultra-cheap money, there has been a great sucking of funds from emerging markets. Currencies and shares have tumbled, from Brazil to Indonesia, but one country has been particularly badly hit.

Not so long ago India was celebrated as an economic miracle. In 2008 Manmohan Singh, the prime minister, said growth of 8%~9% was India's new cruising speed. He even predicted the end of the「chronic poverty, ignorance and disease, which has been the fate of millions of our countrymen for centuries」. Today he admits the outlook is difficult. The rupee has tumbled by 13% in three months. The stockmarket is down by a quarter in dollar terms. Borrowing rates are at levels last seen after Lehman Brothers' demise. Bank shares have sunk.

On August 14th jumpy officials tightened capital controls in an attempt to stop locals taking money out of the country (seearticle). That scared foreign investors, who worry that India may freeze their funds too. The risk now is of a credit crunch and a self-fulfilling panic that pushes the rupee down much further, fuelling inflation. Policymakers recognise that the country is in its tightest spot since the balance-of-payments crisis of 1991. How to lose friends and alienate people

India's troubles are caused partly by global forces beyond its control. But they are also the consequence of a deadly complacency that has led the country to miss a great opportunity.

During the 2003-2008 boom, when reforms would have been relatively easy to introduce, the government failed to liberalise markets for labour, energy and land. Infrastructure was not improved enough. Graft and red tape got worse. Private companies have slashed investment. Growth has slowed to 4%~5%, half the rate during the boom. Inflation, at 10%, is worse than in any other big economy. Tycoons who used to cheer India's rise as a superpower now warn of civil unrest.

As well as undermining 1.2 billion people's hopes of prosperity, failure to reform dragged down the rupee. Restrictive labour laws and weak infrastructure make it hard for Indian firms to export. Inflation has led people to import gold to protect their savings. Both factors have swollen the current-account deficit, which must be financed by foreign capital. Add in the foreign debt that must be rolled over, and India needs to attract $250 billion in the next year, more than any other vulnerable emerging economy.

A year ago the new finance minister, Palaniappan Chidambaram, tried to kick-start the economy. He has attempted to push key reforms, clear bottlenecks and help foreign investors. But he has lukewarm support within his own party and faces obstructionist opposition. Obstacles to growth, such as fuel shortages for power plants, remain. Foreign firms find nothing has changed. Meanwhile, bad debts have risen at state-run banks: 10%~12% of their loans are dud. With an election due by May 2014, some fear that the Congress-led government will now take a more populist tack. A costly plan to subsidise food hints at this.

Stopping the rot

To prevent a slide into crisis, the government needs first to stop making things worse. Those capital controls backfired, yet the urge to tinker runs deep: on August 19th officials slapped duties on televisions lugged in through airports. The authorities must accept that 2013 is not 1991. Then the state nearly bankrupted itself trying to defend a pegged exchange rate. Now the rupee floats, and the state has no foreign debt to speak of. A weaker currency will break some firms with foreign loans, but poses no direct threat to the government's solvency.

And so the Reserve Bank of India must let the rupee find its own level. The currency has not yet wildly overshot estimates of fundamental value. Raghuram Rajan, the central bank's incoming head, should aim to control inflation, not micromanage one of the world's most traded currencies. Second, the government must get its finances in order. The budget deficit has been as high as 10% of GDP in recent years. This year the government must hold down its deficit (including those of individual states) to 7% of GDP. It is already cutting fuel subsidies, and—notwithstanding the pressures in the run-up to an election—should do so faster. This is not enough to fix the government's finances, though. Only 3% of Indians pay income tax, so the government's tax take is puny. A proposed tax on goods and services, known as GST, would drag more of the economy into the net. It is stuck in endless cross-party talks. If the government can rally itself before the election to push for one long-term reform, this is the one it should go for. Last, the government, with the central bank, should force the zombie public-sector banks to recapitalise. In 2009 America did「stress tests」to repair its banks. India should follow. Injecting funds into banks would widen the deficit, but the surge in confidence would be worth it.

There are glimmers of hope: exports picked up in July, narrowing the trade gap. But India faces a difficult year, with jittery global markets and an election to boot. Even if it scrapes past the election without a full-blown financial crisis, the next government must do much, much more to change India. Over the coming decade tens of millions of young people will have to find jobs where none currently exists. Generating the growth to create them will mean radical deregulation of protected sectors (of which retail is only the most obvious); breaking up state monopolies, from coal to railways; reforming restrictive labour laws; and overhauling India's infrastructure of roads, ports and power.

The calamity of 1991 led to liberalising reforms that ended decades of stagnation and allowed a spurt of fast growth. This latest brush with disaster could produce a positive legacy, too, but only if it persuades voters and the next government of the importance of a new round of reforms that deal with the economy's flaws and unleash its mighty potential. [From *the Economist*, 2014 (7)]

Chapter 2
Management

Chapter Check List

When you have completed your study of this chapter, you will be able to
1. Have a brief comprehension of management and its theory framework.
2. Master the basic rules of management.
3. Have basic quality and skill as a manager.
4. Know the development trend of management.

In this chapter, we will introduce some basic knowledge about management. There are four sections in this chapter, they are
1. An introduction to management.
2. The important theories of management.
3. The functions of management.
4. Current trends and issues of management.

The more important and difficult parts include the basic concept about management, the theory framework, and the functions of management.

Section 1　An Introduction to Management

Lesson 1　What Does Management Mean?

The term management refers to the process of getting things done, effectively and efficiently, through and with other people. The term process represents the primary activities managers perform. Efficiency means doing the task correctly with the least amount of time or effort or materials, and refers to the relationship between inputs and outputs. For instance, you have increased efficiency if you get more output for a given input. Since managers deal with input resources that are scarce, they are concerned with the efficient use of those resources. Management, therefore, is concerned with minimizing resources costs. Although, efficiency is important, it is not enough simply to be efficient. Management is also concerned with effectiveness. Effectiveness means doing the right task. In an organization, that translates into goal attainment.

Although efficiency and effectiveness are different terms, they areinterrelated. For instance, it is easier to be effective if one ignores efficiency. For example, some government agencies have been attacked on the ground that they are effective but extremely inefficient. That is, they accomplish their goals but do so at a very high cost. Can organizations be efficient and yet not effective? Yes, by doing the wrong things well. Therefore, good management is concerned with both attaining goals (effectiveness) and doing so as efficiently as possible.

New Words, Phrases and Expressions
effectively   adv. 有效地，生效地；有力地；實際上
efficiently   adv. 有效地；效率高地
deal with   處理；涉及；做生意
scarce   adj. 缺乏的，不足的；稀有的
effectiveness   n. 效力
efficiency   n. 效率，功效
attainment   n. 達到；成就；學識
interrelated   adj. 相關的；互相聯繫的

Notes

1. Efficiency means doing the task correctly with the least amount of time or effort or materials, and refers to the relationship between inputs and outputs.
效率意味著用最少的時間或努力或材料正確完成任務，並且指的是投入和產出之間的關係。

2. Therefore, good management is concerned with both attaining goals (effectiveness) and doing so as efficiently as possible.
因此，良好的管理與實現目標密切相關，並盡可能有效地這樣做。

Lesson 2   Who Are Managers and What Do Managers Do?

It used to be fairly simple to define who managers were: They were the organizational members who told others what to do and how to do it. It was easy to differentiate managers from non-managerial employees; the latter term described those organizational members who worked directly on a job or task and had no one reporting to them. But it isn't quite that simple anymore! The changing nature of organizations and work has, in many organizations, blurred the clear lines of distinction between managers and non-managerial employees. Many traditional jobs now include managerial activities, especially on teams. For instance, team members often develop plans, make decisions, and monitor their own performance. And as these non-managerial employees assume responsibilities that traditionally were part of management, definitions we've used in the past no longer describe every type of managerial situation.

So, how do we define who managers are? A manager is someone who works with and through other people by coordinating their work activities in order to accomplish organizational goals. That may mean coordinating the work of departmental group, or it might mean supervising a single person. It could involve coordinating the work activities of a team composed of people from several different departments of even people outside the organization such as temporary employees or employees who work for the organization's suppliers. Keep in mind, also, that managers may have other work duties not related to coordinating and integrating the work of others. For example, an insurance claims supervisor may also process claims in addition to coordinating the work activities of other claims clerks.

Is there some way to classify managers in organizations? There is, particularly for traditionally structured organizations—that is, those organizations in which the number of employees is greater at the bottom than at the top. (These types of organizations are often pictured as being shaped like a pyramid.) We typically describe managers as first-line, middle, or top in this type of organizations. Identifying exactly who the managers are in these organizations isn't difficult, although you should be aware that managers may have a variety of titles.

First-line managers are the lowest level of management and manage the work of non-managerial individuals who are involved with the production or creation of the organization's products, they're often called supervisors but may also be called line managers, office managers, or even foremen. Middle managers include all levels of management between the first-line level and the top level of the organization. These managers manage the work of first-line managers and may have titles such as department head, project leader, plant manager, or division manager. At or near the top of the organization are the top managers, who are responsible for making organization-wide decisions and establishing the plans and goals that affect the entire organization. These individuals typically have titles such as executive vice president, president, managing director, chief operating officer, chief executive officer, or chairman of the board.

Not all organizations get work done with a traditional pyramidal form, however. Some organizations, for example, are more loosely configured with work being done by ever-changing teams of employees who move from one project to another as work demands arise. Although it's not as easy to tell who the managers are in these organizations, we do know that someone must fulfill that role—that is, there must be someone who coordinates and oversees the work of others, even if that 「someone」 changes as work tasks or projects change.

Simply speaking, management is what managers do. But that simply statement doesn't tell us much, does it? We already know that the second part of the definition-coordinating the work of others-is what distinguishes a managerial position from a non-managerial one.

In addition, management involves the efficient and effective completion of organizational work activities, or at least that's what managers aspire to do.

Efficiency refers to getting the most output from the least amount of inputs. Because managers deal with scarce inputs-including resources such as people, money, and equipment-they are concerned with the efficient use of those resources. For instance, at the Beiersdorf Inc. factory in Cincinnati, where employees make body braces and supports, canes, walkers, crutches, and other medical assistance products, efficient manufacturing techniques were implemented by cutting inventory levels, decreasing the amount of time to manufacture products, and lowering products, and lowering product reject rates. These efficient work practices paid off as the company was named one of Industry week's best plants. From this perspective, efficiency is often referred to as「doing things right」. Management is also concerned with being effective, completing activities so that organizational goals are attained.

Effectiveness is often described as「doing the right things」-that is, those work activities that will help the organization reach its goals. For instance, at the Beiersdorf Inc. factory, goals included open communication between managers and employees and cutting costs. Through various work programs, these goals were pursued and achieved. Whereas efficiency is concerned with the means of getting things done, effectiveness is concerned with the ends, or attainment of organizational goals. Management is concerned, then, not only with getting activities completed and meeting organizational goals but also with doing so as efficiently as possible. In successful organizations, high effectiveness typically go hand in hand. Poor management is most often due to both inefficiency and ineffectiveness or to effectiveness achieved through inefficiency.

New Words, Phrases and Expressions
fairly   adv. 相當地；公平地；簡直
differentiate   v. 區分，區別
non-managerial   adj. 非管理層的
blurred   v. 玷污；使……模糊不清；使……感覺遲鈍
monitor   v. 監控
coordinating   v. 協調；整合
integrating   v. 整合；積分；集成化
supervisor   n. 監督人，管理人；檢查員
foremen   n. 工長，工頭
executive   adj. 行政的；經營的；執行的，經營管理的
pyramidal   adj. 錐體的；金字塔形的；角錐狀的
configured   v. 使……成形；按特定形式裝配
distinguishes   v. 辨別；區分，區別；分清

inventory   n. 存貨，存貨清單；詳細目錄；財產清冊

Notes

1. The changing nature of organizations and work has, in many organizations, blurred the clear lines of distinction between managers and non-managerial employees.

在許多組織中，組織和工作的變化模糊了管理層和非管理層員工之間的區別。

2. And as these non-managerial employees assume responsibilities that traditionally were part of management, definitions we've used in the past no longer describe every type of managerial situation.

這些非管理層員工承擔責任歷來就是管理的一部分，我們之前使用過的定義已經不能再用來描述每一種類型的管理情況。

3. It could involve coordinating the work activities of a team composed of people from several different departments of even people outside the organization such as temporary employees or employees who work for the organization's suppliers.

它可能包括協調來自幾個不同部門的人組成的一個團隊甚至是組織以外的人，如臨時員工或那些為供應商工作的員工，來開展工作活動。

4. In addition, management involves the efficient and effective completion of organizational work activities, or at least that's what managers aspire to do.

此外，管理還包括有效地完成組織工作活動，或者至少是完成管理者期望去做的事情。

5. Management is concerned, then, not only with getting activities completed and meeting organizational goals but also with doing so as efficiently as possible.

對於管理而言，它不僅與完成活動和實現組織目標有關，而且還應盡可能有效地這樣做。

Lesson 3　Management Skills

As you can see from the preceding discussion, a manager's job is varied and complex. Managers need certain skills to perform the duties and activities associated with being a manager. What types of skills does a manager need? Research by Robert. L. Katz found that managers need three essential skills or competencies. Technical skills include knowledge of and proficiency in a certain specialized field, such as engineering, computers, accounting, or manufacturing. These skills are more important at lower levels of management since these managers are dealing directly with employees doing the organization's work. Human skills involve the ability to work well with other people both individually and in a group. Because managers deal directly with people, this skill is crucial! Managers with good human skills are able to get the best out of their people. They know how to communicate, motivate, lead, and inspire enthusiasm and trust. These skills are equally important at all levels of management. Finally, conceptual skills are the skills managers must have to think and to conceptualize about abstract and complex situations. Using these skills, managers must be able to see the organization as a whole, understand the relationships among various subunits, and visualize how the organization fits into its broader environment. These skills are most important at the top management levels.

Table 2.1 Skills Needed at Different Management Levels

Top Management	Conceptual Skills		
Middle Management		Human Skills	
Lower-level Management			Technical Skills

How relevant are management skills to today's managers? In today's demanding and dynamic workplace, employees who are invaluable to an organization must be willing to constantly upgrade their skills and take on extra work outside their own specific job area. There's no doubt that skills will continue to be an important way of describing what a manager does. In fact, understanding and developing management skills are very important.

New Words, Phrases and Expressions
competency   n. 能力；勝任特徵；職業能力素質
crucial   adj. 重要的；決定性的；定局的；決斷的
get the best out of   發揮……最大功效；最有效地使用
conceptualize   v. 使……概念化；概念化
subunit   n. 子單位；二級單位
visualize   vt. 形象，形象化；想像，設想
upgrade   vt. 使升級；提升；改良品種

Notes
1. Using these skills, managers must be able to see the organization as a whole, understand the relationships among various subunits, and visualize how the organization fits into its broader environment.

使用這些技能，管理者必須將組織看成是一個整體，理解各單元之間的關係，設想組織如何適應更廣泛的環境。

2. In today's demanding and dynamic workplace, employees who are invaluable to an organization must be willing to constantly upgrade their skills and take on extra work outside their own specific job area.

在今天的動態工作環境和要求下，一個對於組織來說具有無限價值的員工，組織一定願意不斷提高他們的技能並且讓他們承擔自己的特定工作區域外的額外工作。

Section 2 The Important Theories of Management

Lesson 1 Historical Background of Management

Organized endeavors directed by people responsible for planning, organizing, leading, and controlling activities have existed for thousands of years. The Egyptian pyramids and the Great Wall of China, for instance, are tangible evidence that projects of tremendous scope, employing tens of thousands of people, were undertaken well before modern times. The pyramid occupied more than 100, 000 workers for 20 years. Who told each worker what to do? Who ensured that there would be enough stones at the site to keep workers busy? The answer to such questions is managers. Regardless of what managers were called at the time, someone had to play what was to be done, organize people and materials to do it, lead and direct the workers, and impose some controls to ensure that everything was done as planned.

Another example of early management can be seen during the 1400s in the city of Venice, Italy, a major economic and trade center. The Venetians developed an early form of business enterprise and engaged in many activities common to today's organizations. For instance, at the arsenal of Venice, warships were floated along the canals and at each stop materials and rigging were added to the ship. Doesn't that sound a lot like a car 「floating」 along an automobile assembly line and component being added to it? In addition to this assembly line, the Venetians also had a warehouse and inventory system to monitor its contents, personnel functions required to manage the labor force, and an accounting system to keep track of revenues and costs.

These examples from the past demonstrate that organizations have been around for thousands of years and that managementhas been practiced for an equivalent period. However, two pre-twentieth century events played particularly significant roles in promoting the study of management.

First, in 1776, Adam Smith published a classical economics doctrine, *The Wealth of Nations*, in which he argued the economic advantages that organizations and society would gain from the division of labor, the breakdown of a job into narrow and repetitive tasks. Using the pin manufacturing industry as an example, Smith claimed that 10 individuals, each doing a specialized task, could produce about 48, 000 pins a day among them. However, if each person worked separately and had to perform each task, it would be quite an accomplishment to produce even 10 pins a day! Smith concluded that division of labor increased productivity by increasing each worker's skill and dexterity, by saving time lost in changing tasks, and by creating labor-saving inventions and machinery. The continued popularity of job specialization—for example, specific tasks performed by members of a hospital surgery team, specific meal preparation tasks done by workers in restaurant kitchen, or specific positions played by players on a football team—is undoubtedly due to the economic advantaged cited by Adam Smith.

The second, and possibly most important, pre-twentieth century influence on management was the Industrial Revolution. Starting in the eighteenth century in Great Britain, the revolution had crossed the Atlantic to America by the end of the Civil War. The major contribution of the Industrial Revolution was the substitution of machine power for human power, which, in turn, made it more economical to manufacture goods in factories rather than at home. These large, efficient factories using power-driven equipment required managerial skills. Why? Managers were needed to forecast demand, ensure that enough materials were on hand to make products, assign tasks to people, direct activities, coordinate the various tasks, ensure that the machines were kept in good working condition and work standards were maintained, find markets for the finished products, and so forth. Planning, organizing, leading, and controlling became necessary, and the development of large corporations would require formal management practices. The need for a formal theory to guide managers in running these organizations had arrived. However, it wasn't until the early 1900s that the first major step toward developing such a theory was taken.

The development of management theories has been characterized by differing beliefs about what managers do and how they should do it. Scientific management looked at management from the perspective of improving the productivity and efficiency of manual workers. General administrative theorists were concerned with the overall organization and applying quantitative models to management practices. Finally, a group of researchers emphasized human behavior in organizations, or the「people」side of management.

New Words, Phrases and Expressions
endeavor　n. 盡力，努力
tremendous　adj. 極大的，巨大的；驚人的
impose　vi. 利用；欺騙；施加影響　vt. 強加；徵稅；以……欺騙
arsenal　n. 兵工廠；軍械庫
keep track of　記錄；與……保持聯繫
doctrine　n. 主義；學說；教義；信條
repetitive　adj. 重複的
dexterity　n. 靈巧；敏捷；機敏
substitution　n. 代替；[數] 置換；代替物
power-driven　adj. 機動，電動；[機] 動力傳動的
manual　adj. 手工的；體力的　n. 手冊，指南
quantitative　adj. 定量的；量的，數量的

Notes
1. The continued popularity of job specialization—for example, specific tasks

performed by members of a hospital surgery team, specific meal preparation tasks done by workers in restaurant kitchen, or specific positions played by players on a football team—is undoubtedly due to the economic advantaged cited by Adam Smith.

專業分工的不斷普及，例如，某醫院外科團隊成員執行特定任務，工人在餐館廚房完成特定餐準備任務，或者由足球隊中某個選手執行特定位置的任務，無疑都應歸因於亞當‧斯密提到的經濟優勢。

2. The major contribution of the Industrial Revolution was the substitution of machine power for human power, which, in turn, made it more economical to manufacture goods in factories rather than at home.

工業革命的主要貢獻是機動力代替了人力，這反過來又使人們在工廠生產商品比在家裡生產更為經濟。

Lesson 2　Scientific Management

If you had to pinpoint the year in which modern management theory was born, 1911 might be a logical choice. That was the year in which Frederick Winslow Taylor's works, *Principles of Scientific Management* was published. Its contents became widely accepted by managers around the world. The book described the theory of scientific management: the use of scientific methods to define the「one best way」for a job to be done.

Important contributions to scientific management theory were made by Frederick W. Taylor and Frank and Lillian G. Let's look at what they did.

Frederick W. Taylor did most of his work at the Midvale and Bethlehem Steel Companies in Pennsylvania. As a mechanical engineer with a Quaker and Puritan background, he was continually appalled by workers' inefficiencies. Employees used vastly different techniques to do the same job. They were inclined to「take it easy」on the job. And Taylor believed that worker output was only about one-third of what was possible. Virtually no work standards existed, and workers were placed in jobs with little or no concern for matching their abilities and aptitudes with the tasks they were required to do. Managers and workers were in continual conflict. He set out to correct the situation by applying the scientific method to shop floor jobs. He spent more than two decades passionately pursuing the「one best way」for each job to be done.

Taylor's experiences at Midvale led him to define clear guidelines for improving production efficiency. He argued that these four principles of management would result in prosperity for both workers and managers.

Table 2.2　The Four Principles

Develop a science for each element of an individual's work.
Scientifically select and then train, teach, and develop the worker.

Table 2.2 (Cont.)

Heartily cooperate with the workers so as to ensure that all work is done in accordance with the principles of the science that has been developed.
Divide work and responsibility almost equally between management and workers.

New Words, Phrases and Expressions
pinpoint   vt. 查明；精確地找到；準確描述
appall   vt. 使……膽寒；使……驚駭
vastly   adv. 極大地；廣大地；深遠地
aptitude   n. 適當；傾向；資質
set out   出發；開始；陳述；陳列
passionately   adv. 熱情地；強烈地；激昂地
prosperity   n. 繁榮，成功

Notes

1. The book described the theory of scientific management: the use of scientific methods to define the「one best way」for a job to be done.

這本書描述了科學管理理論：使用科學的方法來定義完成工作的「最好的方式」。

2. Virtually no work standards existed, and workers were placed in jobs with little or no concern for matching their abilities and aptitudes with the tasks they were required to do.

幾乎沒有工作標準存在，工人們在工作中很少或根本沒有去關心他們的能力和資質是否與他們被要求完成的任務相匹配。

Lesson 3 General Administrative Theorists

Another group of writers looked at the subject of management but focused on the entire organization. We call them the general administrative theorists. They developed more general theories of what managers do and what constituted good management practice. Let's look at some important contributions to this perspective.

Important contributions to scientific management theory were made by Henri Fayol and Max Weber.

Henri Fayol, he described management as auniversal set of functions that included planning, organizing, commanding, coordinating, and controlling. Because his ideas were important, let's look closer at what he had said.

Fayol wrote during the same time period as Taylor. While Taylor was concerned with management at the lowest organizational level and used the scientific method, Fayol's attention was directed at the activities of all managers. He wrote from personal experience as a practitioner since he was the managing director of a large French coal-mining firm.

Fayol described the practice of management as something distinct from accounting, finance, production, distribution, and other typical business functions. He argued that management was an activity common to all human endeavors in business, government, and even in the home. He then proceeded to state 14 principles of management—fundamental rules of management that could be taught in schools and applied in all organizational situations. These principles are shown in Table 2.3.

Table 2.3 14 principles of management

Division of work
Authority
Discipline
Unity of command
Unity of direction
Subordination of individual interests to the general interest
Remuneration
Centralization
Scalar chain
Order
Equity
Stability of tenure of personnel
Initiative
Esprit de corps

Max Weber was a German sociologist who studied organizational activity. Writing in the early 1900s, he developed a theory of authority structures and relations. Weber described an ideal type of organization. He called a bureaucracy—a form of organization characterized by division of labor, a clearly defined hierarchy, detailed rules and regulations, and impersonal relationships. Weber recognized that this 「ideal bureaucracy」didn't exist in reality. Instead he intended it as a basis for theorizing about work and how work could be done in large groups. His theory became the model structural design for many of today's large organizations.

New Words, Phrases and Expressions
perspective   n. 觀點；遠景；透視圖   adj. 透視的
universal   adj. 普遍的；通用的；宇宙的；全世界的；全體的
practitioner   n. 開業者, 從業者
sociologist   n. 社會學家
bureaucracy   n. 官僚主義；官僚機構；官僚政治
intend   v. 打算；準備

theorizing　v. 從理論上闡明

Notes

1. Fayol described the practice of management as something distinct from accounting, finance, production, distribution, and other typical business functions.

法約爾認為管理的實踐是不同於會計、財務、生產、分銷和其他典型業務的功能的。

2. He called a bureaucracy—a form of organization characterized by division of labor, a clearly defined hierarchy, detailed rules and regulations, and impersonal relationships.

他認為官僚機構——是一種組織，具有勞動分工、明確定義的層次結構、詳細的規章制度和個人關係的特徵。

Section 3　The Functions of Management

There are four functions of management, which are planning, organizing, leading, and controlling. The planning function involves the process of defining goals, establishing strategies for achieving those goals, and developing plans to integrate and coordinate activities. Managers are also responsible for arranging work to accomplish the organization's goals. We call this function organizing. It involves the process of determining what tasks are to be done, who is to do them, how the tasks are to be grouped, who reports to whom, and where decisions are to be made. Every organization includes people, and management's job is to work with and through people to accomplish organizational goals. This is the leading function. When managers motivate subordinates, influence individuals or teams as they work, select the most effective communication channel, or deal in any way with employee behavior issues, they are leading. The final management function managers perform is controlling. After the goals are set and the plans are formulated, the structural arrangements determined, and the people hired, trained, and motivated, there has to be some evaluation of whether things are going as planned. To ensure that is going as it should, managers must monitor and evaluate performance.

Table 2.4　4 functions of management

Planning	Defininggoals Establishing strategy Developing sub-plans to coordinate activities
Organizing	Determining what needs to be done, how it will be done, and who is to do it.
Leading	Directing and motivating all involved parties and resolving conflicts.
Controlling	Monitoring activities to ensure that they are accomplished as planned.

Lesson 1 Planning

Planning is the process of deciding what objectives to pursue during a future time period and what to do to achieve those objectives. It is the primary management function and is inherent in everything a manager does. Planning enables a manager or an organization to actively affect rather than passively accept the future. By setting objectives and charting a course of actions, the organization commits itself to 「making it happen」. Without a planned course of action, the organization is much more likely to let things happen, and then react to those happenings in a crisis mode.

Planning involves selecting mission and objective and deciding on the actions to achieve them; it requires decision-making, that is, choosing a course of action from alternatives. Planning encompasses defining the organization's objectives or goals, establishing an overall strategy for achieving those goals, and developing a comprehensive hierarchy of plans to integrate and coordinate activities.

Planning can occur at all levels in an organizations. The fact that most managers plan in some form is ample evidence of planning's important in management. We can identify the following four specific benefits.

Coordinating efforts

Management exists because the work of individuals and groups inorganizations must be coordinated, and planning is one important technique for achieving coordinated effort. An effective plan specifies objectives both for the total organization and each part of the organization. By working toward planned objectives, the behavior of each part contributes to and is compatible with goals for the entire organization. Planning reduces overlapping and wasteful activities. Furthermore, when means and ends are clear, efficiency become obvious.

Preparing for change

By forcing managers to look ahead, anticipate change, consider the impact of change and develop appropriate response, planning reduce uncertainty. An effective plan of action allows room for change. The longer the time between completion of a plan and accomplishment of an objective, the greater the necessity to include contingency plans. Yet, if management has considered the potential effect of he change, it can be better prepared to deal with it. History provides some vivid examples of what can result from failure to prepare for change. The collapse of many banks and airlines in the last few years was caused in large part by lack of preparedness of management.

Developing performance standards

Plans define expected behaviors, and in management terms, expected behaviors are performance standard. As plans are implement throughout an organization, the objectives and course are the bases for standard, which can be used to assess actual performance. A manager's performance can be assessed in terms of how close his or her unit comes to accomplishing its objective; a production worker can be held accountable for doing a job in the prescribed manner. Through planning, management gets a rational, objective basis for developing performance standards. Without planning, performance standards are likely to be irrational and subjective.

Developing managers

Good planning involves the art of making difficult things simple. Planning is the mental exercise required to develop a plan. Many people believe the experience and knowledge gained throughout the development of a plan force managers to think in a future and contingency-oriented manner; this can result in great advantages over managers who are static in their thinking.

In short, there is a positive relationship between the extent of planning activities and the performance of small, as well as large businesses. Specifically, the benefitsderived from engaging in formal planning activities include the ability to develop a more complete knowledge of the environment, choose from and implement a wider range of strategy options, and engage in cooperative agreements with other firms. Thus planning is the essence of management, all other managerial function stem from planning.

New Words, Phrases and Expressions
　　objective　　n. 目標
　　inhere　　adj. 固有的；內在的；與生俱來的, 遺傳的
　　chart　　v. 繪製……的海圖；用圖表表示；制訂計劃
　　commits　　vt. 犯罪, 做錯事；把……交托給；使……承擔義務
　　react　　v. 反應；影響；反抗；起反作用；使……發生相互作用；使……起化學反應
　　alternatives　　n. 替代選擇；可供選擇的事物
　　specify　　vt. 指定；詳細說明；列舉
　　overlap　　v. 與……重疊；蓋過
　　anticipate　　vt. 預期, 期望；占先, 搶先；提前使用
　　collapse　　v 倒塌；瓦解；暴跌；使倒塌, 使崩潰；使塌陷；　n. 倒塌；失敗；衰竭
　　assess　　v. 對……進行評估
　　accountable　　adj. 有責任的；有解釋義務的；可解釋的
　　irrational　　adj. 不合理的；無理性的；荒謬的
　　mental　　adj. 精神的；腦力的；瘋的
　　derived from　　來源於

Notes

1. Planning encompasses defining the organization's objectives or goals, establishing an overall strategy for achieving those goals, and developing a comprehensive hierarchy of

plans to integrate and coordinate activities.

計劃包括確定組織的目標，制定實現這些目標的總體戰略，以及建立一個全面的計劃體系以整合和協調活動。

2. The longer the time between completion of a plan and accomplishment of an objective, the greater the necessity to include contingency plans.

完成目標的計劃和實現目標之間間隔的時間越長，那麼制定應急預案的必要性就越大。

3. Many people believe the experience and knowledge gained throughout the development of a plan force managers to think in a future and contingency-oriented manner, this can result in great advantages over managers who are static in their thinking.

許多人認為在整個開發計劃中獲得的經驗和知識迫使管理者考慮到未來，這是一種應急導向方式，相比於那些靜態思考的管理者，這可能會產生明顯的優勢。

Lesson 2　Organizing

Most work today is accomplished through organization. The term organization refers to a group of people working together in some type of concerned or coordinated effort toattain objective. In this way, an organization provides a vehicle for implement strategy and accomplishing objectives that could not be achieved by individuals working separately.

Many writers on management distinguish between formal andinformal organization. Formal organization means the intentional structure of roles in a formally designed organization, while the informal organization is a network of interpersonal relationship that arise when people working within the formal organization associate with each other. Usually, the informal organization has a structure, but it is not formal and consciously designed. Chester Barnard described an informal organization as any joint results. Managers must pay particular attention to the informal organizations and take full advantage of them as much as possible, because employees tend to ask for help from someone they are familiar with instead of from someone they know only as a name when they encounter an organizational problem.

Organizing is one of the four important management functions. The term organizing refers to the grouping of activities necessary to attain common objectives and the assignment of each grouping to a manager who has the authority required to supervise the people performing the activities. Thus, organizing is basically a process of division of labor accompanies by appropriate delegation of authority. In other words, organizing should clarify who is to do what tasks and who is responsible for what results. As a tool of allocating resources to realize the organization's goals, organizing plays a crucial role in management activities.

There are four principles of organizing as follows:

the parity principle

When one needs something done that one either cannot or chooses not to do oneself, delegation occurs. The parity principle states that authority and responsibility must coincide.

When delegating, management must delegate sufficient authority to enable subordinates to do their jobs. On the same time, subordinates can be expected to accept responsibility only for those areas within their authority.

However, a manager is often reluctant to delegate authority, for the following reasons: fear that subordinates will fail in doing the task; the belief that it is easier to do the task oneself rather than delegate it; fear that subordinates will look「too good」; humans' attraction to power; comfort in doing the tasks of the previous job held. Despite all the above reasons, there exist some strong reasons for a manager to delegate. First, the managers is freed to do other more important tasks. Second, the subordinates gain feelings of belonging and being needed. These feelings will motivate them to make any possible contribution to the organization. Delegation is also one of the best methods for developing subordinates and satisfying customers. Taco Bell is a good example. Since it realized that the way to ultimately satisfy customers was to empower its lower-level employees to make change in operational strategies and tactics, Taco Bell went from a $500 million regional company in 1982 to a $3 billion national company in 1993.

Thus, to be a good manager, one must learn to delegate. The general steps a manager can follow when delegating are as follows:

(1) Analyze how you spend your time
(2) Decide which tasks can be assigned
(3) Decide who can handle each task
(4) Delegate the authority
(5) Great the obligation
(6) Control the delegation

A rule of thumb is to delegate authority and responsibility to the lowest organization level that has the competence to accept them.

unity of command principle

The unity of command principle statesthat an employee should have one, and only one, immediate manager. An employee who has to report to two or more bosses might have to cope with conflicting demands or priorities. For instance, the difficulty of serving more than one superior arises when two managers tell the same employee to do different jobs at the same time. The employee is thus placed in a no-win situation. Regardless of which manager the employee obeys, the other will be dissatisfied. The key to avoiding such problem is to make sure employees including managers clearly understand the lines of authority. Subordinates only report to the superior who do not work directly for them. An organization chart may be of great help.

An organization chart is a chart that indicates how departments are tied together along the principal lines of authority. Most often, an organization chart clarifies lines of authority and the chain of command.

scalar principle

The scalar principle states that authority in the organization flows through the chain of managers one link at the time, ranging from the highest to the lowest ranks. It is also called the chain of command. The implement of the scalar principle in organizing is to guarantee smooth communication and unity of command.

The problem withcircumventing the scalar principle is that the link bypassed in the process many have very pertinent information. For example, suppose Jenny goes directly above her immediate boss, Luther, to Edward for permission to take her lunch break 30 minutes earlier. Edward, believing the request is reasonable, approves it, only to find out later that the other two people in Jenny's department had also rescheduled their lunch breaks. Thus, the department would be totally vacant from 12:30 to 1 o'clock. Luther, the bypassed manager, would have known about the other rescheduled lunch breaks.

However, it doesn't mean every action must process through every link in the chain. As Henri Fayol point out,「It is an error to depart needlessly from authority, but it is an even greater one to keep to it when detriment to the business ensues.」Therefore, in certain instances when strict adherence to the chain of command creates a degree of inflexibility that hinders an organization's performance, one can and should shortcut the scalar chain.

span of management principle

The span of management refers to the number of subordinates a manager can effectively manage. How many employees can a manager efficiently and effectively direct? This question received a great deal of attention from early management writers. Many practitioners and scholars, such as Sir Ian Hamilton, V. A. Graicunas and Lyndall Urwick, tried to determine a specific number, but recent studies show that the number of people who should report directly to any one person depends on many underlying factors.

Today, the span of management is increasingly determined by analyzing contingency variables, including:

(1) Management level in the organization. Usually, top managers need a narrower span than do middle managers, and middle mangers require a smaller span than do supervisors.

(2) The preferred managing style of a manager

(3) The training and experience of employees. The more training and experience employees have, the wider span of management managers can function with.

(4) The physical proximity of employee

(5) The similarity and complexity of employee tasks

(6) The degree to which standardized procedures are in place

(7) The sophistication of the organization's management information system

(8) The strength of the organization's value system

New Words, Phrases and Expressions
attain   v. 達到，實現；獲得；到達
vehicle   n. 車輛；工具；交通工具；運載工具；傳播媒介；媒介物
associate with   v. 聯合；與……聯繫在一起；和……來往
assignment   n. 分配；任務；作業；功課
delegation   n. 代表團；授權；委託
coincide   v. 一致，符合；同時發生
subordinates   n. 下級；屬下
reluctant   adj. 不情願的；勉強的；頑抗的
motivate   v. 刺激；使有動機；激發……的積極性
ultimately   adv. 最後；根本；基本上
priorities   n. 優先順序
ranging from   從……到……變動；從……到……範圍；延伸；綿亘
circumventing   v. 繞行；使陷入圈套
bypass   v. 繞開；忽視；設旁路；迂迴
pertinent   adj. 相關的，相干的；中肯的；切題的
rescheduled   v. 重新安排
adherence   n. 堅持；依附；忠誠
sophistication   n. 複雜；詭辯；老於世故；有教養
proximity   n. 親近，接近；鄰近

Notes

1. Managers must pay particular attention to the informal organizations and take full advantage of them as much as possible, because employees tend to ask for help from someone they are familiar with instead of from someone they know only as a name when they encounter an organizational problem.

管理者們必須特別注意非正式組織，並且盡可能地充分利用它們，因為當員工遇到一個組織問題時，他們傾向於從熟悉的人那裡尋求幫助，而不是從他們只知道名字的那些人那裡。

2. Management level in the organization. Usually, top managers need a narrower span than do middle managers, and middle mangers require a smaller span than do supervisors.

組織中的管理水準。通常來說，相比於中層管理者，高級管理者需要一個比較狹窄的管理幅度，並且中層管理者比監管者（工頭）來說需要一個更小的管理幅度。

Lesson 3   Leading

Leadership is an important aspect of managing. The ability to lead effectively is one of the keys to being an effective manager. The essence of leadership is followership. In other words, it is the willingness of people to follow that makes a person a leader. Moreover, people tend to follow those whom they see as providing a means of achieving their own desires, wants, and needs.

Before undertaking a study of leadership, a clear understanding must be developed of the relationships among power, authority, and leadership.

Power is a measure of a person's potential to get others to do what he orshe wants them to do, as well as to avoid being forced by others to do what he or she does not want to do. Power can have both a positive and a negative form. Positive power results when the exchange is voluntary and both parties feel good about the exchange. Negative power results when the individual is forced to change. Power in organizations can be exercised upward, downward, or horizontally. It does not necessarily follow the organizational hierarchy from top to bottom.

Authority, which is the right to issue directives and expend resources, is related to power but is narrower in scope. Basically, the amount of authority a manager has depends on the amount of coercive, reward, and legitimate power the manager can exert. Authority is a function of position in the organizational hierarchy, flowing from the top to the bottom of the organization. An individual can have power without having formal authority. Furthermore, a manager's authority can be diminished by reducing the coercive and reward power in the position.

Leadership is the ability to influence people so that they will strive willingly and enthusiastically toward the achievement of organization goals. Obtaining followers and influencing them in setting and achieving objectives makes a leader. Leader use power in influencing group behavior. For instance, political leaders often use referent power. Informal leaders in organizations generally combine referent power and expert power. Some managers rely only on authority, while others use different combinations of power.

Leaders act to help a group attain objectives through the maximum application of its capabilities. Leaders do not stand behind a group to push; they place themselves before the group as they facilitate progress and inspire the group toaccomplish organizational goals.

In fact, there is a distinction between managers and leader. People frequently use the two termssynonymously. However, they are not necessarily the same.

Managers are appointed. They havelegitimate power that allows them to reward and punish. Their ability to influence is based on the formal authority inherent in their positions. In contrast, leaders may either be appointed or emerge from within a group. Leaders can influence others to perform beyond the actions dictated by formal authority.

Should all managers be leaders? Conversely, should all leaders be managers?

Because no one yet has been able to demonstrate through research argument that leadership ability is a handicap to a manager, we can state that all managers should have managerial functions, and thus not all should hold managerial positions. The fact that an individual can influence others does not mean that he or she can also plan, organize, and control.

The skill of leading seems to be a compound of at least four major ingredients: (1) the ability to use power effectively and in a responsible manager; (2) the ability to comprehend that human beings have different motivating forces at different times and in different situations; (3) the ability to inspire; (4) the ability to act in a manner that will develop a climate conducive to arousing motivations.

The third ingredient of leadership is the ability to inspire followers to apply their full capabilities to a project. While the use of motivators seems to center on subordinates and their needs, inspiration comes from group heads, who may have qualities of charisma and appeal that give rise to loyalty, devotion, and a strong desire on the part of followers to promote what leaders want. The best example of inspirational leadership come from hopeless and frightening situations: an unprepared nation on the eve of battle, or a defeated leader undeserted by faithful followers.

The fourth ingredient of leadership has to do with the style of the leader and the organization climate the leader develops. The strength of motivation greatly depends on expectancies, perceived rewards, the amount of effort believed to be required, the task to be done, and other factors that are part of an environment, as well as on organizational climate. Awareness of these factors has led to considerable research on leadership behavior and to the development of various pertinent theories. The views of those who have long approached leadership as a psychological study of interpersonal relationships have tended to converge with the viewpoint that the primary tasks of managers are the design and maintenance of an environment for performance.

The fundamental principle of leadership is this: since people tend to follow those who, in their view, offer them a means of satisfying their personal goals, the more managers understand what motivates their subordinates and how these motivators operate, and the more they reflect this understanding in carrying out their managerial actions, the more effective they are likely to be as leaders.

Because of the importance of leadership in all kinds of organizational action, there is a considerable volume of theory and research concerning it. It is difficult to summarize such a large body of research theory. However, trait theory, behavioral theories and contingency theories are the major types of leadership theory and research.

Prior to 1949, studies of leadership were based largely on attempts to identify the traits that leaders possess. Starting with the 「great man」 theory, that leaders are born and not made, a belief dating back to the ancient Greeks and Romans, researchers have tried to identify the physical, mental, and personality traits of various leaders. That theory lost much of it acceptability with the rise of the behaviorist school of psychology.

Various researchers have identified specific traits related to leadership ability: physical traits such as energy, appearance, and height, intelligent and ability traits, personality traits such as adaptability, aggressiveness, enthusiasm, and self-confidence, task-related characteristics such as achievement drive, persistence, and initiative, and social characteristic such as cooperativeness, interpersonal skills, and administrative ability.

The discussion of the importance of traits to leadership goes on. Recently, the following key leadership traits have been identified: (1) Drive: Leaders exhibit a high effort level. They have a relatively high desire for achievement, and they are tirelessly persistent in their activities. (2) Desire to lead: Leaders have build trusting relationships between themselves and followers by being truthful and by showing high consistency between word and deed. (3) Honesty and integrity: Leaders build trusting relationships between themselves and followers by being truthful and by showing high consistency between word and deed. (4) Self-confidence: Followers look to leaders for an absence of self-doubt. Leaders, therefore, need to show self-confidence in order to convince followers of the rightness of goals and decisions. (5) Intelligence: Leaders need to be intelligent enough to gather, synthesize, and interpret large amounts of information and to be able to create visions, solve problems, and make correct decisions. (6) Job-relevant knowledge: Effective leaders have a high degree of knowledge about the company, industry, and technical matters. In-depth knowledge allows leaders to make well-informed decisions and to understand the implications of those decisions.

In general, the study of leaders' traits has not been a very fruitful approach to explaining leadership. First, not all leaders possess all the traits and many non-leaders may possess most or all of them. Second, the trait approach gives no guidance as to how much of any trait a person should have. Third, explanations based solely on traits ignore situational factors, and what is right in one situation is not necessarily right for a different situation. Last, the dozens of studies that have been made do not agree as to which traits are leadership traits. Most of these so-called traits are really patterns of behavior.

Therefore, a major movement away form trait theories began as early as the 1940s. Leadership research form the late 1940s through the mid-1960s emphasized the preferred behavioral styles that leaders demonstrated.

New Words, Phrases and Expressions
followership　n. 追隨
voluntary　adj. 自願的；志願的；自發的；故意的 n. 志願者；自願行動
exercise　vt. 鍛煉；練習；使用；使忙碌；使驚恐
horizontally　adv. 水準地；地平地
coercive　adj. 強制的；脅迫的；高壓的
legitimate　adj. 合法的；正當的；合理的；正統的；vt. 使合法；認為正當（等於 legitimize）
diminish　vt. 使減少；使變小
referent　n. 指示物；指示對象

synonymous   adj. 同義詞的；同義的，含義相同的
comprehend   vt. 理解；包含；由……組成
expectanc   n. 期望，期待
pertinent   adj. 相關的，相干的；中肯的；切題的
converge with   與……接軌
contingency   n. 偶然性；意外事故；可能性；意外開支
synthesize   v. 合成；綜合
interpret   v. 說明；解釋；翻譯
situational   adj. 環境形成的；情形的

Notes

1. Power is a measure of a person's potential to get others to do what he or she wants them to do, as well as to avoid being forced by others to do what he or she does not want to do.

權力可以衡量一個人的潛力，讓別人做他或她想要他們做的事，以及避免被他人強迫做他或她並不想做的事情。

2. Because no one yet has been able to demonstrate through research argument that leadership ability is a handicap to a manager, we can state that all managers should have managerial functions, and thus not all should hold managerial positions.

因為目前還沒有人能夠通過研究證明領導能力對管理者來說是一個障礙，我們認為所有管理者都應該具備管理功能，但並不是所有的管理者都應擁有管理職位。

3. While the use of motivators seems to center on subordinates and their needs, inspiration comes from group heads, who may have qualities of charisma and appeal that give rise to loyalty, devotion, and a strong desire on the part of followers to promote what leaders want.

看起來激勵的使用往往集中於下屬和他們的需求，鼓舞往往來源於團隊的領導者，這樣的領導者可能具有個人魅力，同時能夠激發下屬的忠誠、奉獻，促使部分下屬去達成領導者想要的東西。

4. The views of those who have long approached leadership as a psychological study of interpersonal relationships have tended to converge with the viewpoint that the primary tasks of managers are the design and maintenance of an environment for performance.

那些長期建議把領導力作為關係心理學來研究的人的觀點已經逐漸與管理者的首要任務是工作環境的構建與維護這一觀點相融合。

Lesson 4 Controlling

It is the process of monitoring, comparing, and correcting work performance. All managers should control even if their units are performing as planned because they can't really know that unless they have evaluated what activities have been done and compared actual performance against the desired standard. Effective controls ensure that activities are completed in ways that lead to the attainment of goals. Whether controls are effective, then, is determined by how well they help employees and managers achieve their goals.

The control process is a three-step process of establishing standards, monitoring results and comparing them to standard, and correcting deviation. The control process assumes that performance standards already exist, and they do. They are the specific goals created during the planning process.

First, establishing standards. Because plans are the yardsticks against which managers devise control, the first step in the control process logically would be to establish plan. However, since plans vary in detail and complexity, and since managers usually can not watch everything, special standards are established. Standards are simply criteria of performance set at the selected points in an entire planning program so that managers can receive signals about how things are going and thus do not have to watch every step in the execution of plans. In practice, management selects some critical points in the planned program, and makes the expected results criteria for control. The number of criteria should not be too many, because large costs of control would have negative effects on the organization's innovation. However, a single criterion is also not a wise choice of management, no matter how important the objective this criterion reflects. The criteria for control should be concise, accurate, and easy to measure.

Second, monitoring results and comparing them to standards. To determine actual performance, a manger must acquire information about it. The second step in control, then is monitoring results or measuring. Four common sources of information frequently used to measure actual performance are personal observation, statistical reports, oral reports, and written reports. Most jobs and activities can be expressed in tangible and measurable terms. When a performance indicator can not be stated in quantifiable terms, mangers should look for and use subjective measures. The comparing step determines the degree of discrepancy between actual performance and the standard. Some variation in performance can be expected in all activities; it is therefore critical to determine the acceptable range of variation, or control tolerances. Deviations in excess of this range become significant and receive the manager's attention. In the comparison stage, managers are particularly concerned with the size and direction of the variation. The major problem in monitoring performance is deciding when, where, and how often to inspect or check. Checks must be made often enough to provide needed information. However, some mangers become obsessed with the checking process. Monitoring can be expensive if overdone, and can result in adverse reactions from employees. Timing is equally important. The manager must recognize a problem in time to correct it.

Third, correcting deviation. The third and final step in the control process is managerial action. Managers can choose among three actions: they can do nothing, then can correct the actual performance, or they can revise the standard. A manger who decides to correct actual performance has to make another decision between immediate corrective action and once and basic corrective action. Immediate corrective action corrects problems at once and gets performance back on track. Basic corrective action asks how and why performance has deviated and then proceeds to correct the source of deviation. It is not unusual for managers to rationalize that they do not have the time to take basic corrective action and therefore must be content to perpetually put out fires with immediate corrective action. Effective managers, however, analyze deviations, and, when the benefits justify it, take the time to permanently correct significant variances between standard and actual performance. It is also possible that the variance was a result of an unrealistic standard-that it, the goal may have been too high or too low. In such cases the standard needs corrective attention, not the performance. Most approaches are based on good common sense. However, they require effort by the manager. In a book on management, Rue and Byars (2005) list 3 guidelines to help managers lessen negative reactions to control. They are as follows: (1) Make sure the standards and associated controls are realistic. Standards set too high or too low turn people off. Make sure the standards are attainable yet challenging. (2) Involve employees in the control-setting process. Many of the behavioral problems associated with controls result from a lack of understanding of the nature and purpose of the controls. People naturally resist anything new, especially if they do not understand why it is being used. Solicit and listen to suggestions from employees. (3) Use controls only where needed. As the Gouldner model suggests, there is a strong tendency to overcontrol. Periodically evaluate the need for different controls. A good rule of thumb is to evaluate every control at least annually. Changing conditions can make certain controls obsolete. Remember, overcontrol can produce negative results.

New Words, *Phrases and Expressions*
essential   adj. 重要的，基本的
derive   v. 起源，來自
variance   n. 變化，分歧
yardstick   n. 尺碼
devise   vt. 設計
complexity   n. 複雜性
execution   n. 實行，執行
critical   adj. 重要的，關鍵的
quantitative   adj. 數量的
qualitative   adj. 質量的
facilitate   vt. 使便利，推動
comparison   n. 對比，比較
in turn   反過來

discrepancy   n. 不一致，差異
unrealistic   adj. 不現實的
deviate   vt. 偏離
proceed   vt. 繼續下去
perpetually   adj. 永恆地，恆久地
in excess of   超過，比……多
over do   做得過分
timing   n. 適時
lessen   vt. 減少，減輕
solicit   vt. 請求
periodically   adv. 定時性地

Notes

1. All managers should control even if their units are performing as planned because they can't really know that unless they have evaluated what activities have been done and compared actual performance against the desired standard.

所有管理人員應進行控制，即使他們的項目正在按計劃執行，因為他們不總是真的知道，除非他們已經評估了那些已經完成的活動，並將實際的表現和期望的標準進行了對比。

2. Standards are simply criteria of performance set at the selected points in an entire planning program so that managers can receive signals about how things are going and thus do not have to watch every step in the execution of plans.

標準是在整個規劃方案設定的點的簡單性質，使管理者可以收到有關事情進展的信號，因此不必再觀察計劃執行的每一個步驟。

3. It is not unusual for managers to rationalize that they do not have the time to take basic corrective action and therefore must be content to perpetually put out fires with immediate corrective action.

管理者沒有時間採取基本的糾正措施，因此必須立即採取果斷措施來「滅火」。這對於管理者來說實在是太平常不過了。

Section 4   Current Trends and Issues of Management

Where are we today? What current management concepts and practices are shaping 「tomorrow's history」? In this section, we'll attempt to answer those questions by introducing several trends and issues that we believe are changing the way managers do their jobs: globalization, workforce diversity, entrepreneurship, managing in an e-business world, need for innovation and flexibility, quality management, learning organizations and knowledge management, and work place spirituality.

Lesson 1 Globalization

Contemporary globalization is mainly embodied in the internationalization and liberalization, generalization and planet these four aspects. Internationalization mainly refers to across national boundaries, describe the different ethnic groups and political, economic and other aspects of the differences between countries. Liberalization is often used by economists, and generalization more researchers used to culture, mainly related to the specific values: a more global world is culturally similar characteristics. Planet, is involved in the transmission of messages and the cultural security issues.

Management is no longerconstrained by national borders. BMW, a German firm, builds cars in South Carolina. McDonald's, a U. S. firm, sells hamburgers in China. Toyota, a Japanese firm, makes cars in Kentucky. Australia's leading real estate company, etc. The world has definitely become a global village!

Managers in organizations of all sizes and types around the world are faced with the opportunities and challenges of operating in a global market. Globalization is such a significant topic that we should pay more attention to it.

New Words, Phrases and Expressions
entrepreneurship   n. 企業家精神
flexibility   n. 靈活性；彈性；適應性
spirituality   n. 靈性；精神性
constrain   v. 驅使；強迫；束縛
challengers   n. 挑戰者；挑戰賽

Notes

1. Contemporary globalization is mainly embodied in the internationalization and liberalization, generalization and planet these four aspects. Internationalization mainly refers to across national boundaries, describe the different ethnic groups and political, economic and other aspects of the differences between countries. Liberalization is often used by economists, and generalization is more researchers used to culture, mainly related to the specific values: a more global world is culturally similar characteristics. Planet, is involved in the transmission of messages and the cultural security issues.

當代的全球化主要體現在國際化、自由化、普遍化和星球化這四個方面。國際化主要是指跨越國界的，描述不同民族和國家之間政治、經濟等方面的差異。自由化常常被經濟學家使用，而普遍化則更多地為文化研究者所使用，主要涉及特定的價值觀念：一個更加全球性的世界在文化上趨於同質化。星球化則涉及消息的傳播與文化安全問題。

2. Managers in organizations of all sizes and types around the world are faced with the opportunities and challenges of operating in a global market.

世界各地的各種規模和類型組織的管理者在全球市場營運下，都面臨著機遇和挑戰。

Lesson 2　Workforce Diversity

One of the major challengers facing managers in the twenty-first century will be coordinating work efforts of diverse organizations are characterized by workforce diversity: a workforce that's more heterogeneous in terms of gender, race, ethnicity, age, and other characteristics that reflect differences. How diverse is the workforce? A report on work and workers in the twenty-first century, called workforce 2020, stated that the U. S. labor force would continue its ethnic diversification, although at a fairly slow pace. Throughout the early years of the twenty-first century, minorities will account for slightly more than one-half of net new entrants to the U. S. workforce.

The fastest growth will be Asian and Hispanic workers. However, this report also stated that a more significantdemographic force affecting workforce diversity during the next decade will be the aging of the population. This trend will significantly affect the U. S. work-force in three ways. First, these aging individuals may choose to continue full-time work, work part-time, or retire completely. Think of the implications for an organization when longtime employees with their vast wealth of knowledge, experience, and skills choose to retire, or imagine workers who refuse to retire and block opportunities for younger and higher-potential employees. Second, these aging individuals typically begin to receive public entitlements. Having sufficient tax rates to sustain these programs has serious implications for organizations and younger workers since there will be more individuals demanding entitlements and a smaller base of workers contributing dollars to the program budgets. Finally, the aging population will become a powerful consumer force driving demand for certain types of products and services. Organizations in industries of potentially high market demand (such as entertainment, travel, and other leisure-time pursuits; specialized health care; financial planning, home repair, and other professional services; etc.) will require larger workforces to meet that demand whereas organizations in industries in which market demand faces potential declines may have to make adjustments in their workforces through layoffs and downsizing.

New Words, Phrases and Expressions
heterogeneous　adj. 多樣的；異種的；不均勻的
ethnicity　n. 種族劃分
entrant　n. 進入者；新會員；參加競賽者
demographic　adj. 人口統計學的；人口學
implication　n. 含義；暗示；牽連，捲入
entitlemen　n. 權利；津貼

sustain   v. 維持；支撐，承擔；忍受
layoffs   n. 裁員；解雇
downsizing   n. 精簡；縮小規模

Notes

1. One of the major challengers facing managers in the twenty-first century will be co-ordinating work efforts of diverse organizations are characterized by workforce diversity.

在 21 世紀，管理者面臨的主要挑戰之一是協調以勞動力多樣化為特徵的組織的工作。

2. Having sufficient tax rates to sustain these programs has serious implications for organizations and younger workers since there will be more individuals demanding entitlements and a smaller base of workers contributing dollars to the program budgets.

以足夠（高）的稅率來維持這些項目已經嚴重影響了組織和年輕員工，因為將會有更多的個人權利要求和較小的工人基數從而導致項目預算增加。

Lesson 3 Entrepreneurship

Practically everywhere you turn these days you'll read or hear about entrepreneurs. If you pick up a current newspaper or general news magazine or log on to one of the Internet's news sites, chances are you'll find at least one story about an entrepreneur or an entrepreneurial business. Entrepreneurship is a popular topic! But what exactly is it?

Entrepreneurship is the process whereby an individual or a group of individuals uses organized efforts and means to pursue opportunities to create value and grow by fulfilling wants and needs through innovation and uniqueness, no matter what resources are currently controlled. It involves the discovery of opportunities and the resources to exploit them. Three important themes stick out in this definition of entrepreneurship. First is the pursuit of opportunities. Entrepreneurship is about pursuing environmental trends and changes that no one else has seen or paid attention to. For example, Jeff Bezos, founder of Amazon.com, was a successful programmer at an investment firm on Wall Street in the mid-1990s. However, statistics on the explosive growth in the use of the Internet and World Wide Web (at that time, it was growing about 2,300 percent a month) kept nagging at him. He decided to quit his job and pursue what he felt were going to be enormous retailing opportunities on the Internet. And the rest, as they say, is history. Today, Amazon sells books, music, home improvement products, cameras, cars, furniture, jewelry, and numerous other items from its popular Web site.

The second important theme in entrepreneurship is innovation. Entrepreneurship involves changing, revolutionizing, transforming, and introducing new approaches—that is, new products or services or new ways of doing business. Dineh Mohajer is a prime example of this facet of entrepreneurship. As a fashion conscious young woman, she hated the brilliant and bright nail polishes that were for sale in stores. The bright colors clashed with her trendy pastel-colored clothing. She wanted pastel nail colors that would match what she was wearing. When she couldn't find the nail polish colors she was looking for, Mohajer decided to mix her own. When her friends raved over her homemade colors, she decided to take samples of her nail polish to exclusive stores in Los Angeles. They were an instant hit! Today, her company, Hard Candy, sells a whole line of cosmetics in trendy and fashionable stores across the United States—all the result of Mohajer's innovative ideas.

The final important theme in entrepreneurship is growth. Entrepreneurs pursue growth. They are not content to stay small or to stay the same in size. Entrepreneurs want their businesses to grow and work very hard to pursue growth as they continually look for trends and continue to innovate new products and new approaches.

Entrepreneurship will continue to be important to societies around the world, For-profit and even not-for-profit organizations will need to be entrepreneurial - that is, pursuing opportunities, innovations, and growth—if they want to be successful. We think that an understanding of entrepreneurship is so important.

New Words, Phrases and Expressions
exploit　v. 開發，開拓；剝削；開採
stick out　伸出；突出
explosive　adj. 爆炸的，爆炸性的；爆發性的
enormous　adj. 龐大的，巨大的；殘暴的，凶惡的
facet　n. 面；方面；小平面
innovate　v. 創新，革新；改變；創立

Notes

1. If you pick up a current newspaper or general news magazine or log on to one of the Internet's news sites, chances are you'll find at least one story about an entrepreneur or an entrepreneurial business.

如果你拿起報紙或一般的新聞雜誌或登錄互聯網的新聞網站之一，可能你至少會找到一個企業家或創業企業的故事。

2. Entrepreneurship is the process whereby an individual or a group of individuals uses organized efforts and means to pursue opportunities to create value and grow by fulfilling wants and needs through innovation and uniqueness, no matter what resources are currently controlled.

創業的過程是一個個人或一群人使用組織的努力和手段創造價值，通過創新和獨特性滿足需求而獲得成長的機會，不管目前控制的是哪些資源。

Exercises

Ⅰ. Define the following concepts

1. management
2. planning

3. organizing

4. leading

5. controlling

6. trait theory

II. True or false

1. Managers have to perform planning, organizing, leading, and controlling at the same time.

2. Managers at different levels almost spend the same time for each managerial function.

3. The classical approach to management can be broken into two subcategories: scientific management and general administrative theory.

4. Management approaches keep changing due to dynamics of the environment.

5. Planning enables a manager or an organization to passively accept rather than actively affect the future.

6. Organizing is basically a process of division of labor accompanied by appropriate delegation of authority.

7. Subordinates can be expected to accept responsibility only for those areas within their authority.

8. Managers often believe that it iseasier to delegate the task rather than do it oneself.

9. The scalar principle indicates that every action must progress through every link in the chain in all instances.

10. Leaders should have the ability to comprehend the nature of human beings.

11. An effective leader should concentrate all his attention on the organizational goals.

12. Controlling as a separate function of the management system is independent of other functions such as planning and organizing.

13. Controlling involves three basicsteps: monitoring results, comparing them to standards, and correcting deviations.

14. The more the criteria set up by the management in the control process, the better the performance of the organization is.

15. Common sources of information frequentlyused to measure actual performance are personal observation, statistical reports, oral reports, and written reports.

III. Questions and problems

1. What is an organization? Why are managers important to an organization's success?

2. What was Henri Fayol's major contribution to the management movement?

3. What are the functions of planning?

4. Define organization, and differentiate between a formal and an informal organization.

5. Explain the importance of the organizing function.

6. What are the strengths and weaknesses of the trait theory of leadership?

7. Discuss the following statement: Leaders are born and cannot be developed.

8. What is control? When is it in the management system?

9. What are the three steps involved in the control process?

10. What are the guidelines to help managers lessen negative reactions to control? Explain them.

Further Reading Comprehension

Evaluation of Strategic Management in Business

1. Introduction

The field of Strategy has evolved substantially in the past twenty-five years. Firms have learned to analyze their competitive environment, define their position, develop competitive and corporate advantages, and understand threats to sustaining advantage in the face of challenging competitive threats. Developments in the global economy have changed the traditional balance between customer and supplier. New communications and computing technology, and the establishment of reasonably open global trading regimes, mean that customers have more choices, and supply alternatives are more transparent. Businesses, therefore, need to be more customer-centric, especially since technology has evolved to allow the lower cost provision of information and customer solutions.

The understanding of demands could solve some of the problems such as physical factors and personal and social human factors relentlessly businesses are dealing with (Bucki and Suchanek 2012) (Enis 1980) (McCarthy and Perreault 1993). there are more complex aspects on the social, economic, geography and culture factors (Schiffman and Kanuk 2007) brought marketing mix and environment into the types of factors.

The complexity of the factors affecting consumer behavior and their changes in the time shows relations between external stimuli, consumer's features, the course of decision-making process and reaction expressed in choices. As a result, the survey of consumer's behavior is complicated for traditional analytical approaches.

Accordingly, in this paper, we explain and review the business strategies and demonstrate the strategies that an organization can consider. Then, we could choose the best strategy in the organization with AHP method. household industry was chosen for this investigation and the result of AHP process in PARS house appliance was Cost leadership.

2. Strategies in business

(1) Market Penetration

Some of organizations invest on increasing quantity of their products or services that give them more benefits. In this strategy, managers focus on advertising and improve product capacity and other factors to sell their products. It increases the amount of annual sales and consequently, will improve the position of organization.

In the past decade, the scope of utility planning processes has broadened. Traditionally, electric utility planning consisted mainly of matching expected customer load growth with new capacity or energy purchases that is made. Little attention was given to demand-side management or to supply resources than utility owned generating facilities. integrated resource planning (IRP) which considers demand - side factors, transmission and distribution, and a wide scale of supply-side factors (Berry 1993). With the demand from the consumers, the products will seem markeTable. However, the organizations should actively monitor their stock so that the service stations would never have empty stock. After knowing that their sales are increasing permanently, stock quantity can be increased. The expectation that the market share can be high could be met if the consumer is very sensitive toward the price production and distribution cost will decrease if the sales (production) increases and the competitor cannot compete in a very low price.

(2) Market Development

Market development is a strategy that seeks new groups of customers with a little deviation from present and actual customers. According to the Terms of References (ToRs), Market Development Approaches (MDA) are defined as 「any intervention undertaken by donors or implementing agencies that lead to an higher level of financial sustainability for market. The MDA Working Group finds it necessary to improve access and choice of reproductive health supplies for low and moderate income consumers in public, private and commercial sectors」. Thus, MDAs can be seen as initiatives which work towards the outcomes of higher financial sustainability, developed access and developted choice (Gardiner, Schwanenflugel et al. 2006).

A marketing plans lead organizations to achieve one or more marketing objectives. It can be for a productor service, a brand, or a product line. Development in marketing plan, objectives and marketing activities should be clear, quantifiable, focused and realistic.

(3) Product Development

Organization that focuses on product or service development, intendsto improve its products or produce new products. Using this strategy, organization can sell more products to its customer and can make new chance in market for itself. Ocean spray paid attention to this strategy and created innovation in the can of juice and made mix poTable of some concentrate and nuts.

The new product development (NPD) process consists of the activities carried out by firms when developing and begin to produce new products. A new product that is introduced on the market will improve over a sequence of stages, beginning with an initial product concept or idea that is considered, reformed, tested and started to produce in the market (Booz, Allen et al. 1982). NPD focuses on the importance of introducing new products on the market for lasting business success. Its contribution to the growth of the companies, its affect on profit performance, and its role as a key factor in business planning (Urban and Hauser 1993, Cooper 2001, Ulrich 2003).

(4) Horizontal Integration

According to horizontal strategy, factories will be brought to increase the chance for organizations to sell their product and develop the geographies' scope for sell and increase the product lines and services. In this strategy, organizations focus on dominant other competitors and they try to buy other factories instead of improving their product and abilities.

The most important threat to competition from horizontal integrator between existing firms in a market is reducing the number of competitors for marketing.

In Horizontal integration, organizations focus on overcome other competitors and they intend to merge other factories instead of improving their product and abilities.

Horizontal integration is a part of combination and integration that many industries use for evolve. For example, when Hewlett-packard and Compaq were integrated, the new factory could increase in their products in computer's market.

(5) Strategic Alliance

A strategic alliance is an contract between firms to do business together in order to goes beyond normal company-to-company dealings, but fall short of a integrators or a full partnership (Wheelen and Hunger 2011). Organizations do not have much time to establish new market one-by one (Ohmae 1992). In today's economy forming an alliance with an existing company is very appealing. For a company, partnering with an international company can make expansion into unfamiliar territory a lot easier and stressful.

There are four potential useful aspect that international business may realize from strategicalliance that their topics are Ease of market entry, Shared risks, Shared knowledge and expertise and synergy and competitive advantage (Soares 2007).

(6) Stability Strategies

Stability strategy is used for organizations that they are satisfy of their chance in the market. In this situation, they do not choose any development strategies and they prefer to implement their normal strategies in marketing, producing and services that they used to use before.

Stability can be the best strategy for an organization because some situation require slow trend of development. When the market is saturate or it is in the low ebb, payment for implement strategy and effort for product is more than their benefits and incomes of strategies. This situation will be happen in the industries with low income and without development and with a lot of exit barriers.

Chapter 3
International Trade

Chapter Check List

When you have completed your study of this chapter, you will be able to
1. Have a brief comprehension of international trade and its theory framework.
2. Understand what the economic globalization and international organization is.
3. Have some common knowledge about international payment system.
4. Be clear about how people conduct trade between nations.

In this chapter, we will introduce some basic knowledge about international trade. There are five sections in this chapter, they are
1. Introduction to international trade.
2. Theory framework of international trade and basic concept.
3. Economic globalization and international organization.
4. International payment system.
5. Introduction to international practice.

The more important and difficult parts include the theory framework, basic concept about international trade and specific practice of international trade.

Section 1 An Introduction to International Trade

Lesson 1 What Is International Trade?

People always say that 「the world is getting smaller every day」. It means not only the convenient communication and transportation but also the growing international market to buy and sell goods, services and financial assets. It is believed that trade can be tied directly to an improved quality of life for the people of all the participate nations. It's hard to find a person on earth who has not been take part in the trade among nations whether directly nor indirectly.

Why do people choose to do business cross the borders? Can't a powerful country such as USA produce all products it needs? Why USA still has the impulsion to import goods that it can produce with full competent? Here are the reasons:

On national level, international trade can promote economic growth. Different countries have different natural, human, and capital resources and different ways of combining these resources, based on that nations are not equally efficient at producing what their residents demand. They have to trade off the opportunity cost of producing one product with others. Given a choice of producing one product or another, it is more efficient to produce the product with lower opportunity cost. Thus, if a country that can produce goods or services with same resources but lower opportunity cost, it is said to have absolute advantage in produce this kind of product. Different countries with different resources have absolute advantage in different products, if they trade, they will both gain from the transaction. Even if one country has absolute advantage in product both goods that are to be traded. One nation with comparative advantage in producing goods, it can also benefit from the specialization and trade. That is to say, international trade can greatly expend the market and specialization can help suppliers achieve economic of scale. The fast development of economic scale nowadays promotes one nation's economic growth.

For customers in different countries, they can buy goods and services with same quality but lower price, since international trade increases the competition with larger market and more suppliers and increasing competition would lower the price. Moreover, international trade provides each country a chance to obtain a wider variety of products for their consumers by trading with each other. Certainly, it will improve the living standards of the people.

As for producers, international trade will help manufacturers and distributors seek out products and services as well as components and finished goods produced in foreign countries because of the different distribution of the world's resources. In open economy, it is easier for producers to form scale economy because of specialization. Producers can sell products with lower cost to a larger market so as to expend sales and then pursuit more profits. Further, to avoid wild swings in their sales and profits, firms seek out foreign markets and procurement as a means to this end. Many firms take advantage of the fact that the timing of business cycles differs among countries. Thus while sales decrease in one country that is experiencing recession, they increase in another that is undergoing recovery.

International trade is the exchange of goods and services across national borders. It means the activity between buyer and seller in the purchase, sale and or exchange of commodities or goods in overseas markets. It includes wild range of things, not only international goods trading and foreign manufacturing, but also encompasses the growing services industry in areas such as transportation, tourism, banking, advertising, construction, retailing, wholesaling and mass communications. It includes all business transactions that involve two or more countries. In most countries, it represent a significant share of GDP. It's pretty certain that not only the larger absolute level of international trade but also the relative importance of trade has been growing for nearly every country and for all countries as a group. To measure the relative size of trade, we often use the index which compares the size of a country's exports with its gross domestic product (GDP). The higher the index is, the higher percentage of the output of final goods and services to be sold abroad is. Its increase reflects a greater international interdependence and a more complex international trade network encompassing not only final consumption goods but capital goods, intermediate goods, primary goods, and commercial services.

New Words, Phrases and Expressions
impulsion   n. 衝動；衝擊；原動力
competent   adj. 勝任的；有能力的；能幹的；足夠的
residents   n. 居民；房客
trade off   權衡；賣掉；交替使用；交替換位
specialization   n. 專門化；特殊化
procurement   n. 採購；獲得，取得
participate   v. 參與，參加；分享
encompass   v. 包含；包圍，環繞；完成
absolute level   絕對水準；絕對水準
relative importance   相對重要性

Notes

1. It includes wild range of things, not only international goods trading and foreign manufacturing, but also encompasses the growing services industry in areas such as transportation, tourism, banking, advertising, construction, retailing, wholesaling and mass communications.

國際貿易包括的範圍很廣泛，不僅包括國際貨物貿易和外國製造，還包括如交通、旅遊、銀行、廣告、建築、零售、批發和大眾傳媒等不斷增長的服務業領域。

2. On national level, international trade can promote economic growth. Different countries have different natural, human, and capital resources and different ways of combining these resources, based on that nations are not equally efficient at producing what their residents demand.

在國家層面上，國際貿易可以促進經濟增長。基於各個國家並不是同樣有效地生產他們的居民所需產品，不同的國家有不同的自然資源、人力資源和資本資源，也有不同的資源結合方式。

Lesson 2 Some Problems About International Trade

Although countries can gain a lot from international trade, trading aboard is much difficult than trading in domestic. Trading abroad means exchange of commodities with countries that have different economic structure, production conditions, level of productivity, economic policy, industrial policy and trade policy. Doing business with foreigners must overcome language barriers. Different customs and religion will lead to differences in consumption habits. At the meantime, different currency increases the complexity of the transaction. In international trade, at least one of the two parties will be denominated in foreign currencies. It would take a long period of time from conclusion of contract to settlement. If a larger change occurs in foreign exchange rates during this period, there will be exchange risk. Moreover, countries always concerned about improving their own welfare rather than the wellbeing of the universe. Thus, they tend to set up tariff barriers or non-tariff barriers to protect domestic market. In general, there still exist other problems, but the following three are the most common problems in international trade.

Trade barriers

It is generally assumed, as the famous economist David Ricardo stated in the nineteenth century, that the free flow of international trade benefits all who participate. In actual practice, however, the world has never had a completely free trading system. This is because every individual country puts controls on trade for the following reasons: (1) To make up for a deficit on balance of payment. (2) To guarantee national security. (3) To protect their national industry.

Trade barriers can be divided into tariff barriers and non-tariff barriers. Tariff barriers are the most common form of trade barriers which restrain foreign goods by means of customs duties, import surcharge and import variable duties. Tariff increases the prices of imported goods and therefore makes them less competitive with domestic goods.

With the development of international organizations which try to guide countries to conduct free trade with each other, tariffs have been lowered gradually. However, almost countries tend to use other devices to protect their own interests named non-tariff barriers (NTBs). Non-tariff barriers as defined by General Agreement on Tariff and Trade (GATT) allow an importing country to introduce measures which are necessary to protect human, animal or plant life or health. These include quota, voluntary export subsidies and a variety of other regulations and restrictions covering international trade.

Culture differences

When trade with foreign companies, it becomes more complex than promoting businesses in domestic since culture differences faced to a businessman. In fact, the transactions across national borders highlight the differences between domestic and international trade. If traders didn't pay enough attention to it, it will bring about a lot of trade conflicts.

When firms do business overseas, they come to contact with people from different cultures. These people speak different languages and they follow different customs and manners to conduct businesses. People from different culture backgrounds tend to judge the world from their own ways of looking at things. Thus, when trade with foreign people, businessman should be on alert against different local culture and business norms.

Currency conversion

Another problem in international trade is currency conversion. It would be much easier if every country in the world use the same currency. Things go athwart, doing businesses across borders always need exchange money. Different currencies represent same value but different amounts. That is to say, each currency's value is stated in terms of other currencies. The rate at which one currency will be exchanged for another is so called Exchange rate. It is also regarded as the value of one country's currency in terms of another currency. These exchange rates change every day and are constantly updated in banks and foreign exchange offices around the world. Importing and exporting firms to whom the payment is made in foreign currency can be involved in significant foreign exchange risk because of the fluctuation in exchange rate.

New Words, Phrases and Expressions
surcharge   n. 附加費；超載
bring about   引起；使調頭
on alert   警戒狀態
things go athwart   事與願違
be involved in   被捲入……中

Notes

1. Trading abroad means exchange of commodities with countries that have different economic structure, production conditions, level of productivity, economic policy, industrial policy and trade policy.
同外國交易意味著，與有不同的經濟結構、生產條件、生產力水準、經濟政策、產業政策和貿易政策的國家進行商品交換。

2. Importing and exporting firms to whom the payment is made in foreign currency can be involved in significant foreign exchange risk because of the fluctuation in exchange rate.
以外幣付款的進出口公司，有可能因為匯率的波動而被捲入嚴重的外匯風險中。

Section 2   Theory Framework of International Trade and Basic Concepts

For a long time, traditional trade that advocates free trade theory occupied the mainstream. The core of free trade is that free trade allows both parties that get involved in trading to obtain more trade interests. International trade, not only can expand the total production, but also can improve the total national consumption of both sides, so that the level of both parties' welfare improves. Therefore, countries should cancel the import and export trade restrictions and barriers, call off grants to its own exports of various privileges and benefits to achieve free competition at home and abroad.

Free trade theory began in Adam Smith's theory of absolute advantage. Smith believed that countries should specialize in the production and exports the product in which the country have 「absolute advantage」, not produces but imports goods in which it does not have 「absolute advantage」. But this theory has many basic assumptions which leads to many flaws. One flaw is that, in real life, one country with no absolute advantage of the product will still trade and benefit from international trade. It can't be explained by absolute advantage theory. In 1817, David Ricardo made it clear. According to the theory of comparative advantage, the basis of international trade is the relative differences (rather than absolute differences) of production technology, as well as differences in the relative costs arising therefrom. The less productive nation should specialize in the production and export of the commodity in which it has a comparative advantage.

Absolute advantage theory and the theory of comparative advantage is part of the classical trade theory system, known as the 「classical trade theory」. Overall, the classical trade theory suggests that differences in production technology is the cause of international trade, and demonstrates that the free trade benefits all participants, which laid the foundation for the future development of international trade theory.

Heckscher – Ohlin model, that is, factor endowment theory is called 「neoclassical trade theory」. Theory of factor endowments improved and revised the Ricardian model with single element, labor. It explained the cause of international trade from the different factor endowments between nations. Division according to their respective factor endowment can make the most effective use of the factors of production, trade on the basis of this, all countries can benefit.

After World War II, with the development of the productive forces, International trade becomes more and more large, commodity structure and regional distribution have undergone great changes. Faced with the new complex situation, it is difficult to make avery persuasive interpretation by classical theory. Further, classical and non-classical theory assume that the market is perfectly competitive, which is not coincide with the reality of contemporary international trade. Early 1950s, Leontief used input – output analysis to test H-O model by using US data and draw a totally opposite conclusion with H-O model. This test is the famous 「Leontief Puzzle」. Under the impetus of the Leontief paradox, it gave birth to a wide variety of modern trade theory. Representatively, intra – industry trade theory, technology gap theory and product cycle theory, etc.

International trade theory is an important part of international economics, also is the basis for countries to develop trade policies. International trade theory illustrates the following three main issues: the causes of trade, trade pattern and income distribution. The differences between classical, neoclassical and modern trade theory are showed in the following table.

Table 3.1 The differences between classical, neoclassical and modern trade theory

	Cause of international trade	Trade pattern	Income distribution
Classical trade theory	Different production technology	inter-industry	defy explanation
Neoclassical trade theory	Different factor endowment	inter-industry	The incomes of the abundant factor owners will increase and the owners of the scarce factor will find their incomes falling
Modern trade theory	Product differentiation / Similar demand preference / Economies of scale	intra-industry / intra-product	The rich get richer and the poor poorer

Lesson 1 Classical Theory of International Trade

1. Trade Based On Absolute Advantage

Adam Smith, a classical economist, was a leading advocate of free trade on the grounds that it promoted the international division of labor. With free trade, nations could concentrate their production on goods they could make most cheaply, with all the consequent benefits of the division of labor. According to labor theory of value, it assumes that labor is the only factor of production and is homogeneous (of one quality) and the cost or price of a good only depends on the amount of labor required to produce it. For example, if China uses less labor to produce a pair of shoes than USA, then the cost of China's production will be lower.

Smith applies his trading principle, the principle of absolute advantage: in a 2-nation, 2-product world, international specialization and trade will be beneficial when one nation has an absolute cost advantage (use less labor to produce one unit of output.) in one good. For world to benefit from specialization, each country should have a good that it is absolutely more efficient than any other traders. A nation will import those goods in which it has an absolute disadvantage; it will export those goods in which it has an absolute cost advantage. In the case of Zambia, it has an absolute advantage over many nations in the production of copper. This occurs because of the existence of reserves of copper ore or bauxite. We can see that in terms of the production of goods, there are obvious gains from specialization and trade, if Zambia produces copper and exports it to those nations which specialized in other products or services.

2. Trade Based on Comparative Advantage

Absolute advantage theory requires each nation to be the least-cost producer of at least one good that can export to its trade partner. But what if a nation is more efficient than its trading partner in the production of all goods. David Ricardo developed a principle to show trade can benefit both two nations even when one nation is absolutely more efficient in all goods.

Ricardo assumed that there are only two countries in the world and only produce two kinds of goods. Each country has a fixed endowment of resources, and all units of each particular resource are identical. These two countries are in different production techniques, which means differences on labor productivity. Also he supposed that production factor markets and product markets are perfectly competitive market.

Unlike Smith, who emphasized the importance of absolute cost differences among nations, Ricardo emphasized comparative (relative) cost differences. Ricardo's trade theory thus became known as the principle of comparative advantage. According to Ricardo's comparative-advantage principle, even if a nation has an absolute cost disadvantage in the production of both goods, a basis for mutually beneficial trade may still exist. The less efficient nation should specialize in and export the good in which it is relatively less inefficient (where its absolute disadvantage is least). The more efficient nation should specialize in and export that good in which it is relatively more efficient (where its absolute advantage is greatest).

We'll use an example to illustrate Ricardo's model. Table 3.2 reflects the production ability in each country and imply the relative value of each commodity.

Table 3.2 production ability in each country and imply the relative value of each commodity

Nation	Labor hours per unit		Price Ratio in Autarky
	Wine	Cloth	
United States	10 hr/bbl	20 hr/yd	$1W : \frac{1}{2} C$ (or $1C : 2W$)
United Kingdom	60 hr/bbl	40 hr/yd	$1W : \frac{3}{2} C$ (or $1C : \frac{2}{3} W$)

Assume that producing per unit of good, U.S. workers use 10 hours to produce wine or 20 hours to produce cloth, while U.K. workers use 60 hours to produce wine or 40 yards of cloth. In this example, U.S. has an absolute advantage in the production of both commodities. From Adam Smith's perspective, there is no basis for trade between these countries because U.S. is more efficient in the production of both goods. U.K. has an absolute disadvantage in both goods. Ricardo, however, pointed out that each nation can specialize in and export that good in which it has a comparative advantage—the United States in wine, the United Kingdom in cloth. Consider the autarky price ratios (i.e., the price ratios when country has no international trade). Within United States, 1 barrel of wine can exchange for $\frac{1}{2}$ yards of cloth, while within United Kingdom, 1 barrel of wine can exchange

$\frac{3}{2}$ yards of cloth. Thus United States stands to gain if it can specialize in wine and import cloth from United Kingdom at a ratio of 1 barrel: $\frac{3}{2}$ yards. Similarly, the United Kingdom would benefit by specializing in cloth production and exporting cloth to the United States, where it could receive 2 barrels of wine per yard of cloth instead of $\frac{2}{3}$ barrel of wine per yard at home.

Exploring the price ratios further, we will find that within the United States, 1 barrel of wine can exchange for $\frac{1}{2}$ yards of cloth, so any relative price under $\frac{1}{2}$ C have to be given up for 1W is desirable for U. K. Relevantly, U. K. will gain if its wine charge in trade for more than $\frac{3}{2}$ C. In a word, when the international price between the autarky price ratios, both countries will gain. It's important to examine the precise determination of the international price ratio or terms of trade, which Ricardo didn't do. After trade, there will be a common price between two countries since wine is imported into U. K. (new supply from U. S.) and U. S. is now demanding English cloth (new demand), the relative price of English cloth in terms of wine will rise. In U. S. the relative price of wine will rise because cloth is arriving from U. K and the English is demanding for American wine. Thus the autarky price ratios converge toward each other through trade. With trade, prices are no longer only determined by the labor theory of value but also the relative demands in the two countries.

The terms of trade reflect relative demand of two goods trading by these two nations and are important for the distribution of the gains between two countries. It can be proved that the closer the terms of trade are to a country's autarky price ratio, the smaller the gain for that country from international trade. When the terms of trade is equilibrium, it means that trade is balanced for each country, that is, import is equal to export in total value.

It can also use production-possibilities frontier (PPF) to present the Ricardian model. The production-possibilities frontier reflects all combinations of two products that a country can produce at a given point in time given its resource base, level of technology, full utilization of resources, and economically efficient production. Because all of these conditions are showed in the list of assumptions, it is clear that the Classical model assumes the countries to produce and consume on their production possibilities frontiers before trade. Furthermore, the constant-cost assumption indicates that the opportunity cost of production is the same at the various levels of production. So the PPF is a straight line whose slope represents the opportunity cost of economic production.

We use the same example, U. S has the absolute advantage in both wine and cloth production. Thus all points in its PPF curve fallin the right of that of U. K.

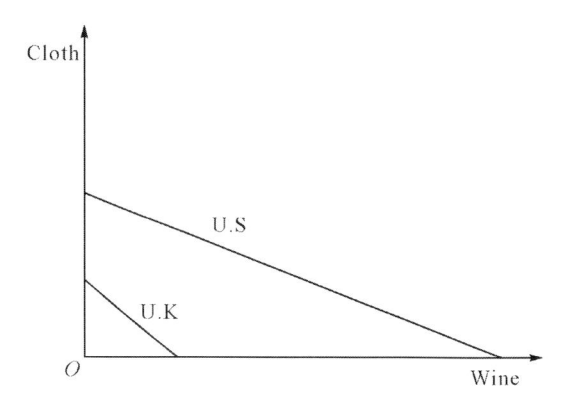

Figure 3.1 PPF of U. S. and U. K.

Consider the opportunity cost of U. S, if U. S produces one more unit of wine, half of a unit of cloth has been foregone. If U. K produces one more unit of wine, one unit and a half of cloth has been foregone. Hence, the opportunity cost will be lower if U. S specializes in wine production.

As for U. K, if U. K produces one more unit of cloth, two-third unit of wine has been foregone. When U. S produces one more unit of cloth, two units of wine has been foregone. Thus if U. K concentrates in the production of cloth, it'll gain from trade.

In conclude, U. S should produce wine and U. K cloth. The surpluses should then be traded.

New Words, Phrases and Expressions
coincide   vi. 一致；符合
impetus   n. 推動；動力
homogeneous   adj. 同種的；均勻的
copper ore   銅礦
bauxite   n. 礬土；鋁礦土
endowment of resources   資源稟賦
identical   adj. 同一的；完全相同的
specialize in   專門研究……
autarky   n. 自給自足
equilibrium   n. 均衡；平靜；保持平衡的能力
Notes
1. Therefore, countries should cancel the import and export trade restrictions and barriers, call off grants to its own exports of various privileges and benefits to achieve free competition at home and abroad.
因此，各國應該取消對進出口貿易的限制和障礙，取消給予本國出口商品的各種特權和優待，使商品自由進出口，在國內外市場上實現自由競爭。

2. Smith applies his trading principle, the principle of absolute advantage: in a 2-nation, 2-product world, international specialization and trade will be beneficial when one nation has an absolute cost advantage (use less labor to produce one unit of output) in one good.

斯密應用他的絕對優勢貿易原則：在只有兩個國家、兩種產品的世界中，如果一個國家擁有一個絕對的成本優勢（使用更少的勞動力生產一單位產出），國際分工和貿易將是有益的。

3. Heckscher - Ohlin model, that is, factor endowment theory is called「neoclassical trade theory.」Theory of factor endowments improved and revised the Ricardian model with single element, labor.

赫克歇爾-俄林模型，也就是要素稟賦理論，被稱為「新古典貿易理論」。要素稟賦理論修正和完善了李嘉圖的單一要素（勞動）模型。

Lesson 2 New Classical Theory of International Trade

1. Factor Endowment Theory

In the 1920s and 1930s, the Swedish economists Eli Heckscher and Bertil Ohlin formulated a theory addressing two questions left largely unexplained by Ricardo: (1) what determines comparative advantage? (2) What effect does international trade have on the earnings of various factors of production (distribution of income) in the trading nations? Because Heckscher and Ohlin maintained the factor endowments underlie a nation's comparative advantage, their theory became known as factor-endowment theory. It is also known as the Hecksher-Ohlin theory.

Many assumptions of Hecksher-Ohlin theory is similar to that of Ricardian model. However, there are two assumptions are critical to the H-O. That is, the factor endowments are different in each country and commodities are always intensive in a given factor regardless of relative factor prices. It has to be noticed that different factor endowments refer to different relative factor endowments, not absolute amounts of it. The definition of relative factor endowments can be divided into physical definition and price definition. The physical one explains factor abundance in terms of the physical units of the two factors, for example, capital/labor ratio. For example, two countries both have labor and capital. Relative factor endowment means that even if a country with fewer absolute units of physical capital than a larger country could still be capital intensive country as long as the amount of capital relative to labor was larger than that of big sized country. On another prospect, the price definition relies on the relative prices of capital and labor to determine the type of factor abundance characterizing the two countries. The relationship between relative factor abundance and relative price is that the greater the relative abundance of a factor, the lower its relative price. For example, Table 3.3 shows the factor conditions of U.S and China.

Table 3.3 Factor Abundance in U. S. and China

Resource	United States	China
Capital	500 machines	100 machines
Labor	1,000 workers	5,000 workers

In this table, the capital/labor ratio of U. S. is 0.5 (500 machines/1,000workers). However, the same ratio of China is only 0.02 which far less than that of United States. Thus, China is a labor–intensive country while U. S. is a capital–intensive country. Relative abundance of a resource indicates that its relative price is lower than in countries where it is relatively scarce. So U. S. has much cheaper capital and China has cheaper labor. Hence, U. S. will produce more products that use more capital and less labor. Meanwhile, China will have a lower opportunity cost if China produce goods with more labor and less capital such as shoes.

Based on the assumptions of the H–O, we can conclude that a country will export that commodity for which a large amount of the relatively abundance input is used. It will import that commodity in the production of which the relatively scarce input is used.

2. Factor Price Equalization Theory

In last lesson, we mentioned that free trade tends to converge commodity prices among trading countries. Could this the same with factor prices? Based on former theory we introduced, we know that countries will expand in its comparative advantage industries, which uses a lot of the cheap, abundant factor. As a result of the increase of demand for the abundance factor, its price will rise. Meanwhile, the scare factor price will decrease. Because this process occurs at the same time in both countries, each nation experiences a rise in the price of the abundant factor and a fall in the price of the scarce factor. Trade therefore leads to an equalization of the relative factor prices in the two trading countries.

Considering that, China imports airplanes from America which are capital intensive results in an increased in American demand for capital; the price of it therefore rises in USA. Meanwhile, China produce less airplane, hence capital price falls. Hence, the effect of trade is to equalize the price of capital in the two nations. Similarly, America demand cloth from China which is labor intensive results in an increased in Chinese demand for labor; thus wages of Chinese workers will rise and wages of American workers will fall. We conclude that by redirecting demand away from the scarce factor and toward the abundant factor in each nation, trade leads to factor–price equalization. In each country, the cheap factor becomes more expensive, and the expensive factor becomes cheaper.

We should note that factor–price equalization only happened if all the assumption of the H–O model have to hold perfectly. Especially, the most important two are the assumptions of no barriers to trade and of access to identical technology.

3. Leontief Paradox

Following the development of the Factor endowment theory (H-O), little empirical evidence was brought to bear about its validity. The first attempt to investigate the factor endowment theory empirically was undertaken by Wassily Leontief in 1954. It had been wildly accepted that in America, capital was relatively abundant and labor was relatively scarce. Based on H-O model, America should export capital-intensive goods and import labor intensive goods.

Leontief made use of his own invention-an input-output table-to test the H-O model. The table is very useful for calculating the aggregate country requirements of capital and labor for producing a bundle of goods such as exports and imports. Leontief used 1947 data and developed the Leontief statistic, which is defined as

$$\frac{(K/L)_M}{(K/L)_X}$$

$(K/L)_M$ refers to the capital (K) /labor (L) ratio used in a country to produce import-competing goods and $(K/L)_X$ refers to the capital/labor used to produce exports. Because America was thought to be a relative capital-intensive country, thus, the Leontief statistic would less than 1. However Leontief found that the capital/labor ratio for American export industries was about \$14,000 per labor year and the capital/labor ratio of its import-competing industries was about \$18,200 per labor year. Hence, the Leontief statistic for American was 1.3 ($= \frac{\$18,200}{\$14,000}$). It was unexpected for a relative capital abundant country. These findings, which contradicted the predictions of the H-O model, became known as the Leontief paradox.

New Words, Phrases and Expressions

converge　v. 使匯聚；聚集，靠攏
redirect　v. 使改方向
bring to bear　施加
validity　n. 有效性
calculate　v. 計算，預測
aggregate　adj. 聚合的，合集的

Notes

1. However, there are two assumptions are critical to the H-O. That is, the factor endowments are different in each country and commodities are always intensive in a given factor regardless of relative factor prices.

然而，H-O 有兩個關鍵假設，即一個假設是不同國家要素稟賦不同和商品總是給定資源稟賦，與相關要素價格無關。

2. Relative factor endowment means that even if a country with fewer absolute units of

physical capital than a larger country could still be capital intensive country as long as the amount of capital relative to labor was larger than that of big sized country.

相對要素禀賦意味著，即使一個國家的物質資本的絕對單位比一個大國的要少，仍然可能是資本密集型國家，只要其資本相對於勞動力的數量比大於大國的資本相對於勞動的數量比。

Lesson 3　Modern Trade Theories

1. Intra-industry Trade Theory

When a country is both exporting and importing items in the same product classification category, intra-industry trade (IIT) occurs. It is different from inter-industry trade, where a country's exports and imports are in different product classification categories. The existence of intra-industry trade appears to be incompatible with the traditional trade theory. In Ricardian theory and the factor endowment theory, a country would not simultaneously export and import the same product. However, computers manufactured by IBM are sold to Japan, meanwhile, America imports computers from Japanese companies. Nations that are net exporters of manufactured goods embodying sophisticated technology also purchase such goods from other nations. In fact, much of the intra-industry trade is conducted between countries whose factor endowments are similar. Then we look at some explanations for the occurrence of IIT.

Product differentiation.

Suppliers are willing to distinguish their products in the minds of consumers in order to buildbrand loyalty and product differentiation could satisfy consumer's numerable different taste. Some intra-industry trade emerges for this reason.

Transport costs.

Sometimes transport cost will be much cheaper if producers conduct business with neighboring country rather than domestic partners in long distance, especially if the product has large bulk relative to its value.

Degree of product aggregation.

IIT can result because of the way trade data are record and analyzed. If the category is broad, therewill be greater IIT than would be the case if a narrow category is examined. Different factor endowments and product variety.

Trading countries with different factor endowments may produce different quality of the same products. Capital-intensive countries will produce higher-quality products and labor-intensive countries will produce lower-quality products.

Intra-industry trade is an economic phenomenon that reflects the complexity of production and trade patterns in the modern world. IIT provides consumers with wider range of goods. That is another kind of gain from trade.

2. Technology Gap Theory

Rposner, an American economist, put forward the technological gap theories in 1961 and tried to explain the international trade problems from the perspective of technology changes.

Different countries didn't confirm with each other in technology development. Countries with advanced technology monopolize the new industry and enjoy the monopolistic profit. When a country develop a new product through technology innovation, it may export such product to other countries by the means of comparative advantage formed by the technology gap. The technology gap will continue until foreign countries can grasp the advanced technology gradually through the import of this new product or technology cooperation. And the technology gap will not disappear until the importing country can use their native cheaper abundant factor to imitate production and export. The innovating country will invent new products, and export in quantity to earn monopolistic profit. Then it can produce a new technology gap. At the same time, the foreign producers continue to imitate, produce and export. With the circles on, trade continues steadily. Thus, technology gap is the real cause of international trade and determines the flow of the international trade.

3. Product Cycle Theory

The product cycle theory (PCT) of trade builds on the imitation lag hypothesis in its treatment of delay in the diffusion of technology. In this theory technology innovation plays an important role as a key to determinate of trade patterns in manufactured products.

PCT was developed by Raymond Vernon in 1966. It's concerned with the life cycle of a typical「new product」and its impact on international trade. This「new product」has two basic characteristics: first is that it will cater to high-income demand; the second is that it will be labor-saving and capital-using in nature.

PCT divides the life cycle of the new product into three stages. The first stage is the new-product stage, the product is produced and consumed only in American market. As the producers familiar with the characteristics of the product and the market, the international trade will take place. The second stage is maturing-product stage, in this stage, the product's characteristics begin to emerge. The production becomes standardized and gradually economic of scale start to be realized. The European producers begin to produce it with lower labor costs and sale to American market. The third stage is standardization-product stage, by this time, the characteristics of the product and the production process are well known; the product is familiar to consumers and the production process to producers. Production may shift to developing countries. The trade pattern changes to be that the U. S. and other countries import it from developing countries.

In a word, PCT postulates a dynamic comparative advantage. At the first, the innovating country export the good but then displaced by other developed countries—which in turn are ultimately displaced by the developing countries.

New Words, Phrases and Expressions
sophisticated technology　工藝精良
monopolize　v. 壟斷，獨占
imitate　v. 模仿；效仿
postulate　v. 假定，要求
brand loyalty　品牌忠誠度
Notes
1. Nations that are net exporters of manufactured goods embodying sophisticated technology also purchase such goods from other nations.
那些複雜技術製成品的淨出口國家，也會從其他國家購買這類商品。
2. Suppliers are willing to distinguish their products in the minds of consumers in order to build brand loyalty and product differentiation could satisfy consumer's numerable different taste.
為了建立品牌忠誠度，供應商就要讓消費者在心裡區分它們的產品，而產品差異化能夠滿足消費者多種不同的偏好。
3. Sometimes transport cost will be much cheaper if producers conduct business with neighboring country rather than domestic partners in long distance, especially if the product has large bulk relative to its value.
有時候如果生產者與周邊國家開展業務，而不是與遠距離的國內夥伴合作，運輸成本將會更便宜，特別是如果產品的價值與它的體積相關時。

Section 3　Economic Globalization and International Organization

Lesson 1　Economic Globalization

When listening to the news, we often hear about economic globalization. What does this term mean?

Economic globalization is the increasing economic integration and interdependence of national, regional and local economies across the world through an intensification of cross-border movement of goods, services, technologies and capital. Economic globalization primarily comprises the globalization of production and finance, markets and technology, organizational regimes and institutions, corporations and labor. In terms of people's daily lives, economic globalization means that the residents of one country are more likely now than they were 50 years ago to consume the products of another country, to invest in another country, to earn income from other countries.

While economic globalization has been expanding since the emergence of trans-national trade, it has grown at an increased rate over the last 20—30 years under the framework of General Agreement on Tariffs and Trade and World Trade Organization, which made countries gradually cut down trade barriers and open up their current accounts and capital accounts. This recent boom has been largely accounted by developed economies integrating with less developed economies, by means of foreign direct investment, the reduction of trade barriers.

Where as globalization is a broad set of processes concerning multiple networks of economic, political and cultural interchange, contemporary economic globalization is propelled by the rapid growing significance of information in all types of productive activities and marketization, and by developments in science and technology.

Economic globalization, as all other significant social changes, has dual features, advantages as well as disadvantages.

There are many benefits that accrue from taking firms into world markets and joining the process of globalization, which include the following 5 aspects. (1) Productivity increases faster when countries produce goods and services in which they have a comparative advantage. Living standards can increase more rapidly. (2) Global competition and cheap imports keep a constraint on prices, so inflation is less likely to disrupt economic growth. (3) An open economy promotes technological development and innovation, with fresh ideas from abroad. (4) Jobs in export industries tend to pay about 15 percent more than jobs in import-competing industries. (5) Unfettered capital movements provide the United States access to foreign investment and maintain low interest rates.

Economic globalization will not present a rosy picture for all countries and firms. There are disadvantages and problems. (1) Millions of Americans have lost jobs because of imports or shifts in production abroad. Most find new jobs that pay less. (2) Millions of other Americans fear getting laid off, especially at those firms operating in import-competing industries. (3) Workers face demands of wage concessions from their employers, which often threaten to export jobs abroad if wage concessions are not agreed to. (4) Besides blue-collar jobs, service and white-collar jobs are increasingly vulnerable to operations being sent overseas. (5) American employees can lose their competitiveness when companies build state-of-the-art factories in low-wage countries, making them as productive as those in the United States.

New Words, Phrases and Expressions

integration　n. 集成；綜合
intensification　n. 強化；加劇；劇烈化
contemporary　adj. 當代的；同時代的
marketization　n. 市場化；自由市場經濟化
dual　adj. 雙重的
rosy　adj. 美好的；樂觀的

concession　n. 讓步；特權

white-collar　白領

Notes

1. While economic globalization has been expanding since the emergence of trans-national trade, it has grown at an increased rate over the last 20-30 years under the framework of General Agreement on Tariffs and Trade and World Trade Organization, which made countries gradually cut down trade barriers and open up their current accounts and capital accounts.

然而經濟全球化自從跨國貿易出現以來一直在擴大，比在關稅與貿易總協定和世界貿易組織框架下的過去20~30年間，以更快的速度增長，這使許多國家逐步減少貿易壁壘並且開放它們的存款帳戶和資本帳戶。

2. Where as globalization is a broad set of processes concerning multiple networks of economic, political and cultural interchange, contemporary economic globalization is propelled by the rapid growing significance of information in all types of productive activities and marketization, and by developments in science and technology.

然而全球化是一組廣泛的過程，涉及多個網路的經濟、政治和文化交流，在所有類型的生產活動和市場化的信息的重要性的快速增長，以及科學技術的發展，推動了當代經濟全球化的快速發展。

Lesson 2　International Organization

International trade is the trade across the borders of countries which have many differences in terms of political system, law, culture, history, economic level and so on. Different nations have their own policies to do international trade. Unfortunately, the policies of different nations differ from one other. In order to promote international trade, international organizations like the WTO emerged. They focus on trade and transnational business transactions and provide useful resources. Listed below are some of the major international organizations involved in trade and related issues.

WTO and GPA

The World Trade Organization (WTO), which came into being on January 1, 1995. The WTO provides a forum for negotiating additional reductions of trade barriers and for settling policy disputes, and enforces trade rules. Until the establishment of the WTO, GATT functioned de facto as an organization, conducting rounds of talks addressing various trade issues and resolving international trade disputes. Since the end of the Uruguay Round, there have been negotiations on topics such as telecommunications services, information technology products, and financial services.

The WTO functions as the principal international body concerned with multilateral negotiations on the reduction of trade barriers and other measures that distort competition. The WTO also serves as a platform for members to raise their concerns regarding the trade policies of their trading partners. The basic aim of the WTO is to liberalize world trade and place it on a secure basis, thereby contributing to economic growth and development.

GPA emphasizes on its high openness, which is similar with the WTO. Joining GPA also needs one-on-one negotiations with members and it is not less complicated with WTO.

GPA is a single trade agreement under the jurisdiction of the world trade organization. The participants open their government procurement market to the outside world, in order to realize the internationalization and liberalization of government procurement legal documents. GPA's goal is for their member to open its government procurement market to other members. And GPA is a plurilateral trade agreement under the WTO agreement. There is no mandatory obligation for WTO members to join the pact.

When China joined the WTO in 2001, it did not sign the GPA at the same time, and the only commitment is willing to accept the GPA, but negotiations as arranged. At the end of 2006, China officially announced that it will start GPA accession negotiations at the end of 2007.

International Chamber of Commerce (ICC)

For most of this century, the International Chamber of Commerce (ICC) has been the world's leading organization in the field of international commercial dispute resolution. Established in 1923 as the arbitration body of ICC, the International Court of Arbitration has pioneered international commercial arbitration as it is known today. The Court took the lead in securing the worldwide acceptance of arbitration as the most effective way of resolving international commercial dispute.

The dispute resolution mechanisms developed by ICC have been conceived specifically for business disputes in an international context. ICC has always led the way in providing international business with alternatives to court litigation. The ICC's Rules of Arbitration, which provide a means for settling disputes in a final and binding manner, are the best known of the services ICC affords international business for the resolution of commercial disputes anywhere in the world.

World Bank

The International Bank for Reconstruction and Development (IBRD), commonly referred to as the World Bank, is an intergovernmental financial institution located in Washington, D. C. Its objectives are to help raise productivity and incomes and reduce poverty in developing countries. It was established in December 1945 on the basis of a plan developed at the Bretton Woods Conference of 1994. The Bank loans financial resources to credit worthy developing countries. It raises most of it funds by selling bonds in the world's major capital markets. Its bonds have, over the years, earned a quality rating enjoyed only by sound governments and leading corporations. Projects supported by the World Bank normally receive high priority within recipient governments and are usually well planned and supervised. The World Bank earns a profit, which is plowed back into its capital.

IMF

International Monetary Fund (IMF) is an international financial institution proposed at the 1944 Bretton Woods Conference and established in 1946. IMF seeks to stabilize the international monetary system as a sound basis for the orderly expansion of international trade. Specifically, among other things, the fund monitors exchange rate policies of members when they experience balance of payments difficulties, and provides, financial assistance through a special 「compensatory financing facility」 when they experience temporary shortfalls in commodity export earnings.

New Words, Phrases and Expressions
preamble   n. 序文；電報報頭
de facto   實際上的
arbitration   n. 公斷，仲裁
litigation   n. 訴訟，起訴
monetary system   貨幣制度

Notes

1. The WTO functions as the principal international body concerned with multilateral negotiations on the reduction of trade barriers and other measures that distort competition.

WTO 的作用就像是旨在減少貿易壁壘的多方談判以及其他扭曲競爭方式的國際機構的委託人。

2. Specifically, among other things, the fund monitors exchange rate policies of members when they experience balance of payments difficulties, and provides financial assistance through a special 「compensatory financing facility」 when they experience temporary shortfalls in commodity export earnings.

特別地，在其他方面，當成員的國際收支平衡出現困難時，基金組織將變動成員的匯率政策；當成員經歷暫時的商品出口收入不足的狀況時，會通過一個特殊的「補償性融資貸款」提供金融援助。

Section 4　International Payment System

「Payment」 is a transmission of an instruction to transfer value that results from a transaction in the economy, and 「settlement」 is the final and unconditional transfer of the value specified in a payment instruction. Thus, if a customer pays a department store bill by check. 「Payment」 occurs when the check is placed in the hands of the department store, and 「settlement」 occurs when the check clears and the department store's bank account is credited. If the customer pays the bill with cash, payment and settlement are simultaneous. Just as each nation has its own national currency, each nation also has its own payment and settlement system. In this part, we will describe international payment system and major payment systems.

Lesson 1 Definition of International Payment System

International payments are financial transaction conducted among different countries/regions in which payments are affected or funds are transferred from one to another in order to settle accounts, debts, claims, etc. emerged in the political, economic or culture activities among them. International payment consists of trade payment and non-trade payment. To be more precise, international payments may arise from: (1) Commercial settlements, that is, trade payment. In international trade, importers in one country must make payment to exporters in another country for their imported goods. (2) Payments for the services rendered. Services rendered by individuals or enterprises in one country to those in another country must also be paid, for example, insurance premium, freight, postage, cable charges, bank commissions, etc. (3) Payments between governments. The government of one country may make payments to that of another country for political, military, or economic reasons, such as extending loans, giving aids and grants, providing disaster relief, etc. (4) Transfer of funds among countries. Following the general trend of capital internationalization in the world, capital is usually exported or imported among developed countries, among developed and developing countries, or even among developing countries by way of making investments, issuing loans, etc. (5) Others. Other international payments such as overseas remittances, educational expenses, inheritance, etc. should also be settled among countries.

Since most of the international payments originate from transaction in trade, in this section, we meanly talk about the international payment system.

Apayment system is the means whereby cash value is transferred between a payer's bank account and a payee's bank account. It includes: Policies and procedures, including rules for crediting and debiting balances; A medium for strong and transmitting payment information; Financial intermediaries for organizing information flow, carrying out value transfer instruction, and generally administering payment activities.

There are four types of value transfer – coin, check, wire, and automated clearing house. They are classified as primary payment systems. There are some other different types of secondary payment systems that convey payment information but ultimately require one of the primary payment systems to transfer value.

New Words, Phrases and Expressions

simultaneous   adj. 同時的

insurance premium   保險費

overseas remittances   海外匯款

inheritance　n. 繼承；遺傳；遺產
automated clearing house　自動清算所
ultimately　adv. 最後；根本；基本上

Notes

1.「Payment」is a transmission of an instruction to transfer value that results from a transaction in the economy, and「settlement」is the final and unconditional transfer of the value specified in a payment instruction.

「付款」是在經濟交易中轉移價值的命令,「結算」是在支付指令中最終的和無條件的價值轉移。

2. International payments are financial transaction conducted among different countries/regions in which payments are affected or funds are transferred from one to another in order to settle accounts, debts, claims, etc. emerged in the political, economic or culture activities among them.

國際收支是在不同國家/地區進行的金融交易,它的支付被影響或資金從一方轉移到另一方是為了在政治、經濟或文化活動中進行結算、債務、索賠等活動。

Lesson 2　Major Payment Systems

In the past, most of debits and credits to be cleared were in the form of paper instruments, which were cleared manually, taking several days, or even weeks, accordingly slowing down turnover of banks and exposing them to higher foreign exchange risks. Along with the ever-changing business environment and rapidly developing information technology, fund transfers between banks and mainly done through electronic fund transfer systems. Funds debited will be transferred and settled from the sending participant's (say, remitting bank's) account to the recipient's (e.g. the paying bank's) account, in real-time, almost without time interruption, and without manual operation. And the recipient bank can use the funds transferred to its account immediately. Final clearing can be done in one day interval. Nowadays, there are four vital electronic transfer systems, they are SWIFT, Fed Wire, CHIPS and CHAPS.

SWIFT (Society for Worldwide Inter-bank Financial Telecommunication) is a computerized international telecommunications which provides a network that enables financial institutions worldwide to send and receive information about financial transactions in a secure, standardized and reliable environment.

The majority of international inter-bank messages use the SWIFT network. As of September 2010, SWIFT linked more than 9,000 financial institutions in 209 countries and territories, who were exchanging an average of over 15 million messages per day (compared to an average of 2.4 million daily messages in 1995). SWIFT transports financial messages in a highly secure way but does not hold accounts for its members and does not perform any form of clearing or settlement.

SWIFT does not facilitate funds transfer; rather, it sends payment orders, which must be settled by correspondent accounts that the institutions have with each other. Each financial institution, to exchange banking transactions, must have a banking relationship by either being a bank or affiliating itself with one (or more) so as to enjoy those particular business features.

SWIFT is an international communications network connecting the world major banks, enabling its members from different countries to get in touch with each other. The system enables member banks to transmit between themselves international payments, statements and other transaction associated with international banking. The use of the network is more convenient and reliable than past methods of communication such as mail, telex and cable, and enables the banks to offer a better service to their customers.

Here are some of the merits of SWIFT as follows: (1) Faster, more Reliable Communication. SWIFT messages can be transmitted worldwide within seconds. SWIFT service is available to its members 24 hours a day. (2) Reduced transaction errors. Standardized formats SWIFT used can decrease the potential wrong information. (3) Lower transmission costs. The cost of a SWIFT transaction is much less than the cost of other methods of electronic communication. (4) Greater efficiency. SWIFT's standardized formats permit automated handling of transactions, eliminating the need for operator intervention. (5) Better statistics. Detail transaction statements provided auditable records of incoming and outgoing transactions. (6) Increased security. All SWIFT transmissions are encrypted to insure their integrity. In addition, funds transfer-type messages are verified a SWIFT「Authenticator」, thus providing further protection from fraud.

SWIFT massages are transmitted from country to country through central, interconnected to operating center located in Brussels. In turn, operating centers are connected by international data-transmission lines to regional processors in the most members.

Here is the operation process: In the first step, the appropriate SWIFT message type is selected, prepared, addressed and released by the sender to the SWIFT network through the bank's SWIFT interface. The massage is then sent to the sender's local SWIFT Regional Processor. The input Regional Processor forwards the SWIFT massage to a Slice Processor. If the message is properly formatted, it is sent to the Receiver's local SWIFT Regional Processor. (Massages that are improperly formatted are returned to the sender along with an explanation of the error.) The output Regional Processor then sends the message to the Receiver. Usually the massage is received a few seconds after it has been transmitted by the Sender. If the receiver is unable to accept the message, it is periodically retransmitted.

In order to ensure error-free identification of parties in automated systems, SWIFT developed the Bank Identifier Code (BIC), which identifies precisely the financial institution involved in financial transactions. The BIC is made of eight or eleven consecutive alphanumeric characters without any spaces of other characters, which the first four-character code being called the Bank Code. It is unique to each financial institution and can only be made up of letters. The 5th and 6th code are country in which the financial or numeral providing geographical distinction within a country, e. g. cities, states, provinces and time zones. Only a SWIFT BIC can appear in the header of a SWIFT message as Sender or Receiver.

FED WIRE stands for the Federal Reserve Wire Transfer System. It is a fund transfer system operated nationwide in the USA by the Federal System. This fund-transfer network handles transfers from one financial institution to another with an account balance held with the Fed. The transfer of reserve account balances is used for the buying and selling of the Fed funds, and credit transfers on behalf of bank customers.

Fund transfer operation under FED WIRE is as follows: (1) In marking a FED WIRE transfer, a payer gives instructions to a bank in which the payer has an available balance. (2) The paying bank passes instructions on to the Fed to move value from the bank's reserve balance account to the reserve balance account of another bank in which the payee has an account. (3) Tough initially generated by voice, paper instructions, or electronic means, the actual transfer of value is merely a bookkeeping entry at the Fed. The Fed credits the reserve account of the payee's bank and debits the reserve account of the payer's bank. (4) Wires also provide a confirmation number to the payer so that the transaction can be traced. (5) When a bank receives an incoming wire, the receiving firm is given notification that value has been received. This is also an important for many types of transactions. No other mechanism provides confirmation help and notification to the parties in a payment transaction.

However, FED WIRE does not permit much additional information to be carried along with the wire information. It is not uncommon for a firm to receive a wire payment and to be in the dark concerning the purpose of the payment.

CHIPS stands for the Clearing House Inter-bank Payment System. It was developed in 1979 by the New York Clearing House Association for transfer of international dollar payments. Operated real-time, it is a final payment system for business-to-business transactions, linking about 140 depository institutions, called also settling banks, that have offices or affiliates in New York City. About 90% of all international inter-bank dollar-denominated transactions. Most international dollar-based transactions flow through New York correspondent banks that clear transactions through CHIPS.

CHIPS works as a netting system. This means that only information, and not values, is transferred during a specified period. At the end of the period (for CHIPS the period is one business day) only the net amount is actually transferred from one party to the other.

The core of CHIPS is the Universal Identifier Database (UID), in which all corporate customers of the member's bank has its own Universal Identification Number (UIN), a unique identifier that tells the CHIPS what private account and bank information to use for processing payments. It allows straight-through processing without manual lookups or costly delays.

CHIPS differs from the Fed wire payment system in three key ways. First, it is privately owned, whereas the Fed is part of a regulatory body. Second, it has 47 member participants (with some merged banks constituting separate participants), compared with 9,289 banking institutions (as of March 19, 2009) eligible to make and receive funds via Fed wire. Third, it is a netting engine.

CHAPS stands for the Clearing House Automated Payment System. It is an electronic transfer system for sending same-day value payments from bank to bank. CHAPS offer its member banks and their participants an efficient, risk-free, reliable same-day payment mechanism. Every CHAPS payment is unconditional, irrevocable and guaranteed. CHAPS is available nationwide in Britain and is operated by a number of settlement banks who communicate directly through computers.

A CHAPS transfer is initiated by the sender to move money to the recipient's account (at another banking institution) where the funds need to be available (cleared) the same working day. Unlike with a bank giro credit, no pre-printed slip specifying the recipient's details is required. Unlike checks, the funds transfer is performed in real-time removing the issue of float or the potential for payments to be purposefully stopped by the sender, or returned due to insufficient funds, even after they appear to have arrived in the destination account.

As well as making transfers originated by banks themselves, CHAPS is frequently used by businesses for high-value payments to suppliers, by mortgage lenders issuing advances, and by solicitors and conveyancers on behalf of individuals buying houses.

CHAPS operates in the following way: Each bank sendsinstructions of the payments to the other banks through a so-called 「gateway」. This uses common software and there is no central executive on computer. Once the messages have been accept in the system as being authentic, they cannot be revoked. This allows same-day transfer throughout the clearing system. Larger companies may be able to link their own technology into the system and operate payments and receipts electronically. Small companies can instruct their banks to make these payments in the usual way—by telephone call. CHAPS means that the expensive telegraphic transfer no longer needs to be used. At the same time, treasurers of companies are able to monitor their cash positions more easily. In 2000. CHAPS Sterling processed an average daily volume of 81,738 with a daily average value of USD 189 billion.

Each clearing system introduced previously has its own characteristics as well as limitations. CHIPS and CHAPS are connected with their own central bank respectively, easily clearing the net position for its settling members, yet limiting the number of participating members on the basis of qualification or location and by operating only in the local business time. Working in different currencies, in more than 10 languages, 24 hours a day and 7 days a week, SWIFT is the most popular international payment system in the world, even though it cannot yet make the ultimate net debit or credit settlement at present.

New Words, Phrases and Expressions
territory   n. 地區；領土；邊疆區
correspondent   n. 代理商行；客戶；通訊記者
integrityn   n. 完整；正直；誠實
Fed wire   聯邦結算系統
depository   adj. 保管的，存儲的
eligible   adj. 合格的，合適的；符合條件的
consolidate   v. 鞏固，加強；聯合
bank giro credit   銀行轉帳信貸
mortgage   v. 抵押
conveyancer   n. 運輸者

Notes

1. Each financial institution, to exchange banking transactions, must have a banking relationship by either being a bank or affiliating itself with one (or more) so as to enjoy those particular business features.

每一個進行外匯交易的金融機構，必須有一種銀行業務聯繫，要麼成為一家銀行，要麼隸屬於一個或多個銀行機構，以便享受那些特定的業務功能。

2. Funds debited will be transferred and settled from the sending participant's (say, remitting bank's) account to the recipient's (e.g. the paying bank's) account, in real-time, almost without time interruption, and without manual operation.

資金將即時地從參與方的帳戶（也就是匯出行），轉移到接收方的（例如付款銀行）帳戶，幾乎在不需要人工操作的情況下，無時間中斷地完成。

3. In order to ensure error-free identification of parties in automated systems, SWIFT developed the Bank Identifier Code (BIC), which identifies precisely the financial institution involved in financial transactions.

為了確保沒有錯誤識別的自動化系統，SWIFT 發展了銀行識別碼（BIC），可以精確地識別涉及金融交易的金融機制。

Section 5　International Trade Practice

Lesson 1　A Brief Introduction to International Trade Practices

International Trade Practices, mainly talking about the specific process of international commodity exchange, generally include preparation for exporting, business negotiation and performance of contract.

Any exporter wants to sell his products in a foreign market must conduct plenty of market researches firstly. Only has an eye for the needs of foreign market, the methods by which the products can be supplied, the market structures and the general economic conditions in those places, can the exporter decide whether there are good prospects for his goods in the foreign market, After that, exporters need to develop a series of marketing initiatives to make his products known to others.

If a foreign company is interested in buying exporter's goods, negotiation should be organized. Business negotiation in international trade is the process by which the seller and buyers negotiate about the trade terms of a transaction with the intent of reaching an agreement about the sales of goods. Business negotiation usually follow such a procedure: enquiry, offer, counter-offer, acceptance and conclusion of sales contract. Among which offer and acceptance are two indispensable links for reaching an agreement and concluding a contract.

Should the business negotiation process between the two sides carried out smoothly, parties to the transaction signed a contract. Next is the part of the exporter to fulfill the contract. The implementation of export contract usually goes through the steps of goods preparation, inspection application, reminding of L/C, examination and modification of L/C, chartering and booking space shipping, shipment, insurance, documents preparation for bank negotiation and the settlement of claims, etc. As it can be seen from the above, the procedures of an export or import business is so complicated that it may take a long time to conclude a transaction. In this process, a substantial start point of trading is the conclusion of the contract. When a contract is formed, which creates legal obligations enforceable by law. Parties to the transaction must be in strict accordance with the contract and fulfill their obligations to ensure the successful completion of the transaction. It is the best way to avoid dispute may occur.

Since the importance of a contract can't be underestimated in an international sale, in this section, we are going to meanly talk about the clause stipulated in sales contract. It includes: e, offer, c.

Lesson 3   Main Elements of Sales Contract

The international sales contract is a legal document to bound rights and obligations of buyers and sellers. Terms of the contract is a core part of the contract, meanly include commodity description, quality clause, package clause, shipping clause, insurance clause, payment clause. These terms must be explicitly stipulated, try to exclude the impact of language and culture. Certainly, conclusion of the contract must not ignore the responsibilities and obligations of each parties, in particular, after the default, the responsibility shall bear. If some terms of the contract to write very vague or in a very general way, the defaulters cannot be held responsible, even the responsibilities and obligations of each parties are provided.

Commodity description

Specific goods must be accurately described and adopt widely accept names. This is sometimes done with standard quality designation. For Santos coffee, for instance, the following specialist terms are used: minimum, regular, good, superior, prime and extra prime. Everyone in the coffee trade knows what it means.

Quality clause

The quality of goods refers to the intrinsic attributes and outer form or shape the goods, such as modeling, structure, color, luster, taste, chemical composition, mechanical performance, biological feature, etc. The quality of commodity is not only one of the major terms of sales contract, but also the first item which should be agreed upon by the exporter and importer while the business is being negotiated.

In international trade, there are two ways to indicate the quality of the goods either by description or by sample. Most goods are sold by the method of sale by specification, grade or standard in international trade. The specification of goods refers to certain major indicators which indicate the quality of goods, such as composition, content, purity, size, length, etc. The grade of goods refers to the classifications of the commodity of one kind which is indicated by words, numbers or symbols. Other methods include sale by brand name or trade mark, sale by place of origin, sale by description and illustration.

Sale by sample refers to the transaction is done by the sample agreed by both parties. A sample is a small quantity of a product, often taken out a whole lot or specially designed and processed. If goods are sold by sample, the goods must be in fully conformity with the sample.

Sale by sample includes three cases: Sale by the seller's sample, Sale by the buyer's sample, Sale by the counter sample.

According to international trade laws and practices, the seller must deliver goods which are of the quality and description required by the contract. If the seller fails to, the buyers has the right to lodge a claim against the seller or even cancel the contract. However international trade usually has a long delivery process both in time and distance. Thus, the quality of the goods may be effected by many factors. Usually, the delivered goods can't fully conformity with the stipulation in contract. Therefore, while concluding the contract, it is necessary to add some flexible conditions such as quality latitude, quality tolerance. Quality latitude means that the permissible range within which the quality of the goods delivered by the seller may be flexibly controlled. Quality tolerance refers to the quality deviation recognized, which allows the quality of the goods delivered to have certain difference within a range.

Quantity clause

Quantity of goods refers to the weight, number, length, volume, area, capacity, etc. which are indicated by different measuring units. Quantity clause is one of the essential terms and conditions for the conclusion of sales contract. The main quantity clause includes units of measurement, which varies with type of goods. Some common units of measurement are weight, number, length, area, volume and capacity.

Weight is one of wildly used units of measurement in international business. Agricultural and sideline products, minerals, manufactured goods such as steel, rice, oil, etc. use weight to measure. Common measurement units are kilogram, gram, ton, pound, ounce, long ton, short ton, etc. As for number, mostly applied to industrial finished goods, especially light industry goods, engineering goods, native produce & animal by-products, etc. Common measurement units are piece, pair, set, dozen, gross, ream, roll, unit, head, etc. case, carton, bundle, barrel, bottle, bag are used as package units. In the trade of goods like textile and metal cord, etc. it usually uses length units such as meter, foot, mile, yard, inch etc. Area is often used in trade of glass, textile product such as carpets, etc. like square meter, square yard, square foot. Goods traded by volume is limited, it is generally used in timber, chemical gas, etc. it includes cubic meter, cubic foot, cubic yard and so on. Capacity is mostly used for grain, corn, wheat, gas, gasoline, petroleum. The commonly used capacity units are liter, gallon, bushel and so on.

While concluding the sales contract, the quantity clause should be clearly and definitely stipulated so as not to give rise to disputes. However, some goods cannot be in conformity with the quantity stipulated in contract accurately since the influence of natural conditions, limitations of packaging and transportation conditions. Thus, a more or less clause is necessary added in sales contract. The quantity clause therefore consists of three parts: the quantity of the deal+ the measurement unit+ a more or less clause. For example, 500 M/T with 5% more or less at seller's option.

Package clause

No matter what kind of transport mode is used, the packaging of goods is necessary. According to CISG, the seller should deliver goods which are contained or packaged in the manner required by contract. In today's world, the significant of packing has been increasingly recognized for the reason that packing has a lot of merits: Reasonable packing can prevent the goods from being stolen or damaged. Packing make it convenient to store, transport, load, unload and count the goods. It also can optimize the use of shipping space, thus reduce the transportation cost. Sales packing provides retailers easier distribution and end customers convenient purchase and delivery. Package clauses include packing, marking.

There are some kinds of packing such as nude cargo, bulk cargo, packed cargo. Nude cargoes refers to those kinds of cargoes whose qualities are stable and to be shipped without any package or in simple bundle. Outside circumstance can hardly influence them. They are not easy to be packed or do not need to be packed. For instance, goods like steel, timber, rubber, automobile, etc. Cargo in bulk refers to goods which are shipped or even sold without packages on the conveyance in bulk, such as oil, grain, coal, etc. Most of commodities in international trade need certain degree of packing during shipment and sales process. According to the functions of the goods in the process of circulating and the packing materials and methods, packing can be divided into transport or shipping packing (also called outer packing), sales packing and neutral packing. It worth mentioning that neutral packing means that there is neither a name of origin, nor a name and address of manufacture, nor a trade mark, a brand, or even any words on the (outer or inner) packing of the commodity and the commodity itself.

Besides packing, marking is also an important package clause. Marking refers to different diagrams, words and figures which are written, printed, or brushed on the outside of the shipping packing. It can divided into shipping marks, indicative marks, warning marks. Among them, shipping marks are much more important. Shipping marks are marks printed on the shipping packing. It quickens the identification and transportation of the goods and help avoid shipping errors. Shipping mark consists of: consignee's code, destination, reference No. i. e. contract number, license number and number of packages.

Shipping clause

The shipment of goods stipulated in contract is a work of wide involved aspects, long distance and strong timeliness. Therefore, sellers and buyers need to protocol the shipping clause with fully attention. Especially the details such as shipment date, port of shipment and destination, partial shipment, transshipment, etc.

As to methods of delivery, there are many types of transportation methods, such as ocean transport, railway transport, air transport, river and lake transport, postal transport, road transport, pipelines transport, land bridge transport and international multimode transport and so on. We are going to introduce ocean transport in details.

Ocean transport is the most wildly used in international trade. It is quite cheap for delivery large amount of goods over long distance. According to ways of operation, ocean transport can be divided into liner transport and charter transport.

Liner transport is a passenger or cargo vesselthat operates over a regular route according to an advertised timetable. It has the following characteristics: (1) Fixed route, ports, schedule and relatively fixed freight; (2) Loading and unloading charges included in the freight; (3) Simple procedures and ideal for cargoes of small quantity.

Liner freight is the remuneration payable to the carrier for the carriage of goods. Normally, the freight of liner service is made up of two parts. One is the basic freight; the other is the surcharges and additionals. Basic freight is charged for carriage of goods from the port of shipment to port of destination. It constitutes the main part of the transportation cost. Payment of surcharged and additional fees may be required on some special occasions or for some cargoes that need special care in handling. Total freight can be calculating using the following formula:

Total freight = Total quantity× Basic rate × (1+ Surcharged/ Additionals rate)

Charter shipment refers to a freight-carrying vessel which has no regular route or fixed schedule of sailing. The shipper charters the ship from the ship-owner and uses it to carry goods. The charter rate is determined by quantities of the goods carried and fluctuates with market conditions of supply and demand.

Insurance clause

Because of perils of the sea, losses and charges, the insurance clauses are important in international trade. As for perils of the sea, marine risk refers to the dangers existing during the course of the ocean transport. It includes three types of risks: (1) Natural calamities are caused by the forces resulting from the changes of nature, e. g., vile weather, thunder, lightning, Tsunami, earthquake, flood, etc. (2) Fortuitous accidents include accidents resulting from unexpected causes, the carrying conveyance being grounded, stranded, or in collision with floating ice of other objects, as well as fire of explosion. (3) Extraneous risks include theft, fresh, or rainwater damage, shortage, leakage, breakage, sweating and heating, intermixture and contamination, odor, hook damage, breakage of packing, rusting, etc.

As for loss, which refers to any loss or damage due to natural calamities and fortuitous accidents and related costs incurred in the process of transit. Loss can be classified into two types: total loss and partial loss. Total loss is usually classified into two kinds: one is actual total loss, which means the insured subject matter is totally and irretrievable lost. The other is constructive total loss and it is found in the case where an actual total loss appears to be unavoidable or the cost to be incurred in recovering or reconditioning the goods together with the forwarding cost to the destination named in the policy would exceed their value on arrival. Partial loss refers to the loss of part of a consignment. According to different causes, partial loss can be either general average or particular average. General average means a certain special sacrifice and extra expense intentionally incurred for the general interests of the ship owner, the insurer, and the owners of various cargoes aboard the ship. Particular average means that a particular cargo is damaged by any cause and the degree of damage dose not reach a total loss.

As for charges, maritime charges are usually occurred in saving the insured cargoes. It can be classified into two kinds: (a) Sue and labor charges. (b) Salvage charges.

As showed in figure 3.2, insurance coverage includes basic insurance coverange and additional insurance coverage. The former includes (1) Free From Particular Average (F. P. A.) (2) With particular average (W. P. A. or W. A.) and (3) All risks (ALL RI). The later inclucds general additional risks and special additional risks.

F.P.A is a kind of limited form of cargo insurance coverage. F.P.A. provides coverage

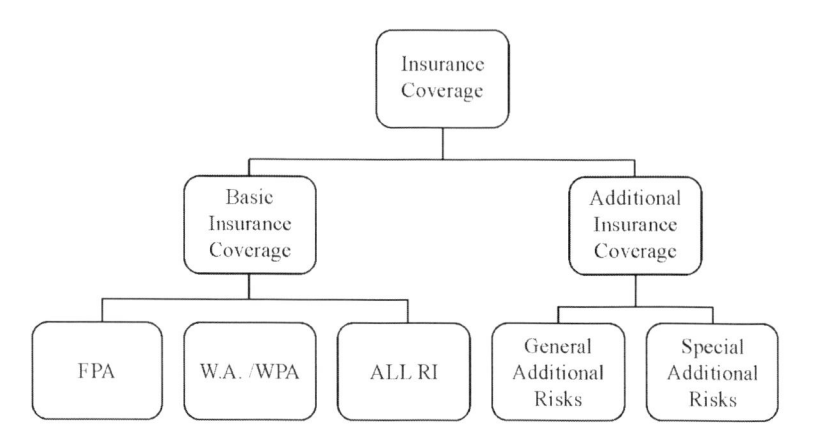

```
                    ┌─────────────┐
                    │  Insurance  │
                    │  Coverage   │
                    └──────┬──────┘
              ┌────────────┴────────────┐
        ┌─────┴─────┐              ┌────┴──────┐
        │   Basic   │              │ Additional│
        │ Insurance │              │ Insurance │
        │  Coverage │              │  Coverage │
        └─────┬─────┘              └────┬──────┘
    ┌─────┬───┴────┐              ┌─────┴─────┐
  ┌─┴─┐ ┌─┴──┐  ┌──┴──┐       ┌───┴───┐   ┌───┴───┐
  │FPA│ │W.A.│  │ALL  │       │General│   │Special│
  │   │ │/WPA│  │ RI  │       │Addit. │   │Addit. │
  └───┘ └────┘  └─────┘       │ Risks │   │ Risks │
                              └───────┘   └───────┘
```

Figure 3.2   Insurance coverage in China

for total losses and general average emerging from actual 「marine perils」. According to PICC's Ocean Marine Cargo Clauses revised in January 1, 1981, the scope of F. P. A. includes: (1) Total loss or constructive total loss of the whole consignment hereby insured caused in the cause of transit by natural calamities, such as vile weather, thunder and lightning, Tsunami, earthquake and flood. (2) Total or partial loss caused by accidents---the carrying vessel being grounded, sanded, sunk, or in collision with floating ice or other objects, as well as fire or explosion. (3) Partial loss caused by vile weather, lightening and /or Tsunami, etc. Where the conveyance has been grounded, stranded, sunk or burnt, irrespective of whether the event or events take place before or after such accidents. (4) Total or partial loss consequent on failing of entire package or package into sea during the process of loading, unloading or transshipment.

Reasonable cost incurred by the insured in salvaging the goods or averting or minimizing a loss recoverable under the policy, provided that cost shall not exceed the sum insured of the consignment so save.

Losses attributable to discharge of the insured goods at a port of distress following a sea peril as well as special charges arising from loading, warehousing and forwarding of the goods at an immediate port of call or refuge.

Sacrifice in and contribution to general average, and salvage charges resulting from the above mentioned accidents as well as other reasonable expenses in salvaging the cargoes from perils.

Such proportion of losses sustained by the ship owners as is to be reimbursed by the cargo owner under the contract of Affreightment 「Both to blame collision」 clause.

With particularaverage (W. P. A. or W. A.) covers partial loss due to vile weather, lightning, Tsunami, earthquake and/or flood as well as the risks covered under F. P. A. condition as mentioned above.

Besides the risks covered under the F. P. A. and W. P. A. conditions as above, this insurance also cover all risks of losses or damage to the insured goods whether partial or total, arising external causes in the course of transit. Depending on the quality of goods, particular transport conditions and the cost of insurance, the exporter has to decide in each case what kind of form is sufficient.

Additional risks cannot be purchased to insure goods independently. They shall be underwritten depending on one kind of basic risks. Meanwhile, it is not necessary for the goods to be insured by additional risks if it is insured by All Risks.

General additional risks have the following eleven kinds: theft, pilferage and non-delivery clause; fresh water and rain damage; shortage; intermixture and contamination; leakage; clash and breakage; odor; sweating and heating; hook and damage; loss or damage caused by breakage of packing; rust.

Special additional risk covers loss or damage caused by some special extraneous reasons such as politics, law, regulations and war.

Insurance value and insurance premium

Insurance value is the actual value of insurable cargo. It is generally calculated as cost of goods + amount of freight + insurance premium + percentage of the total sum to represent a reasonable profit for the buyer.

Insurance premium is the amount of money paid by the insured in order to make the insurance contract come into force and the basic proceeds cared by the insurer. It is usually calculated as insurance value * premium rate.

Insurance document

An insurance policy is the contract made between the insurer and the insured, which is insured by the insurer and confirmed by the insured. Insurance certificate is similar with insurance policy, however insurance policy is more comprehensive and commonly used than it. An insurance policy or certificate forms part of the chief document of transactions. In addition, in case of loss or damage, the insurance policy or certificate is the essential basis for claim and claim settlement.

Overall, insurance clause is an indispensable part of a sales contract, it should be stipulated explicitly and reasonably. The insurance clause may vary with the different trade terms. Generally, it should stipulate the insured amount, basic risk and additional risks, the insurance document required and standard of insurance coverage, etc.

In international trade practice, the buyer is obliged to pay the seller the agreed amount, in the agreed currency, within the agreed period of time and by the agreed method of payment. Thus, it is important to stipulate payment clause clearly in sales contract. Normally, in sales contract, payment clause mainly consists of four elements: time, mode, place, currency of payment.

New Words, Phrases and Expressions
intrinsic attributes　固有屬性

cubic meter　立方米
bushel　n. 蒲式耳
protocol　v. 擬定
remuneration　n. 報酬；酬勞，賠償
natural calamities　自然災害
vile weather　惡劣天氣
fortuitous　adj. 偶然的；意外的
conveyance　n. 運輸；運輸工具
strand　v. 擱淺
collision　n. 碰撞；衝突
extraneous　adj. 外來的；沒有關聯的
leakage　n. 洩露
contamination　n. 污染
pilferage　n. 偷盜，行竊；贓物

Notes

1. If some terms of the contract to write very vague or in a very general way, the defaulters cannot be held responsible, even the responsibilities and obligations of each parties are provided.

如果某些合同條款寫得非常模糊或者是非常普遍的方式，違約者將不會負責，所以每一方的責任和義務是專門列出的。

2. According to the functions of the goods in the process of circulating and the packing materials and methods, packing can be divided into transport or shipping packing (also called outer packing), sales packing and neutral packing.

根據貨物在循環中的功能以及包裝的材料和方法，包裝可分為陸運或海運包裝（也稱為外包裝）、銷售包裝、中性包裝。

3. Constructive total loss: it is found in the case where an actual total loss appears to be unavoidable or the cost to be incurred in recovering or reconditioning the goods together with the forwarding cost to the destination named in the policy would exceed their value on arrival.

推定全損：它存在於一個實際全損似乎不可避免的情況，或者註明在保險單中的恢復調整貨物的費用與運輸到目的地的費用一起，將超過它們到達時的價值的情況。

4. In international trade practice, the buyer is obliged to pay the seller the agreed amount, in the agreed currency, within the agreed period of time and by the agreed method of payment.

在國際貿易實踐中，買方有義務遵守支付賣方議定數額、議定的貨幣種類、在約定期限內和約定的付款方法的要求。

Exercise

Ⅰ. Define the following concepts

1. absolute advantage

2. factor endowment

3. intra-industry trade

4. international payment system

5. shipping mark

6. marine risk

7. total loss

8. insurance premium

II. Multiple-choice question

1. Which of the following sentences is NOT true?

A. Trade is beneficial to all the participants.

B. As long as a country has comparative advantage in something, it can gain in the international trade.

C. If a country has absolute advantage in everything, it can gain nothing in the international trade.

D. Even if a country has no absolute advantage in everything, it still can gain in international trade.

2. Which of the following statements of WTO is NOT true?

A. WTO and GATT are the current existing two big international trade agencies.

B. WTO came after GATT in 1995.

C. WTO helps to solve the international trade disputes between countries.

D. WTO is short for World Trade Organization.

3. Which of the following is considered as international trade?

A. Selling grains to Hong Kong.

B. Traveling to Japan.

C. Buying an iphone in American and taking it back to China.

D. Exporting tea to England.

4. A payment system includes policies and procedures a medium for storing and transmitting payment information and

A. Rules for crediting and debiting balance.

B. Financial intermediaries.

C. Financial transactions and information

D. Financial institution transfer

5. Each primary payment system alone is sufficient to transfer value from one party to another. Which of the following is not a primary payment system?

A. Coin and currency

B. Check

C. Wire and automated clearing house

D. Charge card

6. Which is the following does not belong to the main elements of package clause?

A. The package method

B. The mode of transportation

C. The package materials

D. The package specification

7. According to UCP600,「about」allows the quantity to be _____ more or less.

A. 3%

B. 5%

C. 10%

D. 15%

8. _____ is composed of a specific geometric figure, abbreviations of consignee, the port of destination and the package number.

A. A shipping mark

B. A warning mark

C. An indicative mark

D. A subsidiary mark

Ⅲ. Questions and problems

1. What is international trade and what benefits does international trade bring to us?

2. Use the theory of comparative advantage to explain why it pays for:

■ U. S. to export grains and import oil.

■ Russia to export oil and import grains.

3.「The factor price equalization theory indicates that with free trade the real wage earned by workers equal to the real rental rate earned by landowners.」Is this correct or not, why?

4. Name a product and analyze its product life cycle.

5. What kind of challenges are faced with China in economic globalization?

6. What is the functions of packing?

7. What are the differences among F. P. A., W. P. A. and All risks?

Further Reading Comprehension

An Export Economy that Fails to Import Jobs

This month, brace yourself to hear plenty of rhetoric coming out of Washington about 「exports」and「jobs」. As a new Republican-dominated Congress starts work, energy companies are lobbying to drop a decades-old ban on exports of crude oil, arguing that such sales will create thousands of American jobs.

As pitches go, it is a powerful one. But there is another question about exports and jobs that Congress should be debating more urgently: the fact that US businesses arebecoming so efficient that they require fewer workers than ever before to deliver growth, even — or especially — for exports.

Take a look at some fascinating data compiled by the Commerce Department, and quietly released last year. This shows that in thepast few years, the number of American jobs supported by exports has risen as overseas sales have grown. In 2009, exports created 9.7m jobs; by 2013 the tally was 11.3m.

This is cheering. And since overseas sales rose further in 2014, amid a wider economic recovery, there is every reason to think that when the commerce department publishes the 2014 tally the number of export-linked jobs will have grown again.

But there is a billion dollar catch. Look at how many jobs are being generated per dollar of sales and the graph steadily slopes down. Back in 2009, each billion dollars' worth of exports was creating 6,763 jobs. In 2013, it was 5,590 jobs. That is a fall of 17 per cent — in just four years.

There are two ways to interpret this trend. If members of Congress want to feel cheery at the start of a new year, they could celebrate the fact that American companies are becoming more innovative and competitive on the world stage. This partly reflects lower energy costs. But another factor is that as recovery has taken hold in the US, levels of automation and digitization are rising sharply too.

That is prompting more US companies to develop production inside America, since it holds down labor costs. While Chinese workers might be cheaper than their American-counterparts, robots are more cost effective than both — and often more competent.

Take Alcoa, the world's third-largest aluminum company. Having previously expanded production in places such as Mexico and Asia, it is now focusing heavily on the US. This winter, for example, it is remodeling a plant in Savannah, Georgia, that uses pioneering processes to forge metal to withstand ultra-high temperatures.

A decade ago, such work might have been placed outside America. But Klaus Kleinfeld, Alcoa chief executive, says that it now makes sense to keep it in Savannah, to be closer to customers and research units. 「The pressure to automate is huge,」 he says. 「We are not doing it for cost, but for innovation reasons, a precision is required that no human hand or eye can deliver.」

Companies such as Whirlpool, General Electric and Ford are taking similar steps. Even clothing companies are doing the same. Levi Strauss, for example, still makes its jeans outside America. But it recently brought its innovation center —and those jobs — from Turkey to San Francisco. Boston Consulting Group reckons that over half of all large US manufacturing companies are either actively re-shoring activity, or considering this, as America becomes a more competitive destination.

There is a more pessimistic twist to all this: what will the surplus workers do? An optimistic answer is that the economy will eventually adapt to generate new jobs, as it did 150 years ago when farm workers were forced off the land. A downbeat scenario is that this trend will exacerbate the bifurcation that has developed in the jobs market in the past decade, as mid-tier manufacturing jobs have disappeared. In a world of hyper-efficient companies there is swelling demand for a highly trained elite; indeed, Thursday's jobless claims data suggest that companies are actually finding it hard to hire enough skilled workers. Alcoa needs plenty of computer programmers. But what it does not need (as much) are traditional metal-bashers. Exports can boom — but with fewer blue-collar workers.

Mr. Kleinfeld has recently started working with community colleges on retraining programs, in a bid to help workers to adapt. Other companies are doing the same. But what is lamentably missing is any coherent policy from Washington to support such endeavor. Indeed, Congress seems to be paying woefully little attention to the issue, compared with the focus it is devoting to other topics, such as those energy exports. That needs to change — well before the current recovery loses steam. (From the financial times columnist Gillian Tett)

Chapter 4
International Business

Chapter Checklist

When you have completed your study of this chapter, you will be able to
1. Have a brief comprehension of international business.
2. Understand the environment of international business.
3. Be clear about international business strategies.
4. Have some common knowledge about international business operations.

In this chapter, we will introduce some basic knowledge about international business. There are four sections in this chapter, they are
1. Introduction to international business.
2. The environment of international business.
3. International business strategies.
4. International business operations.

The more important and difficult parts include the environment about international business and specific practice of international business strategy.

Section 1 An introduction to International Business

Lesson 1 What is Globalization and Global Institutions?

As used in this book, globalization refers to the shift toward a more integrated and interdependent world economy. Globalization has several facets, including the globalization of markets and the globalization of production.

The globalization of markets refers to the merging of historically distinct and separate national markets into one huge global marketplace. Falling barriers to cross-border trade have made it easier to sell internationally. It has been argued for some time that the tastes and preferences of consumers in different nations are beginning to converge on some global norm, thereby helping to create a global market. Consumer products such as Citigroup credit cards, Coca-Cola soft drinks, Sony Play Station video games, Starbucks coffee, and

IKEA furniture are frequently held up as prototypical examples of this trend. Firms such as those just cited are more than just benefactors of this trend; they are also facilitators of it. By offering the same basic product worldwide, they help to create a global market.

The globalization of production refers to sourcing goods and services from locations around the globe to take advantage of national differences in the cost and quality of factors of production (such as labor, energy, land, and capital). By using global sourcing, companies hope to lower their overall cost structure or improve the quality or functionality of their product offering, thereby allowing them to compete more effectively.

As markets globalize and an increasing proportion of business activity transcends national borders, institutions are needed to help manage, regulate, and police the global marketplace, and to promote the establishment of multinational treaties to govern the global business system.

Over the past half century, a number of important global institutions have been created to help perform these functions, including the General Agreement on Tariffs and Trade (GATT) and its successor, the World Trade Organization (WTO); the International Monetary Fund (IMF) and its sister institution, the World Bank; and the United Nations (UN). All these institutions were created by voluntary agreement between individual nation-states, and their functions are enshrined in international treaties.

The World Trade Organization (like the GATT before it) is primarily responsible for policing the world trading system and making sure nation-states adhere to the rules laid down in trade treaties signed by WTO members. The WTO is also responsible for facilitating the establishment of additional multinational agreements between WTO members. Over its entire history, and that of the GATT before it, the WTO has promoted lowering barriers to cross-border trade and investment. In doing so, the WTO has been the instrument of its members, which have sought to create a more open global business system unencumbered by barriers to trade and investment between members. Without an institution such as the WTO, it is unlikely that the globalization of markets and production could have proceeded as far as it has. The IMF was established to maintain order in the international monetary system; the World Bank was set up to promote economic development. The World Bank is the less controversial of the two sister institutions. It has focused on making low-interest loans to cash-strapped governments in poor nations that wish to undertake significant infrastructure investments.

New Words, Phrases and Expressions
barrier　　n. 障礙；屏障；界限
preference　n. 偏愛；傾向；優先權
prototypical　adj. 典型的；原型的
benefactor　n. 捐助者；施主

facilitators   n. 輔助商；促進者
transcends   vt. 超過，超越

Notes

1. Firms such as those just cited are more than just benefactors of this trend; they are also facilitators of it. By offering the same basic product worldwide, they help to create a global market.

這些公司不僅僅是潮流的倡導者，更應該是潮流的推動者。通過向世界範圍內提供同樣的基礎產品，它們正在幫助創建一個全球性市場。

2. It has focused on making low-interest loans to cash-strapped governments in poor nations that wish to undertake significant infrastructure investments.

它的重點在於向一些資金短缺的貧窮國家發放低利息貸款，用來資助它們希望從事的重大基礎設施投資項目。

Lesson 2   World Business

There are thousands of multinational enterprises that collectively perform a wide range of operations and services. However, if we were to examine what these companies are doing, we would discover that much of their activity could be classified into two major categories: (1) exports and imports and (2) foreign direct investment.

International trade is the exchange of goods and services across international borders and is also known as exports and imports. Exports are goods and services produced by a firm in one country and then sent to another country. For example, many companies in China export textile products to the United States. Imports are goods and services produced in one country and brought in by another country. UK, the city of London generates exports of financial services that are「invisibles」in the British balance of payments. In most cases people think of exports and imports as physical goods, but they also include services such as those provided by international airlines, cruise lines, reservation agencies, and hotels. Indeed, many international business experts now recognize that one of the major US exports is its entertainment and pop culture such as movies, television, and related offerings.

Information on exports and imports is important to the study of international business for a number of reasons. First, trade is the historical basis of international business and trade activities help us understand MNE practices and strategies. Second, trade helps us better understand the impact of international business on world economies. For example, Japan imports all of its oil. So when the price of oil in the world market rises sharply, we can readily predict that the cost of manufacturing cars in Japan will rise and Japanese auto exports will decline. Conversely, if oil prices decline, we can predict that world imports of Japanese cars are likely to increase. A third reason why exports and imports are important in the study of international business is that they are main drivers of international trade (FDI is another driver, which we discuss below). When worldwide exports and imports begin to slow down, this is a pretty good sign that world economies are going into a slump.

Foreign direct investment (FDI) is equity funds invested in other nations. It is different from portfolio investment in that FDI is undertaken by multinational enterprises (MNEs) who exercise control of their foreign affiliates. Like exports and imports, FDI is a driver of international business and many companies use FDI to establish footholds in the world marketplace by setting up operations in foreign markets or by acquiring businesses there. Firms engage in FDI for a number of reasons: to increase sales and profits, to enter rapidly growing markets, to reduce costs, to gain a foothold in economic unions, to protect domestic markets, to protect foreign markets, and to acquire technological and managerial know-how.

Foreign direct investment (FDI) occurs when a firm invests directly in facilities to produce or market a product in a foreign country. According to the U. S. Department of Commerce, in the United States FDI occurs whenever a U. S. citizen, organization, or affiliated group takes an interest of 10 percent or more in a foreign business entity. Once a firm undertakes FDI, it becomes a multinational enterprise.

FDI takes on two main forms. The first is a greenfield investment, which involves establishing a new operation in a foreign country. The second involves acquiring or merging with an existing firm in the foreign country. Acquisitions can involve a minority stake (in which the foreign firm takes a 10 percent to 49 percent interest in the firm's voting stock), majority stake (a foreign interest of 50 percent to 99 percent), or full outright stake (foreign interest of 100percent). FDI is more and more important to international business activities and we will discuss FDI environment later in next section.

New Words, Phrases and Expressions
multinational   n. 跨國公司；   adj. 跨國公司的；多國的
textile   n. 紡織物，織物；   adj. 紡織的
conversely   adv. 相反的
affiliate   n. 附屬公司；聯播電臺   v. 加入，與……有關；為……工作
foothold   n. 據點；立足處
acquisition   n. 兼併，購並；獲得，所獲之物

Notes

1. Indeed, many international business experts now recognize that one of the major US exports is its entertainment and pop culture such as movies, television, and related offerings.

事實上，許多國際商務方面的專家認為美國主要的出口品之一就是它的娛樂和流行文化，例如電影、電視節目以及相關產品。

2. According to the U. S. Department of Commerce, in the United States FDI occurs whenever a U. S. citizen, organization, or affiliated group takes an interest of 10 percent or

more in a foreign business entity. Once a firm undertakes FDI, it becomes a multinational enterprise.

根據美國商務部的定義，在美國，當美國公民、組織或者是其分支機構擁有外國公司的股份達 10%或以上，國際直接投資就產生了。一旦一家公司開展國際直接投資之後，它就變成了跨國公司。

Section 2 Environment to International Business

International business is much more complicated than domestic business because countries differ in many ways. Countries have different political, economic, and legal systems. Cultural practices can vary dramatically, as can the education and skill level of the population, and countries are at different stages of economic development. All these differences can and do have major implications for the practice of international business. They have a profound impact on the benefits, costs, and risks associated with doing business in different countries; the way in which operations in different countries should be managed; and the strategy international firms should pursue in different countries. A main function of this section is to develop an awareness of and appreciation for the significance of country differences in political systems, economic systems, legal systems, and national culture.

Lesson 1 National System's Differences

Political systems

There are two types of political systems: democracy and totalitarianism. Democracy is a system of government in which the people, either directly or through their elected officials, decide what is to be done. Good examples of democratic governments include the United States, Canada, England, and Australia. Common features of democratic governments include (1) the right to express opinions freely, (2) election of representatives for limited terms of office, (3) an independent court system that protects individual property and rights, and (4) a relatively non-political bureaucracy and defense infrastructure that ensure the continued operation of the system.

Totalitarianism is a system of government in which one individual or political party maintains complete control and either refuses to recognize other parties or suppresses them. A number of types of totalitarianism currently exist. Communism is an economic system in which the government owns all property and makes all decisions regarding the production and distribution of goods and services. To date, all attempts at national communism have led to totalitarianism. The best example of communism is Cuba. Another form of totalitarianism is theocratic totalitarianism, in which a religious group exercises total power and represses or persecutes non-orthodox factions. Iran and some of the sheikdoms of the Middle East are good examples. A third form is secular totalitarianism, in which the military controls the government and makes decisions that it deems to be in the best interests of the country. Examples of this are communist North Korea and Chile under Pinochet. Political systems typically create the infrastructure within which the economic system functions; in order to change the economic system, there often needs to be a change in the way the country is governed.

Economic systems

The three basic economic systems are capitalism, socialism, and mixed. However, for the purposes of our analysis it is more helpful to classify these systems in terms of resource allocation (market-driven versus centrally determined) and property ownership (private versus public). In a market-driven economy, goods and services are allocated on the basis of demand and supply. If consumers express a preference for cellular telephones, more of these products will be offered for sale. If consumers refuse to buy dot-matrix printers, these goods will cease to be offered. The US and EU nations have market-driven economies. In a centrally-determined economy, goods and services are allocated based on a plan formulated by a committee that decides what is to be offered. In these economies people are able to purchase only what the government determines should be sold. Cuba is the best example.

Market-driven economies are characterized by private ownership. Most of the assets of production are in the hands of privately-owned companies that compete for market share by offering the best-quality goods and services at competitive prices. Centrally-determined economies are characterized by public ownership. Most of the assets of production are in the hands of the state, and production quotas are set for each organization.

In examining economic systems, it is important to remember that, in a strict sense, most nations of the world have mixed economies, characterized by a combination of market driven forces and centrally-determined planning. Mixed economies include privately owned commercial entities as well as government–owned commercial entities. Governments in mixed economies typically own the utilities and infrastructural industries—railroads, airlines, shipping lines, and industries considered to be of economic and strategic importance—for instance, petroleum and copper. For example, the United States, a leading proponent of market-driven economic policies, provides health care and other social services to many of its citizens through government-regulated agencies, which gives it some aspects of central planning. Other democratic countries with mixed economies include Great Britain, Sweden, and Germany, all of which have even stronger social welfare systems than the United States.

Another example of the role of government in the economy is that of promoting business and ensuring that local firms gain or maintain dominance in certain market areas. The US and EU governments continually pressure the Chinese to open their doors toforeign MNEs, and the Chinese government is very active in helping its local firms do business with the

West.

As a result of such developments, there has been a blurring of the differences between market-driven and centrally-determined economies. The biggest change has been the willingness of the latter to introduce free-market concepts. Examples include Russia and other Eastern European countries, which began introducing aspects of free enterprise such as allowing people to start their own businesses and to keep any profits that they make. At the same time, however, many market-driven economies are increasingly adopting centrally determined ideas, such as using business—government cooperation to fend off external competitors, or the use of political force to limit the ability of overseas firms to do business in their country. In the United States, for example, the government is frequently being urged to play a more active role in monitoring foreign business practices. An instance of this can be seen in the box International Business Strategy in Action: Softwood lumber: not-so-free trade. On balance, however, we are now seeing a move from central planning to market-driven and mixed economies. The privatization movement that is taking place worldwide provides one prominent example.

New Words, Phrases and Expressions
implication　n. 含義；含蓄，含意，言外之意；捲入，牽連，牽涉，糾纏
totalitarianism　n. 極權主義
infrastructure　n. 基礎設施；基礎建設
allocation　n. 配給，分配；分配額（分配量）；劃撥的款項；撥給的場地
aellular　adj. 細胞的；蜂窩狀的；由細胞組成的；多孔的
blurring　n. 模糊，斑點甚多，（圖像的）混亂；　v.（使）變模糊；（使）難以區分

Notes
1. Another form of totalitarianism is theocratic totalitarianism, in which a religious group exercises total power and represses or persecutes non-orthodox factions.
另一種極權主義就是神權極權主義。在神權極權主義下，宗教團體具有絕對的權力，同時鎮壓和迫害著非正統的宗教派系。
2. As a result of such developments, there has been a blurring of the differences between market-driven and centrally-determined economies.
這種發展導致的後果使得市場經濟和計劃經濟之間的差別正在變得模糊。

Lesson 2　Cultural Differences

Culture can be defined as「the sum total of the beliefs, rules, techniques, institutions, and artifacts that characterize human populations」or「the collective programming of the mind」. Sociologists generally talk about the socialization process, referring to the influence of parents, friends, education, and the interaction with other members of a particular society as the basis for one's culture. These influences result in learned patterns of behavior common to members of a given society.

Definitions of culture vary according to the focus of interest, the unit of analysis, and the disciplinary approach (psychology, anthropology, sociology, geography, etc). This is significant in that studies of cultural differences adopt a specific definition and set of measurable criteria, which are always debatable. Research into culture and its impact in business and management studies is highly contentious and should not just be taken at face value, including the studies described below. There is a strong consensus, however, that key elements of culture include language, religion, values, attitudes, customs, and norms of a group or society.

Language is perhaps the most important key to understanding culture in general and the specific values, beliefs, attitudes, and opinions of a particular individual or group. English is widely accepted as the language of business; many global institutions and companies have adopted English as their official language. For many firms, such as Toyota, NEC, Hitachi, and IBM Japan, English-speaking ability is a pre-requisite for promotion. However, any assumption that speaking the same language removes cultural differences is dangerous—it normally just hides them. Moreover, a reliance on English by British and American managers, and a lack of other language skills, can weaken their ability to empathize with and adapt to other cultures.

Religion, linked to both regional characteristics and language, also influences business culture through a set of shared, core values. Protestants hold strong beliefs about the value of delayed gratification, saving, and investment. Rather than spending, consuming, and enjoying life now, their religious beliefs prompted the Protestants to look to longer-term rewards (including those in the after-life). There are parallels with the Confucian and Shinto work ethics, which also view spiritual rewards as tied to hard work and commitment to the fruits of industry. Contrasting this, a more stoic attitude among some African populations partly explains their acceptance of the ways things are, because it is the 「will of God」.

At the most general level culture can refer simply to the lifestyle and behavior of a given group of people, so corporate culture is a term used to characterize how the managers and employees of particular companies tend to behave. But the term is also used by human resource managers and senior management in their attempts to proactively shape the kind of behavior (innovative, open, dynamic, etc.) they hope to nurture in their organizations. Promoting a distinctive corporate culture is also expected to enhance the sense of community and shared identity that underpins effective organizations.

There are traditionally two different approaches to looking at culture: (1) The psychic or psychological level, which focuses on the internalized norms, attitudes, and behavior of individuals from a particular culture (psychic distance is a measure of differences between groups). (2) The institutional level, which looks at national (or group) culture embodied in institutions (government, education, and economic institutions as well as in business organizations). In this chapter we will mainly discuss the first, culture as shared psychology, with a brief reference to national institutional differences at the end.

People who are born in, or grew up in, the same country tend to share similar cultural characteristics. Nationality and culture tend to coincide, although nations encompass a wide variety of institutions, religions, beliefs, and patterns of behavior, and distinctive subcultures can always be found within individual countries. The only way to make sense of this wide diversity is to characterize distinct cultural groups through simplified national stereotypes. Many studies have attempted to create these stereotypes by mapping and comparing the shared characteristics of managers and employees in different countries. Researchers then examine the effects of key differences on business behavior, organization, structure, and ultimately the performance of companies from different countries.

Different styles of communication and interaction result from the cultural differences listed above. These can lead to workplace misunderstandings, poor interpersonal and Inter group relationships, inefficiency, and higher costs. Three examples provide some insights into how we can apply the above typologies. US managers, according to all of the above studies, are highly assertive and performance oriented relative to managers from other parts of the world (they come around the midpoint on all the other dimensions). Their interaction style is characteristically direct and explicit. They tend to use facts, figures, and logic to link specific steps to measurable outcomes, and this is the main focus of workplace interaction. Greeks and Russians are less individualistic, less performance oriented, and show lower levels of uncertainty avoidance (are less driven by procedures) than the Americans. When Russian and Greek managers, employees, customers, suppliers, or public-sector officials interact with US counterparts, they may well find the American approach too direct and results focused. For them communication is likely to be more about mutual learning and an exploration of relevant issues than an explicit agreement about specific expectations and end results. Similarly, the Swedes may find the US style too aggressive and unfriendly, working against the relationship building process that for them is a major objective of workplace interaction. The Koreans and Japanese have highly gender-differentiated societies with males tending to dominate decision making and leading most face-to-face communication. The agenda for discussion is likely set by males, and traditional language forms differ according to whether a man is addressing a woman or an older person talking to a younger person, and vice-versa. Gender- (and age-) related roles, responsibilities, and behaviors are therefore deeply embedded in language and customs. Poland and Denmark lie at the other end of the continuum on the gender-differentiation dimension. Perhaps even more than other Western managers their lack of awareness of this cultural difference runs the risk of both embarrassing female employees and offending and alienating senior Japanese male managers. This kind of clash can make negotiations and interaction of all kinds between these groups that much more difficult. Certain kinds of human resource management techniques are inappropriate for organizations that show high power distance ratings. Companies and management consultancies in the UK, the United States, and northern European countries have developed fairly participative management systems to improve productivity, based on their characteristically low power distance and flat organizational hierarchies. Techniques such as 360-degree feedback systems for developing management-employee relationships are not likely to work, however, in Mexican, Panamanian, Thai, or Russian organizations, which have high Humane orientation Cultures that emphasize helping others, charity, and People's wider social obligations power distance and steep hierarchies. Subordinates are uncomfortable being asked to evaluate senior managers, and managers would not see subordinates as qualified to comment on their performance. More than this, to employees in some countries this kind of consultation can give the impression that senior management do not know what they are doing! They may lose faith in senior management ability and leave! None of the above examples means that international managers should (or ever could) entirely change their behavior to suit local values and practices. Like many of the challenges facing managers, cultural sensitivity and cross-cultural effectiveness comes from striking a balance between one's own norms, values, and principles and those of the「foreigner」. The lesson for multinational firms is that ethnocentric corporate cultures and completely standardized HR systems do not work. The key challenge is to adapt to get the best from local differences.

New Words, Phrases and Expressions
disciplinary   adj. 紀律的；訓練的；懲罰的
empathize   v. 移情，神會
protestant   n. 新教教徒；   adj. 新教的，新教教徒的
gratification   n. 滿足，滿意；喜悅；使人滿意之事
proactively   adv. 主動地；前攝的；有前瞻性的
internalized   v. 使內在化的
ethnocentric   adj. 種族（民族、集團）優越感的
Notes
1. Sociologists generally talk about the socialization process, referring to the influence of parents, friends, education, and the interaction with other members of a particular society as the basis for one's culture.
當討論社會化進程時，社會學家將一個人的父母、朋友和與特定社會中其他成員互動而產生的對他的影響來作為他的文化基礎。
2. Greeks and Russians are less individualistic, less performance oriented, and show lower levels of uncertainty avoidance (are less driven by procedures) than the Americans.
與美國人相比，希臘人和俄羅斯人沒有那麼強的個人主義，並且較少以績效為

導向，同時不確定性規避（較少的程序驅動）的程度也更低。

Lesson 3 International Investment's Environment

We have discussed international trade environment in chapter 3 international trade. In this lesson we will introduce free trade zone.

A free trade zone is a designated area where importers can defer payment of customs duty while products are processed further (same as a foreign trade zone). Thus, the free trade zone serves as an 「offshore assembly plant」, employing local workers and using local financing for a tax-exempt commercial activity. The economic activity takes place in a restricted area such as an industrial park, because the land is often being supplied at a subsidized rate by a local host government that is interested in the zone's potential employment benefits.

To be effective, free trade zones must be strategically located either at or near an international port, on major shipping routes, or with easy access to a major airport. Important factors in the location include the availability of utilities, banking and telecom services, and a commercial infrastructure.

More than 400 free trade zones exist in the world today, often encompassing entire cities, such as Hong Kong and Singapore. More than two-thirds are situated in developing countries, and most of their future growth is expected to occur there.

The advantages offered by free trade zones are numerous and mutually beneficial to all stakeholders. For private firms, the zones offer three major attractions. First, the firm pays the customs duty (tariff) only when the goods are ready to market. Second, manufacturing costs are lower because no taxes are levied. Third, while in the zone the manufacturer has the opportunity to repackage the goods, grade them, and check for spoilage. Secondary benefits to firms take the form of reduced insurance premiums (since these are based on duty-free values), reduced fines for improperly marked merchandise (since the good can be inspected in a zone prior to customers' scrutiny), and added protection against theft (resulting from security measures in the bonded warehouses).

At the state and local levels, advantages can be realized in terms of commercial services. On a more global level, free trade zones enable domestic importing companies to compete more readily with foreign producers or subsidiaries of MNEs, thereby increasing participation in world trade. Favorable effects are felt on the balance of payments because more economic activity occurs and net capital outflow is reduced. Finally, the business climate is improved due to reduced bureaucracy and resultant savings to business capital, currently inaccessible because of the delay in paying duties and tariffs. A free trade zone is a step toward free trade. and can be an important signal by government to business that the economy is opening up. Opportunity replaces regulation, and growth of economic activity should result.

Before the establishment of more free trade zones becomes fully accepted and encouraged, governments must be convinced of their many economic benefits. Free trade zones are a vital necessity if nations are to remain competitive on an international scale. Not only will existing companies benefit from their use, but new industries will be attracted, keeping up the same benefits of world trade.

FDI environment

When discussing foreign direct investment, it is important to distinguish between the flow of FDI and the stock of FDI. The flow of FDI refers to the amount of FDI undertaken over a given time period (normally a year). The stock of FDI refers to the total accumulated value of foreign-owned assets at a given time. We also talk of outflows of FDI, meaning the flow of FDI out of a country, and inflows of FDI, the flow of FDI into a country.

Historically, most FDI has been directed at the developed nations of the world as firms based in advanced countries invested in the markets of other advanced countries. During the 1980s and 1990s, the United States was often the favorite target for FDI inflows. The United States has been an attractive target for FDI because of its large and wealthy domestic markets, its dynamic and stable economy, a favorable political environment, and the openness of the country to FDI. Investors include firms based in Great Britain, Japan, Germany, Holland, and France.

FDI can take the form of a greenfield investment in a new facility or an acquisition of or a merger with an existing local firm. The data suggest the majority of cross-border investments are in the form of mergers and acquisitions rather than greenfield investments. UN estimates indicate that some 40 to 90 percent of all FDI inflows were in the form of mergers and acquisitions between 1998 and 2007. In 2001, for example, mergers and acquisitions accounted for some 78 percent of all FDI inflows. In 2004, the figure was 59 percent, while in 2007 it was back up to 89 percent. However, FDI flows into developed nations differ markedly from those into developing nations. In the case of developing nations, only about one-third of FDI is in the form of cross-border mergers and acquisitions. The lower percentage of mergers and acquisitions may simply reflect the fact that there are fewer target firms to acquire in developing nations. When contemplating FDI, why do firms apparently prefer to acquire existing assets rather than undertake greenfield investments? Now we will make a few basic observations. First, mergers and acquisitions are quicker to execute than greenfield investments. This is an important consideration in the modern business world where markets evolve very rapidly. Many firms apparently believe that if they do not acquire a desirable target firm, then their global rivals will. Second, foreign firms are acquired because those firms have valuable strategic assets, such as brand loyalty, customer relationships, trademarks or patents, distribution systems, production systems, and the like. It is easier and perhaps less risky for a firm to acquire those assets than to build them from the ground up through a greenfield investment. Third, firms make acquisitions because they believe they can increase the efficiency of the acquired unit by transferring capital, technology, or management skills. However, there is evidence that many mergers and acquisitions fail to realize their anticipated gains.

Through their choice of policies, home (source) countries can both encourage and restrict FDI by local firms. We look at policies designed to encourage outward FDI first. These include foreign risk insurance, capital assistance, tax incentives, and political pressure. Then we will look at policies designed to restrict outward FDI.

Many investor nations now have government-backed insurance programs to cover major types of foreign investment risk. The types of risks insurable through these programs include the risks of expropriation (nationalization), war losses, and the inability to transfer profits back home. Such programs are particularly useful in encouraging firms to undertake investments in politically unstable countries. In addition, several advanced countries also have special funds or banks that make government loans to firms wishing to invest in developing countries. As a further incentive to encourage domestic firms to undertake FDI, many countries have eliminated double taxation of foreign income (i. e., taxation of income in both the host country and the home country). Last, and perhaps most significant, a number of investor countries (including the United States) have used their political influence to persuade host countries to relax their restrictions on inbound FDI. For example, in response to direct U. S. pressure, Japan relaxed many of its formal restrictions on inward FDI in the 1980s. Now, in response to further U. S. pressure, Japan has moved toward relaxing its informal barriers to inward FDI. One beneficiary of this trend has been Toys「R」Us, which, after five years of intensive lobbying by company and U. S. government officials, opened its first retail stores in Japan in December 1991. By 2008, Toys「R」Us had more 170 stores in Japan, and its Japanese operation, in which Toys「R」Us retained a controlling stake, had a listing on the Japanese stock market.

Virtually all investor countries, including the United States, have exercised some control over outward FDI from time to time. One policy has been to limit capital outflows out of concern for the country's balance of payments. From the early 1960s until 1979, for example, Britain had exchange-control regulations that limited the amount of capital a firm could take out of the country. Although the main intent of such policies was to improve the British balance of payments, an important secondary intent was to make it more difficult for British firms to undertake FDI.

In addition, countries have occasionally manipulated tax rules to try to encourage their firms to invest at home. The objective behind such policies is to create jobs at home rather than in other nations. At one time, Britain adopted such policies. The British advanced corporation tax system taxed British companies' foreign earnings at a higher rate than their domestic earnings. This tax code created an incentive for British companies to invest at home.

Finally, countries sometimes prohibit national firms from investing in certain countries for political reasons. Such restrictions can be formal or informal. For example, formal U. S. rules prohibited U. S. firms from investing in countries such as Cuba and Iran, whose political ideology and actions are judged to be contrary to U. S. interests. Similarly, during the 1980s, informal pressure was applied to dissuade U. S. firms from investing in South Africa. In this case, the objective was to pressure South Africa to change its apartheid laws, which happened during the early 1990s.

Host countries adopt policies designed both to restrict and to encourage inward FDI. As noted earlier in this chapter, political ideology has determined the type and scope of these policies in the past. Inthe last decade of the 20th century, many countries moved quickly away from some version of the radical stance (which prohibited much FDI), and toward a combination of free market objectives and pragmatic nationalism.

It is common for governments to offer incentives to foreign firms to invest in their countries. Such incentives take many forms, but the most common are tax concessions, low-interest loans, and grants or subsidies.

Incentives are motivated by a desire to gain from the resource-transfer and employment effects of FDI. They are also motivated by a desire to capture FDI away from other potential host countries. For example, in the mid-1990s, the governments of Britain and France competed with each other on the incentives they offered Toyota to invest in their respective countries. In the United States, state governments often compete with each other to attract FDI. For example, Kentucky offered Toyota an incentive package worth $112 million to persuade it to build its U. S. automobile assembly plants there. The package included tax breaks, new state spending on infrastructure, and low-interest loans.

Host governments use a wide range of controls to restrict FDI in one way or another. The two most common are ownership restraints and performance requirements. Ownership restraints can take several forms. In some countries, foreign companies are excluded from specific fields, for example, tobacco and mining in Sweden and the development of certain natural resources in Brazil, Finland, and Morocco. In other industries, foreign ownership may be permitted although a significant proportion of the equity of the subsidiary must be owned by local investors. Foreign ownership is restricted to 25 percent or less of an airline in the United States. In India, foreign firms were prohibited from owning media businesses until 2001, when the rules were relaxed, allowing foreign firms to purchase up to 26 percent of a foreign newspaper.

The rationale underlying ownership restraints seems to be twofold. First, foreign firms are often excluded from certain sectors on the grounds of national security or competition. Particularly in less developed countries, the feeling seems to be that local firms might not be able to develop unless foreign competition is restricted by a combination of import tariffs and controls on FDI.

Second, ownership restraints seem to be based on a belief that local owners can help to maximize the resource-transfer and employment benefits of FDI for the host country. Until the early 1980s, the Japanese government prohibited most FDI but allowed joint ventures between Japanese firms and foreign MNEs if the MNE had a valuable technology. The Japanese government clearly believed such an arrangement would speed up the subsequent diffusion of the MNE's valuable technology throughout the Japanese economy.

Performance requirements can also take several forms. Performance requirements are controls over the behavior of the MNE's local subsidiary. The most common performance requirements are related to local content, exports, technology transfer, and local participation in top management. As with certain ownership restrictions, the logic underlying performance requirements is that such rules help to maximize the benefits and minimize the costs of FDI for the host country. Many countries employ some form of performance requirements when it suits their objectives. However, performance requirements tend to be more common in less developed countries than in advanced industrialized nations.

Until the 1990s, multinational institutions were not consistently involved in governing FDI. This changed with the formation of the World Trade Organization in 1995. The WTO embraces the promotion of international trade in services. Since many services have to be produced where they are sold, exporting is not an option (for example, one cannot export McDonald's hamburgers or consumer banking services). Given this, the WTO has become involved in regulations governing FDI. As might be expected for an institution created to promote free trade, the thrust of the WTO's efforts has been to push for the liberalization of regulations governing FDI, particularly in services. Under the auspices of the WTO, two extensive multinational agreements were reached in 1997 to liberalize trade in telecommunications and financial services. Both these agreements contained detailed clauses that require signatories to liberalize their regulations governing inward FDI, essentially opening their markets to foreign telecommunications and financial services companies.

The WTO has had less success trying to initiate talks aimed at establishing a universal set of rules designed to promote the liberalization of FDI. Led by Malaysia and India, developing nations have so far rejected efforts by the WTO to start such discussions. In an attempt to make some progress on this issue, the Organization for Economic Cooperation and Development (OECD) in 1995 initiated talks between its members. (The OECD is a Paris-based intergovernmental organization of ⌈wealthy⌋ nations whose purpose is to provide its 29 member states with a forum in which governments can compare their experiences, discuss the problems they share, and seek solutions that can then be applied within their own national contexts. The members include most EU countries, the United States, Canada, Japan, and South Korea). The aim of the talks was to draft a multilateral agreement on investment (MAI) that would make it illegal for signatory states to discriminate against foreign investors. This would liberalize rules governing FDI between OECD states.

These talks broke down in early 1998, primarily because the United States refused to sign the agreement. According to the United States, the proposed agreement contained too many exceptions that would weaken its powers. For example, the proposed agreement would not have barred discriminatory taxation of foreign-owned companies, and it would have allowed countries to restrict foreign television programs and music in the name of preserving culture. Environmental and labor groups also campaigned against the MAI, criticizing the proposed agreement because it contained no binding environmental or labor agreements. Despite such setbacks, negotiations on a revised MAI treaty might restart in the future. Moreover, as noted earlier, many individual nations have continued to liberalize their policies governing FDI to encourage foreign firms to invest in their Economies.

New Words, Phrases and Expressions
subsidize   v. 給……津貼或補貼，資助或補助……
encompassing   v. 包括，包含；包圍；圍繞
spoilage   n. 損耗；（食品等的）腐敗，損壞，毀壞
readily   adv. 樂意地；快捷地；輕而易舉地；便利地
inbound   adj. 入境的；回內地的；歸本國的；到達的
rationale   n. 理論的說明；基本原理，基礎理論；根據

Notes
1. Incentives are motivated by a desire to gain from the resource-transfer and employment effects of FDI.
從國際直接投資中的資源轉移效應和就業效應中獲利的期望激發了這一動機。
2. Particularly in less developed countries, the feeling seems to be that local firms might not be able to develop unless foreign competition is restricted by a combination of import tariffs and controls on FDI.
特別是在經濟欠發達的國家中，人們認為除非通過進口關稅和控制國際直接投資相結合的方法來限制來自外國公司的競爭，否則當地企業不會得到發展。

Lesson 4　International Finance Market

International financial markets are relevant to companies, whether or not they become directly involved in international business through exports, direct investment, and the like. Purchasesof imported products or services may require payment in foreign exchange, thus involving exchange risk. A company may choose to invest in a foreign business or security, and face both different interest rates and different risks from those at home. Often companies. find that borrowing funds abroad is less expensive than borrowing domestically, in the United States or in any other home country. The relatively unrestricted 「euromarkets」 generally offer better terms to borrowers (and lenders) than do domestic financial markets in any country. Likewise, in the 2000s, investors are discovering more and more opportunities to diversify their portfolios into holdings of foreign securities. This chapter explores the various foreign and international financial markets, including the foreign exchange market, and examines ways to utilize them for domestic and international business.

Foreign exchange markets

Currencies, like any other products, services, or claims, can be traded for one another. The foreign exchange market is simply a mechanism through which transactions can be made between one country's currency and another's. Or, more broadly, a foreign exchange market is a market for the exchange of financial instruments denominated in different currencies. The most common location for foreign exchange transactions is a commercial bank, which agrees to ⌈make a market⌋ for the purchase and sale of currencies other than the local one. In the United States, hundreds of banks offer foreign exchange markets in dozens of cities. However, over 40 per cent of all foreign exchange business is done through the 10 largest banks in New York.

Foreign exchange is not simply currency printed by a foreign country's central bank; rather, it includes such items as cash, checks (or drafts), wire transfers, telephone transfers, and even contracts to sell or buy currency in the future. Foreign exchange is really any financial instrument that carries out payment from one currency to another. The most common form of foreign exchange in transactions between companies is the draft (denominated in a foreign currency). The most common form of foreign exchange in transactions between banks is the telephone or Internet transfer, generally for transactions of $ 1 million or more. Far down the list of instruments, in terms of the value exchanged, is actual currency, which is often exchanged by tourists and other travelers. Since no currency is necessarily fixed in value relative to others, there must be some means of determining an acceptable price, or exchange rate. How many British pounds (£) should one dollar buy? How many Brazilian reals should be paid to buy a dollar? Since 1973, most of the industrialized countries have allowed their currency values to fluctuate more or less freely (depending on the time and the country), so that simple economics of supply and demand largely determine exchange rates.

The exchange rate of $ 1.2058 per euro is offered by one specific bank. This was the closing exchange rate quoted by Bankers Trust Company of New York for purchases of at least $ 1 million worth of euros on October 6, 2004. Another bank in New York may have quoted a slightly higher or slightly lower rate at the same time, and banks in San Francisco, London, or Tokyo probably quoted rates that differed even more. The rates quoted depend on each bank's interest in being involved in foreign exchange transactions and on the varying strategic positions and risks taken by the banks. Thus, the foreign exchange market among banks may be viewed as one large market or as many small markets with extensive (potential and actual) interrelations. Generally, the interbank market within any country is viewed as one market, and intercountry dealings are somewhat segmented by different rules and risks (though in the United States, the San Francisco market often offers quotations from various banks that differ noticeably from rate quotations in New York, due partly to the difference in time zones).

The differences in rate quotations may appear very small to an outsider. Eurosmay be quoted at $ 1.2055 at Citibank in New York and at $ 1.2060 at Bank of America in San Francisco. This difference seems small, but for every euros 1 million sold, the buyer of dollars in San Francisco receives an extra $ 500. Since the transactions are generally rapid, for values of $ 5 million to $ 10 million such differences may add up to a substantial gain (or cost) to the participant in the foreign exchange market.

The various foreign exchange transactions just mentioned involve large commercial banks and large amounts of money. Most of the companies that do international business also utilize foreign exchange markets, but far less frequently than banks and for transactions that involve less money. Consequently, these companies' access to foreign currencies differs from that of the banks. (Specifically, banks make the market and companies pay for the use of this service.) The next section describes the various participants in the US foreign exchange market.

Foreign money and capital market

In each country the MNE enters, it will be able to obtain some degree of access to local financial markets. The MNE will generally utilize such markets to perform necessary local financial transactions and often to hedge its local asset exposure through local borrowing (or its local liability exposure through local deposits or investments). But the MNE can also utilize local financial markets to obtain additional funding (or place additional funds) for its non-local activities. A Mexican firm, for example, may choose to borrow funds through its

Houston affiliate to finance its local needs in Texas and finance some of its needs in Mexico. Usually national financial markets are not fully accessible to foreign firms; they are largely reserved for domestic borrowers and lenders. In these cases, use of local financial markets must be supplemented by use of the financial markets in the country of the parent company or elsewhere. Typically, such problems arise in countries that impose exchange controls, which limit the MNE's ability to put needed funds in the affiliate.

MNEs and national money markets

The advantages MNEs obtain from entering national money markets as borrowers of funds stem from the portfolio effect of holding liabilities in more than one (fluctuating) currency and from the local risk hedging that occurs as the MNE moves to balance local assets such as plant and equipment with local liabilities, such as local-currency borrowing. Additional benefits may occur if the local government subsidizes such loans through interest rate reduction, through tax breaks, or otherwise; a policy of this kind may be offered to attract the MNE's investment. This may lower the whole MNE's cost of capital if the subsidy does not disappear and if the interest rate remains below the comparable home-currency rate during the borrowing period.

The advantages MNEs obtain from entering national money markets as lenders of funds stem from the portfolio effect of holding financial instruments in more than one currency. There may also be a gain from balancing local-currency assets and liabilities if an excess of local liabilities was held previously. In addition, local financial investments may be a necessary strategy when exchange controls or profit remittance limits exist; such investments offer the firm an outlet for retained earnings that yields interest. Sometimes an exchange control policy may make the local-currency interest rates higher than those available in other currencies (after adjusting for expected exchange rate changes). In these cases, higher-than-normal profits can be earned by investing funds in local instruments, although exchange controls may limit conversion of the profits into the home-country currency, or a government-imposed exchange-rate change may wipe out the advantage.

MNEs and national capital markets

The advantages MNEs obtain from entering national capital markets as borrowers are similar to those obtained from entering national money markets. However, the opportunities are generally much more limited. Very few countries have substantially developed stock markets, and those that do usually restrict the issuance of shares by foreigners. A few exceptions exist, such as the United States, the United Kingdom, Germany, and France, but even they serve only the largest and most creditworthy MNEs. National bond markets, when they exist, are also very restrictive. Thus far, virtually all of the foreign bond issues have taken place in New York (called 「Yankee bonds」), London, Switzerland, Germany, and Japan (called 「Samurai bonds」). The eurobond market, which functions mostly outside the United States, is dominated by dollar issues, which constitute about 40 per cent of that total. In all, dollar-denominated issues account for about 36 per cent of the two types of international bonds. The advantages MNEs obtain from entering national capital markets as lenders come mainly from diversification and from higher returns that may be protected by exchange controls, just as in the short-term market. Since the national capital markets are generally small, opportunities to use them are quite limited for foreign as well as domestic investors. Of course, since the main currencies of interest to MNEs and other international investors are those of the few large industrial countries, opportunities in those national capital markets do exist.

New Words, Phrases and Expressions
  unrestricted adj. 無限制的；不受限制的；自由的；敞開的
  portfolio n. 投資組合；（保險）業務量；（公司或機構提供的）系列產品
  utilize vt. 利用，使用
  interbank adj. 銀行同業
  hedge n. 樹籬；保護手段；防止損失（尤指金錢）的手段

Notes

1. Likewise, in the 2000s, investors are discovering more and more opportunities to diversify their portfolios into holdings of foreign securities.

而且，在21世紀的頭十年裡，投資者發現他們有著越來越多的機會通過持有外國證券的方式來分散他們的投資組合。

2. Houston affiliate to finance its local needs in Texas and finance some of its needs in Mexico. Usually national financial markets are not fully accessible to foreign firms; they are largely reserved for domestic borrowers and lenders.

休斯頓附屬公司來為它們在得克薩斯州本地和墨西哥的需求提供資金。通常情況下，國家資本市場並不完全對外國公司開放。它們是為國內借款人和貸款人服務的。

Lesson 5　Regional Money and Capital Markets

Until recently, the scope of financial markets and instruments was predominantly domestic or fully international, but not regional. In the past two decades, however, the European Union has designed both a regional monetary system (the European Monetary Union, or EMU) and a regional currency unit (the euro) for intergovernmental financial transactions in the Union. The eurocurrency market A eurodollar is a dollar-denominated bank deposit located outside the United States. This simple statement defines the basic instrument called the eurodollar. (Eurodollars are not dollars in the pockets of people who live in Europe! They are deposit liabilities of banks.) The widely discussed eurocurrency market is simply a set of bank deposits located outside the countries whose currencies are used in the deposits. Since over half of the deposits are denominated in US dollars, it is reasonably accurate to call this the eurodollar market. Notice that the eurodollar market is not limited to Europe; very large eurodollar markets exist in Tokyo, Hong Kong, Bahrain, Panama, and other cities throughout the world. For this reason, the eurodollar market is sometimes called the international money market.

What is the significance of the international money market? Since the eurocurrency market rivals domestic financial markets as a funding source for corporate borrowing, it plays a key role in the capital investment decisions of many firms. In addition, since this market also rivals domestic financial markets as a deposit alternative, it absorbs large amounts of savings from lenders (i.e., depositors) in many countries. In fact, the eurocurrency market complements the domestic financial markets, giving greater access to borrowing and lending to financial market participants in each country where it is permitted to function. Overall, the eurocurrency market is now the world's single most important market for international financial intermediation. This market is completely a creation of the regulatory structures placed by national governments on banking or, more precisely, on financial intermediation. If national governments allowed banks to function without reserve requirements, interest rate restrictions, capital controls, and taxes, the eurocurrency market would involve only the transnational deposits and loans made in each country's banking system. Instead, national governments heavily regulate national financial markets in efforts to achieve various monetary policy goals. Thus, the eurocurrency market provides a very important outlet for funds flows that circumvent many of the limitations placed on domestic financial markets. Many national governments have found the impact of the eurocurrency market on their firms and banks to be favorable, so they have allowed this market to operate. Since the eurocurrency market is a creation of the regulatory structure, it may be helpful to think about the key underpinnings of the system that allow eurodollars. Essentially three conditions must be met for a eurocurrency market to exist. First, some national government must allow foreign currency deposits to be made so that, for example, depositors in London can obtain dollar-denominated time deposits there. Second, the country whose currency is being used—in this example, the United States—must allow foreign entities to own and exchange deposits in that currency. Third, there must be some reason, such as low cost or ease of use that prompts people to use this market rather than other financial markets, such as the domestic ones. The eurocurrency market has met these conditions for the past three decades, and its phenomenal growth testifies that the demand for such a market has been very large.

A wide range of countries allow foreign-currency deposits to be held in their banking systems. Many of them impose restrictions (interest rate limits, capital controls, and soon).

New Words, Phrases and Expressions
rival   n. 競爭對手
underpinning   n. 基礎
testify   v. 作證，證明；證明，證實
Notes
Many national governments have found the impact of the eurocurrency market on their firms and banks to be favorable, so they have allowed this market to operate.
許多國家的政府發現了歐洲貨幣對於它們公司和銀行的積極作用，所以它們開展了歐洲貨幣業務。

Section 3   International business strategies

Lesson 1   The Strategy of International Business

Our primary concern thus far in this book has been with aspects of the larger environment in which international businesses compete. As we have described it in the preceding section. Now our focus shifts from the environment to the firm itself and, in particular, to the actions managers can take to compete more effectively as an international business. In this section, we look at how firms can increase their profitability by expanding their operations in foreign markets. We discuss the different strategies that firms pursue when competing internationally. We consider the pros and cons of these strategies. We discuss the various factors that affect a firm's choice of strategy. We also look at why firms often enter into strategic alliances with their global competitors, and we discuss the benefits, costs, and risks of strategic alliances.

We have seen how firms that expand globally can increase their profitability and profit growth by entering new markets where indigenous competitors lack similar competencies, by lowering costs and adding value to their product offering through the attainment of location economies, by exploiting experience curve effects, and by transferring valuable skills between their global network of subsidiaries. For completeness it should be noted that strategies that increase profitability may also expand a firm's business and thus enable it to attain a higher rate of profit growth. For example, by simultaneously realizing location economies and experience effects, a firm may be able to produce a more highly valued product at a lower unit cost, thereby boosting profitability. The increase in the perceived value of the product may also attract more customers, thereby growing revenues and profits as well. Furthermore, rather than raising prices to reflect the higher perceived value of the product, the firm's managers may elect to hold prices low in order to increase global market share and attain greater scale economies (in other words, they may elect to offer consumers better 「value for money」). Such a strategy could increase the firm's rate of profit growth even further since consumers will be attracted by prices that are low relative to value. The strategy might also increase profitability if the scale economies that result from market share gains are substantial. In sum, managers need to keep in mind the complex relationship between profitability and profit growth when making strategic decisions about pricing.

Lesson 2　some international business strategies

Marketing strategies

Every multinational has a marketing strategy designed to help identify opportunities and take advantage of them. This plan of action typically involves consideration of four primary areas: the product or service to be sold, the way in which the output will be promoted, the pricing of the good or service, and the distribution strategy to be used in getting the output to the customer. The primary purpose of this chapter is to examine the fundamentals of international marketing strategy. We will look at five major topics: market assessment, product strategy, promotion strategy, price strategy, and place strategy. We will consider such critical marketing areas as product screening, modification of goods and services in order to adapt to local needs, modified product life cycles, advertising, personal selling, and ways in which MNEs tailor-make their distribution systems. International marketing is the process of identifying the goods and services that customers out side the home country want and then providing them at the right price and location. In the international marketplace, this process is similar to that carried out at home, but with some important modifications that can adapt marketing efforts to the needs of the specific country or geographic locale. For example, some MNEs are able to use the same strategy abroad as they have at home. This is particularly true in promotions where messages can carry a universal theme. Some writing implement firms advertise their pens and pencils as 「the finest writing instruments in the world」, a message that transcends national boundaries and can be used anywhere. Many fast-food franchises apply the same ideas because they have found that people everywhere have the same basic reasons for coming there to eat. In most cases, however, a company must tailor-make its strategy so that it appeals directly to the local customer. These changes fall into five major areas: market assessment, product decisions, promotion strategies, pricing decisions, and place or distribution strategies. The latter four areas—product, promotion, price, and place—are often referred to as the four Ps of marketing, and they constitute the heart of international marketing efforts.

Human resource management strategy

Human resource management strategy provides an MNE with the opportunity to truly outdistance its competition. For example, if HP develops a laser printer that is smaller, lighter, and less expensive than competitive models, other firms in the industry will attempt to reverse engineer this product to see how they can develop their own version. However, when a multinational has personnel who are carefully selected, well trained, and properly compensated, it has a pool of talent that the competition may be unable to beat. For this reason, human resource management (HRM) is a critical element of international management strategy. This paragraph considers the ways in which multinationals prepare their people to take on the challenges of international business. We focus specifically on such critical areas as selection, training, managerial development, compensation, and labor relations. International human resource management (IHRM) is the process of selecting, training, developing, and compensating personnel in overseas positions. This paragraph will examine each of these activities. Before doing so, however, it is important to clarify the general nature of this overall process, which begins with selecting and hiring.

There are three basic sources of personnel talent that MNEs can tap for positions. One is home-country nationals, who reside abroad but are citizens of the multinational's parent country. These individuals are typically called expatriates. An example is a US manager assigned to head an R&D department in Tokyo for IBM Japan. A second is host-country nationals, who are local people hired by the MNE. An example is a British manager working for Ford Motor Company in London. The last is third-country nationals, who are citizens of countries other than the one in which the MNE is headquartered or the one in which it has assigned them to work. An example is a French manager working for Sony in the United States. Staffing patterns may vary depending on the length of time the MNE has been operating. Many MNEs will initially rely on home-country managers to staff their overseas units, gradually putting more host-country nationals into management positions as the firm gains experience. Another approach is to use home-country nationals in less developed countries and employ host-country nationals in more developed regions, a pattern that is fairly prevalent among US and European MNEs. A third pattern is to put a new operation under the supervision of a home-country manager but turn it over to a host-country manager once it is up and running. When an MNE is exporting into a foreign market, host-country nationals will handle everything. As the firm begins initial manufacture in that country, the use of expatriate managers and third-country nationals begins to increase. As the company moves through the ensuing stages of internationalization, the nationality mix of the managers in the overseas unit continues to change to meet the shifting demands of the environment. In some cases staffing decisions are handled uniformly. For example, most Japanese MNEs rely on home-country managers to staff senior-level positions. Similarly, some European MNEs assign home-country managers to overseas units for their entire careers. US MNEs typically view overseas assignments as temporary, so it is more common to find many of these expatriates working under the supervision of host-country managers. The size of the compensation package also plays an important role in personnel selection and placement. As the cost of sending people overseas has increased, there has been a trend toward using host-country or third-country nationals who know the local language and customs. For example, in recent years many US multinationals have hired English or Scottish managers for the top positions at subsidiaries in former British colonies such as Jamaica, India, the West Indies, and Kenya.

Political risk and negotiation strategy

Country risk analysis can build on these approaches by providing insights into the country-level factors that will affect the leveraging of these advantages in a particular country, and help managers balance potential risks against rewards. Country risk analysis examines the chances of non-market events (political, social, and economic) causing financial, strategic, or personnel losses to a firm following FDI in a specific country market. Rather than looking at market, industry, or group of competitors, country risk analysis tries to predict macroeconomic and political sources of change that might undermine a firm's position in a local market. This kind of analysis is used to compare country markets before making investment decisions. Decision makers supplement strategic information about market size, market growth, presence of local suppliers, support industries, and the strengths of local competitors with, for example, estimates of current and future socio-economic and political risks that are country specific.

Political risk is the probability that political forces will negatively affect a firm's profit or impede the attainment of other critical business objectives. The study of political risk addresses changes in the environment that are difficult to anticipate. Common examples include the election of a government that is committed to nationalization of major industries or one that insists on reducing foreign participation in business ventures. Most people believe that political risk is confined to third-world countries or to those with unstable governments. However, the shifting policies of triad region governments illustrate that political risk is also an issue for firms doing business in highly industrialized nations. Governments routinely protect key industries for reasons of national security, self sufficiency, or national culture. Examples include US restrictions on foreign investment in the banking, defense, and commercial airline industries; Japanese protection of its rice producers; and France's limitations on foreign involvement in its film and media businesses.

The EU and US governments have long supported their regional champions in the aerospace industry, partly for security reasons, partly to protect local firms. Most recently Airbus has spent $ 11 billion on building the A380 double-decker jumbo jet, one-third of which has come from European taxpayers. This may well initiate another round of trade battles, refereed by the WTO. Much of this political sparring is predictable. It is unanticipated change stemming from complexity and uncertainty that creates more dangerous kinds of risk. However, these examples indicate the pervasiveness of political risk. Given the large number of international markets in which MNEs operate, political risk is going to remain an area of concern. In dealing with this issue, effective negotiating can help to reduce and contain problem areas.

New Words, Phrases and Expressions
modification　n. 改變；更改；緩和；限制
implement　n. 工具，器械；家具，手段；履行（契約等）
　　　　　　vt. 實施，執行；使生效，實現；落實（政策等）；把……填滿
reside　vi. 住，居住，（官吏）留駐；（性質）存在，具備；（權力等）屬於
prevalent　adj. 流行的，盛行的；普遍存在的，普遍發生的
supervision　n. 監督，管理；監督的行為、過程或作用
personnel　n. 全體員工；（與複數動詞連用）人員，員工；人事部門
Notes
1. However, the shifting policies of triad region governments illustrate that political risk is also an issue for firms doing business in highly industrialized nations.
但是，三區政府的政策轉變說明了即使在高度工業化國家開展商務時，政治風險仍然是一個不可忽略的問題。
2. It is unanticipated change stemming from complexity and uncertainty that creates more dangerous kinds of risk.
來源於複雜性和不確定性的不可預測的改變引起了更多危險類型的風險。

Lesson 3　Choosing a Strategy

Pressures for local responsiveness imply that it may not be possible for a firm to realize the full benefits from economies of scale, learning effects, and location economies. It may not be possible to serve the global marketplace from a single low-cost location, producing a globally standardized product and marketing it worldwide to attain the cost reductions associated with experience effects. The need to customize the product offering to local conditions may work against the implementation of such a strategy. For example, automobile firms have found that Japanese, American, and European consumers demand different kinds of cars, which requires them to produce products customized for local markets. In response, firms such as Honda, Ford, and Toyota are pursuing a strategy of establishing top-to-bottom design and production facilities in each of these regions so theycan better serve local demands. Although such customization brings benefits, it also limits the ability of a firm to realize significant scale economies and location economies.

In addition, pressures for local responsiveness imply that it may not be possible to leverage skills and products associated with a firm's core competencies wholesale from one nation to another. Concessions often have to be made to local conditions. Despite being depicted as a 「poster boy」 for the proliferation of standardized global products, even McDonald's has found that it has to customize its product offerings (its menu) to account for national differences in tastes and preferences.

How do differences in the strength of pressures for cost reductions versus those for local responsiveness affect the firm's choice of strategy? Firms typically choose among four main strategic postures when competing internationally: a global standardization strategy, a localization strategy, a transnational strategy, and an international strategy. The appropriateness of each strategy varies given the extent of pressures for cost reductions and local responsiveness. Figure 4.1 illustrates the conditions under which each of these strategies is most appropriate.

	low pressures for local responsiveness	high pressures for local responsiveness
high pressures for cost reductions	Global Standardization Strategy	Transnational Strategy
low pressures for cost reductions	International Strategy	Localization Strategy

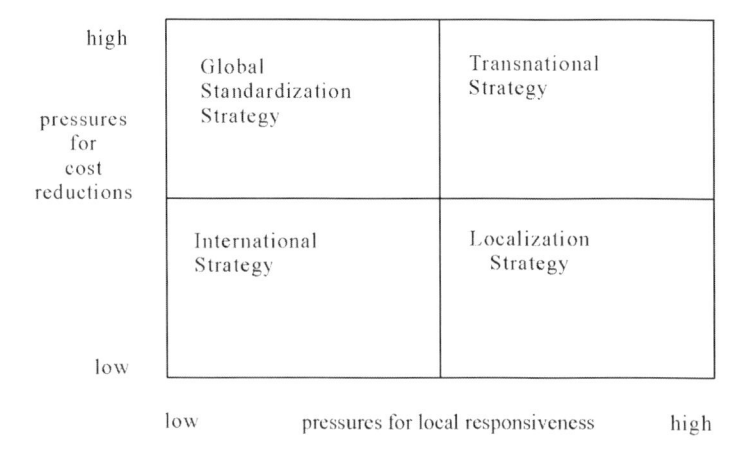

Figure 4.1 Four basic strategies

Global standardization strategy

Firms that pursue a global standardization strategy focus on increasing profitability and profit growth by reaping the cost reductions that come from economies of scale, learning effects, and location economies; that is, their strategic goal is to pursue a low-cost strategy on a global scale. The production, marketing, and R&D activities of firms pursuing a global standardization strategy are concentrated in a few favorable locations. Firms pursuing a global standardization strategy try not to customize their product offering and marketing strategy to local conditions because customization involves shorter production runs and the duplication of functions, which tend to raise costs. Instead, they prefer to market a standardized product worldwide so they can reap the maximum benefits from economies of scale and learning effects. They also tend to use their cost advantage to support aggressive pricing in world markets.

This strategy makes most sense when there are strong pressures for cost reductions and demands for local responsiveness are minimal. Increasingly, these conditions prevail in many industrial goods industries, whose products often serve universal needs. In the semiconductor industry, for example, global standards have emerged, creating enormous demands for standardized global products. Accordingly, companies such as Intel, Texas Instruments, and Motorola all pursue a global standardization strategy. However, these conditions are not yet found in many consumer goods markets, where demands for local responsiveness remain high. The strategy is inappropriate when demands for local responsiveness are high.

Localization Strategy

A localization strategy focuses on increasing profitability by customizing the firm's goods or services so that they provide a good match to tastes and preferences in different national markets. Localization is most appropriate when there are substantial differences across nations with regard to consumer tastes and preferences, and where cost pressures are not too intense. By customizing the product offering to local demands, the firm increases the value of that product in the local market. On the downside, because it involves some duplication of functions and smaller production runs, customization limits the ability of the firm to capture the cost reductions associated with mass-producing a standardized product for global consumption. The strategy may make sense, however, if the added value associated with local customization supports higher pricing, which enables the firm to recoup its higher costs, or if it leads to substantially greater local demand, enabling the firm to reduce costs through the attainment of some scale economies in the local market.

At the same time, firms still have to keep an eye on costs. Firms pursuing a localization strategy still need to be efficient and, whenever possible, to capture some scale economies from their global reach. As noted earlier, many automobile companies have found that they have to customize some of their product offerings to local market demands—for example, producing large pickup trucks for U. S. consumers and small fuel-efficient cars for Europeans and Japanese. At the same time, these multinationals try to get some scale economies from their global volume by using common vehicle platforms and components across many different models and manufacturing those platforms and components at efficiently scaled factories that are optimally located. By designing their products in this way, these companies have been able to localize their product offering, yet simultaneously capture some scale economies, learning effects, and location economies.

Transnational Strategy

We have argued that a global standardization strategy makes most sense when cost pressures areintense and demands for local responsiveness limited. Conversely, a localization strategy makes most sense when demands for local responsiveness are high, but cost pressures are moderate or low. What happens, however, when the firm simultaneously faces both strong cost pressures and strong pressures for local responsiveness?

How can managers balance the competing and inconsistent demands such divergent pressures place on the firm? According to some researchers, the answer is to pursue what has been called a transnational strategy. Two of these researchers, Christopher Bartlett and Sumantra Ghoshal, argue that in today's global environment, competitive conditions are so intense that to survive, firms must do all they can to respond to pressures for cost reductions and local responsiveness. They must try to realize location economies and experience effects, to leverage products internationally, to transfer core competencies and skills within the company, and to simultaneously pay attention to pressures for local responsiveness. 35 Bartlett and Ghoshal note that in the modern multinational enterprise, core competencies and skills do not reside just in the home country but can develop in any of the firm's worldwide operations. Thus, they maintain that the flow of skills and product offerings should not be all one way, from home country to foreign subsidiary. Rather, the flow should also be from foreign subsidiary to home country and from foreign subsidiary to foreign subsidiary. Transnational enterprises, in other words, must also focus on leveraging subsidiary skills.

In essence, firms that pursue a transnational strategy are trying to simultaneously achieve low costs through location economies, economies of scale, and learning effects; differentiate their productoffering across geographic markets to account for local differences; and foster a multidirectional flow of skills between different subsidiaries in the firm's global network of operations. As attractive as this may sound in theory, the strategy is not an easy one to pursue since it places conflicting demands on the company. Differentiating the product to respond to local demands in different geographic markets raises costs, which runs counter to the goal of reducing costs. Companies such as Ford and ABB (one of the world's largest engineering conglomerates) have tried to embrace a transnational strategy and found it difficult to implement.

How best to implement a transnational strategy is one of the most complex questions that large multinationals are grappling with today. Few if any enterprises have perfected this strategic posture. But some clues as to the right approach can be derived from a number of companies. For example, consider the case of Caterpillar. The need to compete with low-cost competitors such as Komatsu of Japan forced Caterpillar to look for greater cost economies. However, variations in construction practices and government regulations across countries mean that Caterpillar also has to be responsive to local demands. Therefore, Caterpillar confronted significant pressures for cost reductions and for local responsiveness. To deal with cost pressures, Caterpillar redesigned its products to use many identical components and invested in a few large-scale component manufacturing facilities, sited at favorable locations, to fill global demand and realize scale economies. At the same time, the company augments the centralized manufacturing of components with assembly plants in each of its major global markets. At these plants, Caterpillar adds local product features, tailoring the finished product to local needs. Thus, Caterpillar is able to realize many of the benefits of global manufacturing while reacting to pressures for local responsiveness by differentiating its product among national markets. Caterpillar started to pursue this strategy in 1979 and by the late 1990s hadsucceeded in doubling output per employee, significantly reducing its overall cost structure in the process. Meanwhile, Komatsu and Hitachi, which are still wedded to a Japan-centered global strategy, have seen their cost advantages evaporate and have been steadily losing market share to Caterpillar. Changing a firm's strategic posture to build an organization capable of supporting a transnational strategy is a complex and challenging task. Some would say it is too complex because the strategy implementation problems of creating a viable organizational structure and control systems to manage this strategy are immense.

International Strategy

Sometimes it is possible to identify multinational firms that find themselves in the fortunate position of being confronted with low cost pressures and low pressures for local responsiveness. Many of these enterprises have pursued an international strategy, taking products first produced for their domestic market and selling them internationally with only minimal local customization. The distinguishing feature of many such firms is that they are selling a product that serves universal needs, but they do not face significant competitors, and thus unlike firms pursuing a global standardization strategy, they are not confronted with pressures to reduce their cost structure. Xerox found itself in this position in the 1960s after its invention and commercialization of the photocopier. The technology underlying the photocopier was protected by strong patents, so for several years Xerox did not face competitors—it had a monopoly. The product serves universal needs, and it was highly valued in most developed nations. Thus, Xerox was able to sell the same basic product the world over, charging a relatively high price for that product. Since Xerox did not face direct competitors, it did not have to deal with strong pressures to minimize its cost structure.

Enterprises pursuing an international strategy have followed a similar developmental pattern as they expanded into foreign markets. They tend to centralize product development functions such as R&D at home. However, they also tend to establish manufacturing and marketing functions in each major country or geographic region in which they do business. The resulting duplication can raise costs, but this is less of an issue if the firm does not face strong pressures for cost reductions. Although they may undertake some local customization of product offering and marketing strategy, this tends to be rather limited in scope. Ultimately, in most firms that pursue an international strategy, the head office retains fairly tight control over marketing and product strategy. Other firms that have pursued this strategy include Procter & Gamble and Microsoft. Historically, Procter & Gamble developed innovative new products in Cincinnati and then transferred them wholesale to local markets. Similarly, the bulk of Microsoft's product development work takes place in Redmond, Washington, where the company is headquartered. Although some localization work is undertaken elsewhere, this is limited to producing foreign-language versions of popular Microsoft programs.

The evolution strategy

The Achilles' heel of the international strategy is that over time, competitors inevitably emerge, and if managers do not take proactive steps to reduce their firm's cost structure, it will be rapidly outflanked by efficient global competitors. This is exactly what happened to Xerox. Japanese companies such as Canon ultimately invented their way around Xerox's patents, produced their own photocopiers in very efficient manufacturing plants, priced them below Xerox's products, and rapidly took global market share from Xerox. In the final analysis, Xerox's demise was not due to the emergence of competitors, for ultimately that was bound to occur, but to its failure to proactively reduce its cost structure in advance of the emergence of efficient global competitors. The message in this story is that an international strategy may not be viable in the long term, and to survive, firms need to shift toward a global standardization strategy or a transnational strategy in advance of competitors. The same can be said about a localization strategy. Localization may give a firm a competitive edge, but if it is simultaneously facing aggressive competitors, the company will also have to reduce its cost structure, and the only way to do that may be to shift toward a transnational strategy. This is what Procter & Gamble has been doing. Thus, as competition intensifies, international and localization strategies tend to become less viable, and managers need to orientate their companies toward either a global standardization strategy or a transnational strategy.

New Words, Phrases and Expressions

customize　vt. 定制, 定做; 按規格改制

concession　n. 特許權; 承認; 減價; (在某地的) 特許經營權

proliferation　n. 增生; 增殖, 分芽繁殖; 再育

variation　n. 變異；變化；變奏（曲）；變量
aggressive　adj. 侵略的，侵犯的，攻勢的；有進取心的；積極行動的

Notes

1. Firms pursuing a global standardization strategy try not to customize their product offering and marketing strategy to local conditions because customization involves shorter production runs and the duplication of functions, which tend to raise costs.

追求全球標準化戰略的公司不會根據當地的情況來制定它們的產品和市場策略。因為量身定制需要更短的生產運行週期和功能的複製，而這些都會提高成本。

2. The Achilles' heel of the international strategy is that over time, competitors inevitably emerge, and if managers do not take proactive steps to reduce their firm's cost structure, it will be rapidly outflanked by efficient global competitors.

國際戰略中的致命的弱點指的是：隨著時間的推移，競爭不可避免地出現。如果管理者不採取積極的措施來降低公司的成本結構，它將迅速被有效的全球競爭對手擊敗。

Section 4　International Business Operations

Lesson 1　Exporting, Importing, and Countertrade

We reviewed exporting from a strategic perspective. We considered exporting as just one of a range of strategic options for profiting from international expansion. This lesson is more concerned with the nuts and bolts of exporting (and importing). Here we look at how to export. As the opening case makes clear, exporting is not just for large enterprises; many small entrepreneurial firms such as MD International have benefited significantly from the money-making opportunities of exporting. The volume of export activity in the world economy has been increasing as exporting has become easier. The gradual decline in trade barriers under the umbrella of GATT and now the WTO, along with regional economic agreements such as the European Union and the North American Free Trade Agreement, have significantly increased export opportunities.

At the same time, modern communication and transportation technologies have alleviated the logistical problems associated with exporting. Firms are increasingly using the World Wide Web, toll-free 800 phone numbers, and international air express services to reduce the costs of exporting. Consequently, it is no longer unusual to find small companies that are thriving as exporters. Nevertheless, exporting remains a challenge for many firms. Smaller enterprises can find the process intimidating. The firm wishing to export must identify foreign market opportunities, avoid a host of unanticipated problems that are often associated with doing business in a foreign market, familiarize itself with the mechanics of export and import financing, learn where it can get financing and export credit insurance, and learn how it should deal with foreign exchange risk. The process can be made more problematic by currencies that are not freely convertible. Arranging payment for exports to countries with weak currencies can be a problem. This brings us to the topic of countertrade, by which payment for exports is received in goods and services rather than money. In this lesson, we will discuss all these issues with the exception of foreign exchange risk. We open the lesson by considering the promise and pitfalls of Exporting.

The Promise and Pitfalls of Exporting

The great promise of exporting is that foreign markets offer most firms in most industries substantial revenue and profit opportunities. The international market is normally so much larger than the firm's domestic market that exporting is nearly always a way to increase the revenue and profit base of a company. By expanding the size of the market, exporting can enable a firm to achieve economies of scale, thereby lowering its unit costs. Firms that do not export often lose out on significant opportunities for growth and cost reduction.

Studies have shown that while many large firms tend to be proactive about seeking opportunities for profitable exporting, systematically scanning foreign markets to see where the opportunities lie for leveraging their technology, products, and marketing skills in foreign countries, many medium-sized and small firms are very reactive. Typically, such reactive firms do not even consider exporting until their domestic market is saturated and the emergence of excess productive capacity at home forces them to look for growth opportunities in foreign markets. Also, many small and medium-sized firms tend to wait for the world to come to them, rather than going out into the world to seek opportunities. Even when the world does come to them, they may not respond.

An example is MMO Music Group, which makes sing-along tapes for karaoke machines. Foreign sales accounted for about 15 percent of MMO's revenues of $ 8 million, but the firm's CEO admits that this figure would probably have been much higher had he paid attention to building international sales. Unanswered faxes and phone messages from Asia and Europe often piled up while he was trying to manage the burgeoning domestic side of the business. By the time MMO did turn its attention to foreign markets, other competitors had stepped into the breach and MMO found it tough going to build export volume. MMO's experience is common, and it suggests a need for firms to become more proactive about seeking export opportunities. One reason more firms are not proactive is that they are unfamiliar with foreign market opportunities; they simply do not know how big the opportunities actually are or where they might lie. Simple ignorance of the potential opportunities is a huge barrier to exporting.

Also, many would-be exporters, particularly smaller firms, are often intimidated by the complexities and mechanics of exporting to countries where business practices, language, culture, legal systems, and currency are very different from the home market.

This combination of unfamiliarity and intimidation probably explains why exporters still account for only a tiny percentage of U. S. firms, less than 5 percent of firms with fewer than 500 employees, according to the Small Business administration.

To make matters worse, many neophyte exporters run into significant problems when first trying to do business abroad and this sours them on future exporting ventures. Common pitfalls include poor market analysis, a poor understanding of competitive conditions in the foreign market, a failure to customize the product offering to the needs of foreign customers, lack of an effective distribution program, a poorly executed promotional campaign, and problems securing financing. 8 Novice exporters tend to underestimate the time and expertise needed to cultivate business in foreign countries. 9 Few realize the amount of management resources that have to be dedicated to this activity. Many foreign customers require face-to-face negotiations on their home turf. An exporter may have to spend months learning about a country's trade regulations, business practices, and more before a deal can be closed. The next Management Focus feature, which documents the experience of FCX Systems in China, suggests that it may take years before foreigners are comfortable enough to purchase in significant quantities. Exporters often face voluminous paperwork, complex formalities, and many potential delays and errors.

According to a UN report on trade and development, a typical international trade transaction may involve 30 parties, 60 original documents, and 360 document copies, all of which have to be checked, transmitted, reentered into various information systems, processed, and filed. The United Nations has calculated that the time involved in preparing documentation, along with the costs of common errors in paperwork, often amounts to 10 percent of the final value of goods exported.

Utilizing export management companies

One way for first-time exporters to identify the opportunities associated with exporting and to avoid many of the associated pitfalls is to hire an export management company (EMC). EMCs are export specialists who act as the export marketing department or international department for their client firms. MD International, which we looked at in the opening case, is one such enterprise. EMCs normally accept two types of export assignments. They start exporting operations for a firm with the understanding that the firm will take over operations after they are well established. In another type, start-up services are performed with the understanding that the EMC will have continuing responsibility for selling the firm's products. Many EMCs specialize in serving firms in particular industries and in particular areas of the world. Thus, one EMC may specialize in selling agricultural products in the Asian market, while another may focus on exporting electronics products to Eastern Europe. MD International, for example, focuses on selling medical equipment to Latin America. In theory, the advantage of EMCs is that they are experienced specialists who can help the neophyte exporter identify opportunities and avoid common pitfalls. A good EMC will have a network of contacts in potential markets, have multilingual employees, have a good knowledge of different business mores, and be fully conversant with the ins and outs of the exporting process and with local business regulations. However, the quality of EMCs varies. While some perform their functions very well, others appear to add little value to the exporting company. Therefore, an exporter should review carefully a number of EMCs and check references. One drawback of relying on EMCs is that the company can fail to develop its own exporting capabilities.

Export strategy

In addition to using EMCs, a firm can reduce the risks associated with exporting if it is careful about its choice of export strategy. A few guidelines can help firms improvetheir odds of success.

The probability of exporting successfully can be increased dramatically by taking a handful of simple strategic steps. First, particularly for the novice exporter, it helps to hire an EMC or at least an experienced export consultant to help identify opportunities and navigate the paperwork and regulations so often involved inexporting. Second, it often makes sense to focus initially on one market or a handful of markets. Learn what is required to succeed in those markets before moving on to other markets. The firm that enters many markets at once runs the risk of spreading its limited management resources too thin. The result of such a shotgun approach to exporting may be a failure to become established in any one market. Third, it often makes sense to enter a foreign market on a small scale to reduce the costs of any subsequent failure. Most important, entering on a small scale provides the time and opportunity to learn about the foreign country before making significant capital commitments to that market. Fourth, the exporter needs to recognize the time and managerial commitment involved in building export sales and should hire additional personnel to oversee this activity. Fifth, in many countries, it is important to devote a lot of attention to building strong and enduring relationships with local distributors and/or customers . Sixth, it is important to hire local personnel to help the firm establish itself in a foreign market. Local people are likely to have a much greater sense of how to do business in a given country than a manager from an exporting firm who has previously never set foot in that country. Seventh, several studies have suggested that the firm needs to be proactive about seeking export opportunities. Armchair exporting does not work! The world will not normally beat a pathway to your door. Finally, it is important for the exporter to retain the option of local production. Once exports reach a sufficient volume to justify cost-efficient local production, the exporting firm should consider establishing production facilities in the foreign market. Such localization helps foster good relations with the foreign country and can lead to greater market acceptance. Exporting is often not an end in itself, but merely a step on the road toward establishing foreign production.

Export and Import Financing

Firms engaged in international trade have to trust someone they may have never seen, who lives in a different country, who speaks a different language, who abides by (or does not abide by) a different legal system, and who could be very difficult to track down if he or she defaults on an obligation. Consider a U. S. Firm exporting to a distributor in France. The U. S. businessman might be concerned that if he ships the products to France before he receives payment from the French businesswoman, she might take delivery of the products and not pay him. Conversely, the French importer might worry that if she pays for the products before they are shipped, the U. S. firm might keep the money and never ship the products or might ship defective products.

Neither party to the exchange completely trusts the other. This lack of trust is exacerbated by the distance between the two parties—in space, language, and culture—and by the problemsof using an underdeveloped international legal system to enforce contractual obligations. Due to the lack of trust between the two parties, each has his or her own preferences as to how the transaction should be configured. To make sure he is paid, the manager of the U. S. firm would prefer the French distributor to pay for the products before he ships them. Alternatively, to ensure she receives the products, the French distributor would prefer not to pay for them until they arrive . Thus, each party has a different set of preferences. Unless there is some way of establishing trust between the parties, the transaction might never occur.

The problem is solved by using a third party trusted by both—normally a reputable bank—to act as an intermediary. What happens can be summarized as follows. First, the French importer obtains the bank's promise to pay on her behalf, knowing the U. S. exporter will trust the bank. This promise is known as a letter of credit. Having seen the letter of credit, the U. S. exporter now ships the products to France. Title to the products is given to the bank in the form of a document called a bill of lading. In return, the U. S. Exporter tells the bank to pay for the products, which the bank does. The document for requesting this payment is referred to as a draft. The bank, having paid for the products, now passes the title on to the French importer, whom the bank trusts. At that time or later, depending on their agreement, the importer reimburses the bank. In the remainder of this section, we examine how this system works in more detail.

A letter of credit, abbreviated as L/C, stands at the center of international commercial transactions. Issued by a bank at the request of an importer, the letter of credit states that the bank willpay a specified sum of money to a beneficiary, normally the exporter, on presentation of particular specified documents. Consider again the example of the U. S. exporter and the French importer. The French importer applies to her local bank, say, the Bank of Paris, for the issuance of a letter of credit. The Bank of Paris then undertakes a credit check of the importer. If the Bank of Paris is satisfied with her creditworthiness, it will issue a letter of credit. However, the Bank of Paris might require a cash deposit or some other form of collateral from her first. In addition, the Bank of Paris will charge the importer a fee for this service. Typically this amounts to between 0.5 percent and 2 percent of the value of the letter of credit, depending on the importer's credit worthiness and the size of the transaction. (As a rule, the larger the transaction, the lower the percentage.)

Assume the Bank of Paris is satisfied with the French importer's creditworthiness and agrees to issue a letter of credit. Theletter states that the Bank of Paris will pay the U.S. exporter for the merchandise as long as it is shipped in accordance with specified instructions and conditions. At this point, the letter of credit becomes a financial contract between the Bank of Paris and the U.S. exporter. The Bank of Paris then sends the letter of credit to the U.S. exporter's bank, say, the Bank of New York. The Bank of New York tells the exporter that it has received a letter of credit and that he can ship the merchandise. After the exporter has shipped the merchandise, he draws a draft against the Bank of Paris in accordance with the terms of the letter of credit, attaches the required documents, and presents the draft to his own bank, the Bank of New York, for payment. The Bank of New York then forwards the letter of credit and associated documents to the Bank of Paris. If all of the terms and conditions contained in the letter of credit have been complied with, the Bank of Paris will honor the draft and will send payment to the Bank of New York. When the Bank of New York receives the funds, it will pay the U.S. Exporter. As for the Bank of Paris, once it has transferred the funds to the Bank of New York, it will collect payment from the French importer. Alternatively, the Bank of Paris may allow the importer some time to resell the merchandise before requiring payment. This is not unusual, particularly when the importer is a distributor and not the final consumer of the merchandise, since it helps the importer's cash flow. The Bank of Paris will treat such an extension of the payment period as a loan to the importer and will charge an appropriate rate of interest. The great advantage of this system is that both the French importer and the U.S. exporter are likely to trust reputable banks, even if they do not trust each other. Once the U.S. exporter has seen a letter of credit, he knows that he is guaranteed payment and will ship the merchandise. Also, an exporter may find that having a letter of credit will facilitate obtaining preexport financing. For example, having seen the letter of credit, the Bank of New York might be willing to lend the exporter funds to process and prepare the merchandise for shipping to France. This loan may not have to be repaid until the exporter has received his payment for the merchandise. As for the French importer, she does not have to pay for the merchandise until the documents have arrived and unless all conditions stated in the letter of credit have been satisfied. The drawback for the importer is the fee she must pay the Bank of Paris for the letter of credit. In addition, since the letter of credit is a financial liability against her, it may reduce her ability to borrow funds for other purposes.

A draft, sometimes referred to as a bill of exchange, is the instrument normally used in international. commerce to effect payment. A draft is simply an order written by an exporter instructing an importer, or an importer's agent, to pay a specified amount of money at a specified time. In the example of the U. S. Exporter and the French importer, the exporter writes a draft that instructs the Bank of Paris, the French importer's agent, to pay for the merchandise shipped to France. The person or business initiating the draft is known as the maker. The party to whom the draft is presented is known as the drawee.

International practice is to use drafts to settle trade transactions. This differs from domestic practice in which a seller usually ships merchandise on an open account, followed by a commercialinvoice that specifies the amount due and the terms of payment. In domestic transactions, the buyer can often obtain possession of the merchandise without signing a formal document acknowledging his or her obligation to pay. In contrast, due to the lack of trust in international transactions, payment or a formal promise to pay is required before the buyer can obtain the merchandise.

Drafts fall into two categories, sight drafts and time drafts. A sight draft is payable on presentation to the drawee. A time draft allows for a delay in payment—normally 30, 60, 90, or 120 days. It is presented to the drawee, who signifies acceptance of it by writing or stamping a notice of acceptance on its face. Once accepted, the time draft becomes a promise to pay by the accepting party. When a bank draws on and accepts the time draft, it is called a banker's acceptance. When a business firm draws on and accepts the draft, it is called a trade acceptance.

Time drafts are negotiable instruments; that is, once the draft is stamped with an acceptance, the maker can sell the draft to an investor at a discount from its face value. Imagine that the agreement between the U. S. exporter and the French importer calls for the exporter to present the Bank of Paris (through the Bank of New York) with a time draft requiring payment 120 days after presentation. The Bank of Paris stamps the time draft with an acceptance. Imagine further that the draft is for $ 100, 000. The exporter can either hold onto the accepted time draft and receive $ 100, 000 in 120 days or he can sell it to an investor, say, the Bank of New York, for a discount from the face value. If the prevailing discount rate is 7 percent, the exporter could receive $ 97, 700 by selling it immediately (7 percent per year discount rate for 120 days for $ 100, 000 equals $ 2, 300, and $ 100, 000 − $ 2, 300 = $ 97, 700). The Bank of New York would then collect the full $ 100, 000 from the Bank of Paris in 120 days. The exporter might sell the accepted time draft immediately if he needed the funds to finance merchandise in transit and/or to cover cash flow shortfalls.

The third key document for financing international trade is the bill of lading. The bill of lading is issued to the exporter by the common carrier transporting the merchandise. It serves three purposes: it is a receipt, a contract, and a document of title. As a receipt, the bill of lading indicates that the carrier has received the merchandise described on the face of the document. As a contract, it specifies that the carrier is obligated to provide a transportation service in return for a certain charge. As a document of title, it can be used to obtain payment or a written promise of payment before the merchandise is released to the importer. The bill of lading can also function as collateral against which funds may be advanced to the exporter by its local bank before or during shipment and before final payment by the importer.

Now that we have reviewed the elements of an international trade transaction, let us see how the process works in a typical case, sticking with the example of the U. S. exporter and the French importer. The typical transaction involves 14 steps as following: (1) The French importer places an order with the U. S. exporter and asks the American if he would be willing to ship under a letter of credit. (2) The U. S. exporter agrees to ship under a letter of credit and specifies relevant information such as prices and delivery terms. (3) The French importer applies to the Bank of Paris for a letter of credit to be issued in favor of the U. S. exporter for the merchandise the importer wishes to buy. (4) The Bank of Paris issues a letter of credit in the French importer's favor and sends it to the U. S. exporter's bank, the Bank of New York. (5) The Bank of New York advises the exporter that a letter of credit has been opened in his favor. (6) The U. S. exporter ships the goods to the French importer on a common carrier. An official of the carrier gives the exporter a bill of lading. (7) The U. S. exporter presents a 90-day time draft drawn on the Bank of Paris in accordance with its letter of credit and the bill of lading to the Bank of New York. The exporter endorses the bill of lading so title to the goods is transferred to the Bank of New York. (8) The Bank of New York sends the draft and bill of lading to the Bank of Paris. The Bank of Paris accepts the draft, taking possession of the documents and promising to pay the now-accepted draft in 90 days. (9) The Bank of Paris returns the accepted draft to the Bank of New York. (10) The Bank of New York tells the U. S. exporter that it has received the accepted bank draft, which is payable in 90 days. (11) The exporter sells the draft to the Bank of New York at a discount from its face value and receives the discounted cash value of the draft in return. (12) The Bank of Paris notifies the French importer of the arrival of the documents. She agrees to pay the Bank of Paris in 90 days. The Bank of Paris releases the documents so the importer can take possession of the shipment. (13) In 90 days, the Bank of Paris receives the importer's payment, so it has funds to pay the maturing draft. (14) In 90 days, the holder of the matured acceptance presents it to the Bank of Paris for payment. The Bank of Paris pays.

New Words, Phrases and Expressions
familiarize   vt. 使熟悉
convertible   n. 有活動折疊篷的汽車；adj. 可改變的；同意義的；可交換的
proactive   adj. 有前瞻性的，先行一步的；積極主動的

transmitted　v. 傳輸；傳送；adj. 透射的
reentered　n. 重進入，再加入；vt. 重新加入；vi. 重新入內，重返
reimburse　vt. 償還；賠償

Notes

1. While some perform their functions very well, others appear to add little value to the exporting company. Therefore, an exporter should review carefully a number of EMCs and check references. One drawback of relying on EMCs is that the company can fail to develop its own exporting capabilities.

一些出口管理公司運行得很好，而另外一些則對出口公司而言價值甚微。所以，對於出口商來說，認真瞭解一些出口管理公司並從其過去的客戶處瞭解實情很重要。依賴出口管理公司的一個弊端是企業可能難以發展自己的出口能力。

2. Time drafts are negotiable instruments; that is, once the draft is stamped with an acceptance, the maker can sell the draft to an investor at a discount from its face value.

遠期匯票是一種可轉讓票據，也就是說，匯票一經承兌，持票人就可以從票面價值中按一定的貼現率將匯票賣給投資者。

Lesson 2　Global Production, Outsourcing, and Logistics

As trade barriers fall and global markets develop, many firms increasingly confront a set of interrelated issues. First, where in the world should production activities be located? Should they be concentrated in a single country, or should they be dispersed around the globe, matching the type of activity with country differences in factor costs, tariff barriers, political risks, and the like to minimize costs and maximize value added? Second, what should be the long-term strategic role of foreign production sites? Should the firm abandon a foreign site if factor costs change, moving production to another more favorable location, or is there value to maintaining an operation at a given location even if underlying economic conditions change? Third, should the firm own foreign production activities, or is it better to outsource those activities to independent vendors? Fourth, how should a globally dispersed supply chain be managed, and what is the role of Internet-based information technology in the management of global logistics? Fifth, should the firm manage global logistics itself, or should it outsource the management to enterprises that specialize in this activity?

We will focus on two of these activities—production and logistics—and attempt to clarify how they might be performed internationally to lower the costs of value creation and add value by better serving customer needs.

We defined production as「the activities involved in creating a product.」We used the term production to denote both service and manufacturing activities, since one can produce a service or a physical product. Although in this lesson we focus more on the production of physical goods, one should not forget that the term can also be applied to services. This has become more evident in recent years with the trend among U. S. firms to outsource the 「production」 of certain service activities to developing nations where labor costs are lower (for example, the trend among many U. S. companies to outsource customer care services to places such as India, where English is widely spoken and labor costs are much lower). Logistics is the activity that controls the transmission of physical materials through the value chain, from procurement through production and into distribution. Production and logistics are closely linked since a firm's ability to perform its production activities efficiently depends on a timely supply of high-quality material inputs, for which logistics is responsible. The production and logistics functions of an international firm have a number of important strategic objectives. One is to lower costs. Dispersing production activities to various locations around the globe where each activity can be performed most efficiently can lower costs. Costs can also be cut by managing the global supply chain efficiently so as to better match supply and demand. Efficient supply chain management reduces the amount of inventory in the system and increases inventory turnover, which means the firm has to invest less working capital in inventory and is less likely to find excess inventory on hand that cannot be sold and has to be written off.

A second strategic objective shared by production and logistics is to increase product quality by eliminating defective products from both the supply chain and the manufacturing process. (In this context, quality means reliability, implying that the product has no defects and performs well.) The objectives of reducing costs and increasing quality are not independent of each other. Improved quality control reduces costs in several ways: Increasing productivity because time is not wasted producing poor-quality products that cannot be sold, leading to a direct reduction in unit costs. Lowering rework and scrap costs associated with defective products. Reducing the warranty costs and time associated with fixing defective products. The effect of improved quality control in these ways is to lower the costs of value creation by reducing both production and after-sales service costs.

New Words, Phrases and Expression
dispersed   adj. 散布的；被分散的；被驅散的 v. 分散；傳播
inventory   n. 存貨，存貨清單；詳細目錄；財產清冊
eliminating adj. 排除的 v. 消除

Notes
1. Logistics is the activity that controls the transmission of physical materials through the value chain, from procurement through production and into distribution.
物流是指控制原料在價值鏈中轉移的活動，包括從獲得原料到生產再到分配的過程。

Exercise
Ⅰ. Define the following concepts

1. International Monetary Fund
2. multinational enterprise
3. power distance
4. green-field investment
5. merger
6. localization
7. uncertainty avoidance
8. strategic alliances

Ⅱ. Please choose the best answer from the following choices of each question.

1. The shift toward a more integrated and interdependent world economy is referred to as (　　).

   A. economic integration
   B. economic interdependency
   C. globalization
   D. global integration

2. The sourcing of good and services from around the world to take advantage of national differences in the cost and quality of factors of production is called (　　).

   A. economies of scale
   B. the globalization of production
   C. global integration
   D. global sourcing

3. The establishment of a wholly new operation in a foreign country is called (　　).

   A. an acquisition
   B. a merger
   C. a greenfield investment
   D. a multinational venture

4. The amount of FDI undertaken over a given time period is known as (　　).

   A. the flow of FDI
   B. the stock of FDI
   C. FDI outflow
   D. FDI inflow

5. Advantages that arise from using resource endowments or assets that are tied to a particular location and that a firm finds valuable to combine with its own unique assets are (　　).

   A. First mover advantages
   B. Location advantages
   C. Externalities
   D. Proprietary advantages

6. Which strategy tries to simultaneously achieve low costs through location economies, economies of scale, and learning effects, and differentiate the product offering across geographic markets to account for local differences? (   ).

A. Internationalization

B. Localization

C. Global standardization

D. Transnational

7. The basic social organization of a society is its (   ).

A. culture

B. social strata

C. social structure

D. caste system

8. _____ refers to the time and effort spent learning the rules of a new market. (   )

A. First mover advantages

B. Strategic commitments

C. Pioneering costs

D. Market entry costs

9. _____ focuses on how society deals with the fact that people are unequal in physical and intellectual capabilities. (   )

A. power distance

B. individualism versus collectivism

C. uncertainty avoidance

D. masculinity versus femininity

10. All of the following except _____ contribute to unethical behavior by international managers. (   )

A. Decision-making processes

B. Leadership

C. Personal ethics

D. National culture

Ⅲ. Questions and problems

1. Describe the shifts in the world economy over the last 30 years. What are the implications of these shifts for international businesses based in Great Britain? North America? Hong Kong?

2. 「Ultimately, the study of international business is not different from the study of domestic business. Thus, there is no point in having a separate course on international business.」Evaluate this statement.

3. You are the CEO of a company that has to choose between making a ＄100 million

investment in either Russia or the Czech Republic. Both investments promise the same long-run return, so your choice of which investment to make is driven by considerations of risk. Assess the various risks of doing business in each of these nations. Which investment would you favor and why?

4. Outline why the culture of a country influences the costs of doing business in that country. Illustrate your answer with examples.

5. Compare and contrast these explanations of vertical FDI: the strategic behavior approach and the market imperfections approach. Which theory do you think offers the better explanation of the historical pattern of vertical FDI? Why?

6. Discuss how the need for control over foreign operations varies with firms' strategies and core competencies. What are the implications for the choice of entry mode?

Further Reading Comprehension

China's environment: Foreign trade and investment and international organizations

Kimbrely Anne Wilhelm

Not only is China's environment in decline, but China's contribution to global environmental deterioration is significant. Because of the global implications of China'senvironmental pollution, international action is needed. Foreign trade and investment should be encouraged to incorporate environmental concerns and technology transfer, especially in the area of energy efficiency. Continued and enhanced international cooperation as well as loans and investment through international and regional organizations should also contribute to improving China's and the global environment. Finally, China should assist in this process by continuing domestic economic reform and the open door policy. Environmental and economic benefits will result from efforts in these three areas.

As national security is redefined in the 1990s, it will include not only strategic military and political concerns but also environmental, economic, energy, and other transnational concerns. Pollution and resulting ecological degradation do not stop at national borders. An international response is needed for the deteriorating global environment. International cooperation is necessary: today's challenges can be dealt with because even though ⌈man is still utterly dependent on the natural world,⌋ he now has for the first time the ability to alter it, rapidly and on a global scale.

China is a major contributor to the deterioration of the global environmentand ecology. Therefore, what China does or does not do with regard to its environment will increasingly affect the environmental condition of other nations and the global community as a whole. The financing of efforts to deal with environmental problems and the transfer of sound environmental technology (including management, education, and infrastructure issues) are obstacles to China's modernization and environmental management (protection, prevention, and clean-up). There are several means by which China and the international community can address the problems of finance and technology, a three of which will be addressed in this article: (1) foreign trade and investment that focus on using and transferring sound environmental technology, (2) cooperation through international organizations and multilateral agency loans, and (3) economic reform with an emphasis on enhancing China's investment environment and instituting price reform.

Background

Due to international pressures and economic considerations, China's leaders have begun refining China's environmental policy; in doing so they have begun to recognize the advantages in emphasizing its environmental policy as an aspect of China's foreign policy. Since the mid 1970s, environmental concerns have been afforded greater awareness and legal recognition and resulted in policy improvements in China. Despite these changes and the formation of policies that stress incorporating environmental concerns into the economic and development plan, the implementation of these policies has been weak. China faces severe domestic environmental problems with potentially enormous global implications. While there is no overlooking the vast and serious environmental problems China faces, it would be a mistake to ignore China's accomplishments and its potential for improvement.

China's focus on modernization has resulted in a growing degree of economic growth and prosperity that has created a demand for greater energy use and consumption. As China's economy and population grow, the increased demands for energy and other basic services result in major pollution of the environment. Improving China's infrastructure, environmental management, and technology transfer to China have costs, but opportunities exist to make efforts that will have significant economic and environmental benefits. Increasingly, it is asserted that more energy-efficient technologies are not necessarily costly, particularly if attributes such as better reliability, product quality, and environmental controls are considered.

The global environmental consequences of China's industrial and economic activities threaten industrialized countries which should realize that it is in 「their own best interest」 to support sustainable development in developing countries, like China. Foreign influence has played a significant role in promoting environmental protection in China and is likely to be greatest when it coincides with China's domestic political interests (such as modernization) or when China is heavily dependent on foreign suppliers and customers. To the extent that foreign trade and investment and international institutions can affect China domestically, China is presented with a means of coping with its environmental problems. In addition to foreign and international endeavors, China must also make efforts to improve its environmental management by continuing economic reform.

Foreign Trade And Investment

International trade and investment are a primary means for China and the global community to address environmental deterioration. The environmental standards of developed and developing countries are linked through foreign trade and investment. The paradox of technology is that 「the industrial economy causes environmental damage, but also offers the main way to repair the damage」. Thus, industrial nations have contributed to environmental deterioration but can also provide the means for improving environmental protection and management in developing and developed nations.

In general, foreign trade and investmentare critical to the growth of China's economy and modernization and China needs foreign technology, capital, and investment to deal with its environmental obstacles. China seeks to increase its foreign exchange reserves and its importation of foreign capital and technology. Foreign investors seek opportunities for joint ventures and investment, but generally feel constrained by China's economic situation. Some investment has been made in environmental technology, management, and equipment in China. Environmentally sound technology and good environmental practices have also been passed to China through general trade and investment.

Ventures in environmental technologies offer risks and opportunities. Investing in environmentally sound technologies is an opportunity to enhance competitiveness. Foreign investors and enterprises can capitalize on and grasp the competitive advantages of 「green」 products and processes. The argument that it is not economically wise or competitive to invest in the environment is increasingly disputed. Opportunities are rapidly expanding to invest and profit from environmental investments; regulations are spurring growth and costs are dropping. 「There is a lot of money to be made. And this is a field in which technologies thrive.」 There is also great potential for technologies innovation. The long-term gains cannot be argued.

Trade and foreign investment provide opportunities for China and foreign investors to address finance and technology issues as they relate to the environment. The value of China's trade increased from U. S. $ 13. 44 billion in 1976 to U. S. $ 73. 80 billion in 1986. Foreign investment in China reached U. S. $ 6, 596 million in 1990. The types of investment have expanded to include wholly foreign-owned enterprises, development projects, joint ventures, and joint management companies. Trade and foreign investment in machine-building, textiles, food, nuclear power plants, motor vehicles, coal-mining, chemicals, and petroleum all have the potential of having a profound negative or positive impact on China's environment.

Kozue Hiraiwa maintains that the type and growth of large-scale investment projects in 1991 indicates that foreign companies are 「getting serious about large-scale investments in the energy and high-tech fields」. Projects have included construction of a chemical plant, development of an electronic industry zone, and establishment of an agrochemical plant in Pudong and an automobile plant in Changchun. Investments in the energy, industrial, and agricultural sectors present opportunities for the use of environmentally sound technology in China and its transfer to China.

Table 4.1 Investment in China During 1990 by Type

Type/Number of Investment	Value (U.S. $ mill.)	Percent Change
4,091 Joint Ventures	$2,703.95	+1.7%
1,317 Joint Management Companies	$1,254.1	+15.8%
1,860 Wholly-Foreign Owned Companies (WFOEs)	$2,443.81	+47.8%
5 Joint Development Projects	$194.25	-4.7%
Total	$6,596.11	

Source: Kozue Hiraiwa, *Foreign Investment in the PRC*, 1990-1991, JETRO: China Newsletter 96 (January/February 1992): pp. 11-12.

Trade, Investment, and Technology Transfer

China can acquire advanced environmental technology, management skills, and equipment through the foreign projects included in its overall modernization program. In general, various forms of investment can be environmentally beneficial in that their advanced technologies would lead to less pollution and improved energy efficiency. If foreign firms focus on integrating 「environmental concerns」 into their contracts, management systems, and technology transfer, there will be mutually beneficial gains—improved efficiency, cost reduction, product innovation, and higher productivity. Several opportunities for environmental trade, investment, and technology transfer to China are as follows:

Environmental Regulations. Due to the new and increasing number of China's environmental laws and their implementation, foreign business opportunities are increasing. For example, the 1986 Provisional Regulations for Environmental Management in Foreign Economic Development Zones require that 「environmental protection clauses be included in all future economic contracts in the (special economic) zones (SEZs)」. Enterprises are also required to conduct environmental impact analyses of construction projects and use zoning to reduce effects on population centers. The SEZs are required to set tighter standards than national standards and to exercise preference for non-polluting or low-polluting import of technology and equip-merit. In order to discourage the importing of outmoded equipment or the stripping of imported machinery, the 1989 Environmental Protection Law (EPL, Article 30) prohibits the 「importation of technology or equipment that does not comply with domestic environmental regulations」.

In addition, foreign investors are accountable to China's environmental laws and are subject to on-site inspections of technology and records, fines, and criminal prosecution. Since 「environmental regulations can apparently override the provisions of any contract」, ma foreign investors should be aware of local as well as national laws. Joint ventures and foreign investors in general are watched more closely for effluent and particle emissions than Chinese factories. These types of regulations and requirements not only indicate a new direction in the importance of environmental protection but also an opportunity for foreign firms to invest in China. As a result, opportunities are increasing for investment and sales in municipal sewage treatment plants, point-source pollution control, monitoring equipment, consulting contracts, and transportation and supply infrastructures. Becoming technology brokers, private sector intermediaries could work at matching partners, arranging finances, and negotiating and managing licensing agreements.

Hainan Island. In 1984, Catherine Enderton noted the importance of environmental quality for the growing tourist industry. This provided a ⌈foreign exchange rationale for pollution control and the preservation of nature⌋ in Hainan Island and other fragile ecologies. More recently, businesses on Hainan Island are beginning to insist that they do not want investment there if it is not environmentally clean. Whether this line of thinking can take hold in other areas or whether it will win out against those who insist on modernization and investment at the expense of the environment of Hainan Island has yet to be seen. Nevertheless, statements such as these show a growing concern for environmental management and pollution prevention and an opportunity for foreign trade and investment.

The Automobile Industry. The development of the automobile industry in China also provides an opportunityfor prevention and foreign investment to work together for an environmental good. Richard Carpenter has commented that in the case of the growing use of internal combustion engines in China, there is time to prevent a potentially enormous source of environmental degradation. He recommends that a motor vehicle research program be established with international cooperation--to study the ⌈emission characteristics for China's engine designs, maintenance practices, operating modes, fuel quality, and vehicle age⌋. Other related issues that should be considered are the discarding of tires and old automobiles. Steps can be taken domestically but also by foreign companies to prepare for and prevent these potentially enormous pollution problems.

Clean environmental technology should be assimilated by foreign investors involved with the automobile industry in China. Additional focus could be given to the development of the Pudong Zone in Shanghai which the Shanghai Automobile Industry Corporation (SAIC) has said will be the home to the largest automobile spare parts manufacturing base in China. The effort to make the automobile industry the ⌈number one backbone industry⌋ of the city will center on three spare parts manufacturing projects that will receive 530 million yuan from SAIC and additional investment from several joint ventures. Foreign companies include the German Eckel and Funke companies, the British FSV company, and the German ZF company.

Joint ventures in the automobile industry such as these deserve attention which Takayama notes ⌈will continue to be a lever of reform⌋ as foreign technology is introduced. The 1990s will be an important period for the Chinese automobile industry as it consolidates the foundation for a modern production system and as it determines the quality of the environmental standards it will select for its development.

Environmental Technology and Equipment Sales. Foreign businesses are already taking advantage of some of the opportunities to sell environmental technology and pollution abatement equipment to China. One U. S. company that produces clean-coal technologies, Joy Environmental Equipment Company, signed a technology transfer agreement with a local company in Harbin to produce electrostatic precipitators and scrubbers for the domestic market. Another U. S. company, Thermo Environmental Instruments, has sold air pollution monitoring equipment to China for the past ten years. Canadian and Japanese firms have also been involved in similar sales to China.

In addition, Canada has recently been actively exploring the opportunities for environmental sales and investment in China. For example, the Canada-China Trade Council (CCTC) has been commissioned by External Affairs and International Trade Canada (EAITC) to undertake a study of the China market potential for Canadian pollution control technology, engineering consulting services, and related equipment and instruments. Such efforts to identify a marketing strategy are expected to「promote and facilitate successful Canadian exports in this key sector, which is receiving increasing attention from both government and industry alike」.

Furthermore, Ralph Klein, Alberta's Environment Minister, led a nine-day trade mission to China which investigated the opportunities to export Alberta's「special waste treatment technology」to Liaoning Province. In addition to having the「potential to generate business in the millions of dollars for Albertans」, the mission also contributed to Alberta's efforts to diversify its economy through strengthening the growing environmental industry.

The United States has also been making efforts in establishing the means and administrative structure to expand environmental technology transfer to China through trade and investment. For example, the recently founded United States-Asia Environmental Partnership (US-AEP) and the International Environment Technology Transfer Advisory Board (IETTAB) were established to explore environmental business opportunities in Asia.

The recently established US-AEP has the potential to be instrumental in assisting U. S. companies in becoming involved with environmental technology business opportunities in China. The US-AEP is a coalition of American and Asian businesses, governments, and community groups working together to「enhance Asia's environment and promote economic programs」. The four components of the program are (1) environmental fellowships, exchange, and training, (2) technology cooperation, (3) environmental and energy infrastructure, and (4) regional biodiversity conservation network.

Opportunities for foreign businesses to engage in trade and investment in areas that will improve China's environment are increasing. China will need to continue to improve the investment incentives for foreign firms as well as continue to pursue domestic economic reforms, if these opportunities are to be fully realized. But at the same time China's leaders will need to give the environment higher priority if as the open-door policy encourages investment in China, it will encourage environmentally positive investment. In order for China's leaders to enforce this prioritizing, China must be able to remain competitive in the international and domestic markets.

International Cooperation and Institutions

A second key means for addressing China's environmental problems, the need for technology transfer, and the financing of such efforts is through China's participation in international and regional organizations, World Bank loans, and international S&T research. There are economic incentives for China to adopt international environmental initiatives. For instance, China fears the imposition of barriers to its exports for environmental reasons. China should consider not「using CFCs if it wants to export to markets that are adopting legal restrictions against their use」. In addition to international pressure focusing on reducing the export of environmentally incorrect goods, international pressure can focus on encouraging China to adopt and abide by international agreements.

International S&T Research

Increased international scientific and academic cooperation and exchanges have also become means for China to acquire advanced technology, including environmental protection technology and research. Chinese scientists seek international assistance with regard to several of China's critical ecological obstacles, including water and food supply. Some consider China an excellent research site for international scientific cooperation on ecological and environmental studies. China will also benefit from Chinese researchers and scientists who have studied in the West and return to China with an understanding of advanced technologies.

Economic Reform and The Open Door Policy

In addition to these foreign and international efforts to protect China's and theglobal environment, China must be required to do its part and shoulder some of the responsibility for environmental management. Two critical means through which China can confront its environmental problems are to continue domestic economic reforms and the open door policy and to improve its investment environment. It can be argued that simply improving the investment environment in and of itself will not necessarily lead to environmentally sound investment in China. However, without improvements to China's investment environment, there are no incentives at all to invest in improving China's environment. If laws and investment requirements can be altered to encourage foreign investment in China, they can be improved to encourage environmentally sound investment. Chinese leaders need to act on the fact that improving the investment environment so that it will encourage not only more foreign investment but also environmentally sound investment will have environmental and economic benefits for China.

Economic Reform

Pollution in China is worse because of economic inefficiency. As He Bochuan has aptly noted, economic difficulties are a major factor in environmental problems because they create inefficiency, waste, lack of innovation, and ineffective use of capital which compound environmental damage and obstruct solutions to it. 48 China is one of the most inefficient users of energy in the world. The incentive for improving efficiency is that「the greater the efficiency, the smaller the volume of residuals」. Fundamentally, without price reform, there is no incentive to improve energy use and efficiency. Intensifying China's domestic economic reforms will generate pres-sure to correct these deficiencies.

China's Investment Environment

Foreign investors experience frustrations in doing business in China. Difficulties arise in repatriating profits, shortages, and limited access to needed materials, inadequate infrastructure, bureaucratic inefficiencies, and uncertainty about consistency in government policies and procedures. China has made efforts to define foreign investment regulations, improve the legal environment, encourage continued reforms, increase the degree of transparency, and increase the acceptance of wholly foreign-owned enterprises. Some efforts have been made, but more are needed to eliminate the limitations to increasing foreign investment, especially if China wants to encourage environmentally sound investments.

For example, effective 1 July 1991,「The Income Tax Laws for Foreign Invested Enterprises and Foreign Enterprises」was passed to replace two previous tax laws in an effort to clarify tax laws and thus improve China's investment environment. The CPC Central Committee's Proposals for the Eighth Five-Year Plan states that「China will, on the principle of equality and mutual benefit, expand its foreign economic and technological exchanges and cooperation and strive for greater progress in foreign trade, utilization of foreign funds, import of foreign technology and exchange of experts」. The government generally desires to improve the investment environment and encourage foreign businesses to launch more projects that use advanced technology.

Conclusion

Even though statements by China's leaders continue to insist that they do pay attention to environmental issues and that they have made unremitting efforts in this area, there is still a gap between stated national policy and actual practice. Li Peng said at UNCED 1992 that China「stands ready to undertake international responsibilities and obligations compatible with our development level and expand international cooperation in the world environmental protection and development」. Despite claims that China is holding its own in protecting the environment, leaders simultaneously seek out foreign trade and investment and international loans and aid to support environmental management efforts. In October 1990, Beijing reported that it could not meet the costs o f the estimated $46.7 billion needed by the year 2000 to control pollution.

Eventhough China is receptive to foreign and international assistance in addressing its environmental problems, the bottom line is, Who will pay for the clean-up, protection, prevention equipment, technology, and expertise, and how will technology be transferred? China's contribution to global environmental deterioration is significant and as such is not only China's concern but also a global concern. Because of the global implications of China's environmental pollution, action needs to be taken on an international level. Three steps will contribute to hurdling the funding and technology obstacles and to cleaning up and protecting China's and the global environment: encouraging (1) foreign trade and investment that incorporate environmental concerns and technology transfer, especially in the area o f energy-efficiency, (2) international cooperation through participation in international organizations and through loans that have environmental dimensions, and (3) economic reforms and the open door policy with emphasis on improving China's investment environment and price reform. As China and the international community act on their national and international responsibility toward the environment by making efforts in these three areas, each can reap environmental and economic benefits.

From Kimberly Anne Wilhelm, East Aisa.

Chapter 5
International Finance

Chapter checklist

When you have completed your study of this chapter, you will be able to
1. Found a brief framework towards finance.
2. Explore the basic approaches and instruments for analyzing international monetary.
3. Explain the core factors of international finance working system.

In this chapter, we will concentrate on the role of international finance plays in the globalization process, pay attention to how the financial markets are linked and integrated and learn about some approaches in measuring international monetary which are used to deal with financial problems in global. There are five sections in this part, which are an overview of finance, the framework of international finance theories, financial instruments and assets, financial institutions, and the international markets.

Section 1 An Introduction to Finance

Lesson 1 What is Finance?

As an important subfield of economics, the basic focus of finance is the working of the capital markets and the pricing of capital assets, its methodology is the use of close substitutes to price financial contracts and instruments, especially applied to value instruments whose characteristics extend across time and whose payoffs depend upon the resolution of uncertainty.

This description, in the above, reveals that finance is not much concerned with barter in objective economy. In addition, time and uncertainty as two core factors result in finance pays less attention to problems of static or certain world.

Indeed, finance as a concept is also dynamical in different historical contexts and academic definitions. Viewing finance as a horizontal and vertical matrix, its essence has experienced a process from simple to complex and from qualification to quantification. In the early phase of economic history, finance is a limited connotation and scope due to the constraint of economic level as a general intermediary. With the proceeding of financialization and monetization, the requirement of finance services is getting more diversified, which promotes the innovation of financial instruments and the improvement of financial markets. After the World War II, the economic globalization and financial deregulation become to be an irreversible tendency, and finance is penetrating to all over the socio-economic and independent to the barter economy. Based on the above discussion, there are different ranges when we refer to finance.

In academic, especially in the modern time, finance is a subtract essence of capital market to consider about the operation and pricing of monetary asset with certain instruments. It is a narrow definition which inherits the neoclassical economics tradition that analyses the optimal allocation of resource in time series under equilibrium. And there is a parallel scope of finance that includes any affairs concern to currency in national or international. This wide range is a part of macroeconomics, focusing on arranging and coordinating the policy portfolios. In this framework, money supply, exchange rates, capital flows among countries and even financial crises are not exogenous and as causes and results of financial activities in the macro-level.

Therefore, we might classify finance in terms of different subject. Generally speaking, asset pricing, risk management and corporate finance describe the definition of finance.

Asset pricing theory focus on the prices or values of claims to uncertain payments. A low price implies a high rate of return, so one can also think of the theory as explaining why some assets pay higher average returns than others. Time and uncertainty offers two branches of valuing asset, we have to account for the delay and the uncertainty of its payments. Time effect is easy to observe, the uncertainty, expressed as risk indeed, is much more important to ensure the reliability of our estimation.

Thus we can derive risk management. In the classic risk management model, we can judge risk and expected return of asset or portfolio at same time, which are denoted by standard variance and mean of historical returns individually. A portfolio is considered efficient if it achieves maximum expected return under a given level of expected risk or it bears minimum risk at a given level of expected return. These goals are two sides of a coin, because we can draw a curve, which we call efficient frontier, based on the correlation between the expected return and risk under a budget restraint. Any points on the curve can realize the two purposes above simultaneously.

Corporate finance is a mixed practice of asset pricing and risk management to deal with the sources of funding, the capital structure of corporations and the actions that managers take to increase the value of the firm to the shareholders. The main concept of study corporate finance is applicable to the financial firm of all kinds of firms. The primary goal of financial management is to maximize or to continually increase shareholder value. Maximizing shareholder value requires managers to be able to balance capital funding between investments in projects that increase the firm's long term profitability and sustainability, along with paying excess cash in the form of dividends to shareholders.

And there are two other subject in finance are international finance and behavioral finance. International finance is much concern with international trade theory and we will discuss in the following section. Behavioral finance is a more recently developed area that tries to analysis the anomalies in financial practices in irrational such as the herd phenomenon or excess volatility. It is new and interdisciplinary with the tools of psychology and brain science as the frontier of modern finance.

New Words, Phrases and Expressions
subfield　n. 子域；分支；子字段；分欄
static　adj. 靜態的
barter　vt. & vi. 作物物交換，以貨換貨；拿……進行易貨貿易
　　　　n. 易貨貿易；換貨，實物交易；交易品，互換品
horizontal　adj. 水準的，臥式的；地平線的
vertical　adj. 垂直的，豎立的
connotation　n. 內涵，含義；言外之意
deregulation　n. 違反規定，反常；解除管制規定
irreversible　adj. 不可逆的；不能翻轉的
penetrate　v. 滲入；穿透，刺入；秘密潛入；洞悉
inherit　v. 繼承
optimal　adj. 最佳的，最優的；最理想的
portfolio　n. 證券投資組合
simultaneous.　adv. 同時地
anomaly　n. 異常現象
volatility　n. 波動性
herd phenomenon　羊群效應，從眾現象

Notes

1. As an important subfield of economics, the basic focus of finance is the working of the capital markets and the pricing of capital assets, its methodology is the use of close substitutes to price financial contracts and instruments, especially applied to value instruments whose characteristics extend across time and whose payoffs depend upon the resolution of uncertainty.

作為經濟學的重要分支，金融依靠替代品對金融合約和工具，特別是那些具備跨期和基於解決不確定性問題來進行結算的工具進行定價，以實現其資本市場營運與資本定價的功能。

2. It is a narrow definition which inherits the neoclassical economics tradition that analyses the optimal allocation of resource in time series under equilibrium.

對金融的狹義定義其實脫胎於新古典經濟學的傳統分析框架，即是在時間序列之下的均衡中尋找對資源的最優配置。

3. A portfolio is considered efficient if it achieves maximum expected return under a given level of expected risk or it bears minimum risk at a given level of expected return.

如果一個投資組合是最有效的，就意味著它能在一定的風險水準下獲得最大化的預期收益，或者在一定的預期收益水準下承擔最小的風險。

4. Maximizing shareholder value requires managers to be able to balance capital funding between investments in projects that increase the firm's long term profitability and sustainability, along with paying excess cash in the form of dividends to shareholders.

想要最大化股東價值，那麼管理者就需要合理規劃，是將剩餘的資金投向可以促進企業長期利潤和可持續發展的項目，還是作為股利發放給股東。

Lesson 2　The Fundamental Insights of Finance

In fact, there are two scopes as the fundamental insight measure finance in different dimensions and connect the micro- and macro-economic theories in overlapped fields.

The first topic is about the efficient markets. Other than the Pareto efficiency in neoclassical equilibrium, efficient markets in financerefer to the utilization of all the available information in setting prices. It means that individual traders could base on their own situations as well as information to take positions in monetary assets, these decisions constitute the market prices which reflect information and also traders' senses to them. If the capital markets are competitive and efficient, the return which an investor expects to get on an investment in certain assets will be equal to the opportunity cost of using the funds. This hypothesis is the backbone of much of financial research and it continues to guide a large body of theoretical and empirical work. And by specifying the information set that is determine prices, efficient markets could be ranked as strong-form efficiency, semi-strong efficiency and weak-form one. Strong-form efficiency indicates that the information set, used by markets to set prices at each date contains all of the available information which not only is the public one embodied, but also the privately held information as well.　The semi-strong one excludes the private information, and the weak-form one, at the bottom of efficiency hierarchy, requires only that the current and past price history be incorporated in the information set. As most economic theories, the efficiency markets hypothesis (EMH) is more intuitive than empirical, because its supportive evidences are as spread as its challenge in tests. Nevertheless, it is the fundamental insight of finance without competitor.

The trade-off between risk and return, as a derivate of EMH, is the second topic of finance. In a competitive market, higher return is accompanied with higher risk is a core intuition of economics which traces back to Adam Smith. In financial theory, the description of risk and return is explained as risk premia, the gap between expected returns and riskless interest rate. It means any traders make decision to alter the proportion of portfolio by investing in an asset will depends on whether the cost of doing so in terms of risk more or less than the benefit of expected return. Hence some scholars created a rigorous micro-model in a mean-variance world where given levels of variance weight the risk of any investing behavior so that investors have to choose the portfolio with highest mean return in different circumstances. And a well-diversified investment will eliminate the unsystematic risk in portfolio that only leaves systematic risk what requires to be priced. And this certain risk is defined with reference to the market portfolio as the residual from a regression of returns on the market portfolio's returns, so this relationship drives us to calculate a factor structure among different portfolios.

As mentioned earlier, close substitutes have the same price so that arbitrage is an undoubted tool in finance. The EMH says that the price of an asset should fully reflect all available information, so we can suggest that if the price has not reflected the information, then there is an arbitrage opportunity to get profit by the inappropriate price. This kind of profit is risk free and attractive to any investors, and their behaviors would change the demand and supply of certain assets. This process pushes the price back to equilibrium without arbitrage opportunity, and such a price is reasonable. So in finance, the absence of arbitrage is an important premise of pricing and as a principle which is applied in investment decision, option pricing and corporation financing.

Based on these micro foundations, capital markets offer a macro-system to merge policies and instruments by specific financial activities, the constant interplay between empirical test and theory consists of the comprehensive concept of modern finance, which has changed the economic and social development profoundly around the world.

New Words, Phrases and Expressions
dimension   n. 尺寸；面積，範圍
overlap   n. 重疊部分；vt. 重疊
equilibrium   n. 均衡
backbone   n. 支柱；脊梁骨，脊椎；分水嶺；毅力
embody   vt. 表現，象徵；包含，收錄
hierarchy   n. ［計］分層，層次；等級制度；統治集團
rigorous   adj. 嚴格的；嚴密的；縝密的；枯燥的
arbitrage   n. 仲裁；套匯，套利
interplay   n. 相互作用
be accompanied with   伴隨著

Notes

1. If the capital markets are competitive and efficient, the return which an investor expects to get on an investment in certain assets will be equal to the opportunity cost of using the funds.

如果資本市場是充分競爭及有效的，那麼投資者對一項資產的投資回報預期與其機會成本是相等的。

2. As most economic theories, the efficiency markets hypothesis (EMH) is more intuitive than empirical, because its supportive evidences are as spread as its challenge in tests. Nevertheless, it is the fundamental insight of finance without competitor.

和大多數經濟理論一樣，有效市場假說的理論意義比實證意義要大，因為它的實證支持結果同時也是對其成立性的挑戰。不過無論如何，有效市場假說都是金融研究中最為深刻的理論基礎。

3. Hence some scholars created a rigorous micro-model in a mean-variance world where given levels of variance weight the risk of any investing behavior so that investors have to choose the portfolio with highest mean return in different circumstances.

因此一些學者基於均值-方差方法建立了一個嚴格的個體模型，在模型裡每個給定水準下的方差值都可以衡量一定的投資行為的風險值，因此投資者需要做的，就是在不同的方差水準下選擇最高回報的資產組合。

4. So in finance, the absence of arbitrage is an important premise of pricing and as a principle which is applied in investment decision, option pricing and corporation financing.

所以在金融領域中，沒有套利機會是進行定價的一個重要前提，同樣也是投資決策、期權定價和公司金融所遵循的原則。

Section 2　The Framework of International Finance Theory

Lesson 1　Exchange Rates and Balance of Payments

The economics of the international economy can be divided into two broad subfields: the study of international trade and the study of international money. The real side and the monetary side are contacted, influenced and cause-and-effect each other. Following the definition of finance in last lesson, international finance is concerned with the determination of real income and the allocation of consumption over time in economies linked to world markets. International monetary analysis focuses on the monetary side of the international economy, that is, on financial transactions such as foreign purchases of certain currencies and capital flows among countries.

Almost every country issues its own currency which is only circulated in domestic. But nowadays to be isolated in the tendency of economic globalization is difficult and unwise, thus there is a requirement of measuring the value of goods and services by different currencies, and the basic and fundamental tool is exchange rates.

Foreign exchange is an acceptable medium of payment includes foreign currencies, foreign securities and foreign currency-based pay orders, and the flow of international capital and trade asks exchange reserves as a guarantee of payment. Then the prices of certain currency towards other currencies in global market are the exchange rates. There are two different ways to express exchange rates: the direct quotation means how many units of native foreign it takes to purchase one unit of a foreign currency, meanwhile the indirect quotation is the other way around. Actually these two ways are the reciprocal of each, and we choose one of them for the convenience and conciseness of research. Moreover, exchange rates are a bilateral concept because their value depends on other currencies, so a currency that has lost value relative to another currency is said to have depreciated, and a currency that has gained value relative to another currency is said to have appreciated.

Exchange rates are the external value of currency, and then the market prices of currency are also determined by the equilibrium of demand and supply as other goods. The first, and oldest, exchange-rate theory used to determine if a currency might be over- or undervalued is known as purchasing power parity (PPP). It states that, ignoring transportation costs, tax differentials, and trade restrictions, trade homogeneous goods and services should have the same price in two countries after converting their prices into a common currency. Thus, purchasing power parity is often called the law of one price. In this model, price differences could be considered a factor of currency demand because relatively lower prices in a nation cause the demand for the nation's currency to increase, and the exchange rates, therefore, shift as well. In words, the domestic price level expressed in the domestic currency should equal the foreign price level expressed in the domestic currency.

So far we have discussed nominal exchange rates. These are exchange rates that do not reflect changes in price levels in the two countries. The nominal exchange rate tells us the purchasing power of our own currency in exchange for a foreign currency. The real exchange rate adjusts the nominal exchange rate for changes in nations' price levels and thereby measures the purchasing power of domestic goods and services in exchange for foreign goods and services. Hence, if we want to know how much of another country's goods and services that a unit of our own nation's goods and services can buy, we need to know the value of the real exchange rate. This rate measures the relative price level of native and foreign currencies by calculating the degree of variability in prices instead of only the difference between them, so real exchange rate explains the change of a nation's competiveness in international trade.

As we discussed earlier, international finance is a subject focus on the external balance of open economy. In this situation, the national income equals domestic product plus net factor payments from abroad plus net international transfer payments, and then it promotes the concept of balance of payment. The balance of payments records all transactions of an economy with other regions in given time as a flow indicator, including the current account and the capital and financial account generally. And here we will concentrate on the current account.

According to the national income identity, which the national income is equivalent to the sum of consumption expenditures, investment expenditures and net exports, the current account equals net exports of goods and services (including all net factor payments) plus net transfers. If national expenditure is defined as the sum of consumption and investment, then national income less national expenditure equals the current account. When in surplus, the current account therefore measures the growth of the economy's external assets; when in deficit, it measures the growth of external debt.

Since balance of payments acts as an independent factor in economic analysis, it has become a key point for policy makers and been established a framework of theory. Generally speaking, there are three essential approaches in practice.

The first approach is the elasticities approach to the balance of payments. This theory focuses on the price elasticities of supply and demand for exports and imports. It holds that the devaluation will be successful if the price elasticities of demand for exports and imports are large enough so that the increase in exports sold to foreigners and the reduction in imports bought by domestic residents together more than offset the terms of trade loss caused by the devaluation. Elasticity measures help us to determine the amount by which the quantity of exports supplied and the quantity of imports demanded adjust in response to a change in the exchange rate. In principle, this can be a useful tool when considering how to reduce a balance-of-payments deficit.

The second approach is the absorption approach to the balance of payments. It states that a country's balance of trade will only improve if the country's output of goods and services increases by more than its absorption, where the term 「absorption」 means expenditure by domestic residents on goods and services. The absorption approach shows us why a nation enjoying an economic expansion often experiences a currency appreciation, while a nation in a recession often experiences a currency depreciation, and offers a view of changing internal expenditures to affect the external balance.

And the last approach to the balance of payments is the monetary one. It is an analytical formulation which emphasizes the interaction between the supply and the demand for money in determining the country's overall balance of payments position. Under this approach, the change of money supply by the policy of monetary authority will affect the cash balance in markets, which leads to the alteration of price and income level. Such a chain reaction would be reflected in the flow of reserve currency and the balance of payments finally.

Altogether, as the most underlying and useful indicators, exchange rates and balance of payments are tools to understand and analyze the mechanism of international finance, and also are targets of international finance activities. We will probe into the policies coordination of international finance next lesson.

New Words, Phrases and Expressions
medium   adj. 中等的, 中级的; 普通的; 平均的; 半生

n. 媒介物，媒質；中間，中庸；［數］中位數，平均；
guarantee　n. 保證，擔保；保證人，保證書；抵押品；vt. 保證，擔保
quotation　n. 引用，引證；行情，行市
reciprocal　adj. 互惠的；倒數的；相互的；n. 倒數；互相關聯的事物
bilateral　adj. 兩側的；雙邊的，雙方的；雙向的
homogeneous　adj. 均勻的；同性質的，同類的
convert　vt. （使）轉變；使皈依；兌換，換算；侵占
nominal　adj. 名義上的；微不足道的
devaluation　n. 貶值
indicator　n. 指示器
alteration　n. 變更；變化，改變
coordination　n. 協調；和諧
probe into　探究；細查，深究

Notes

1. Foreign exchange is an acceptable medium of payment includes foreign currencies, foreign securities and foreign currency-based pay orders, and the flow of international capital and trade asks exchange reserves as a guarantee of payment.

外匯是一種被廣為接受的支付方式，其中包括外國貨幣、外國證券和以外幣為標的的支付命令，此外在國際資本和貨物流動中，也會要求一國以外匯儲備作為支付保證。

2. This rate measures the relative price level of native and foreign currencies by calculating the degree of variability in prices instead of only the difference between them, so real exchange rate explains the change of a nation's competiveness in international trade.

真實匯率通過計算價格的波動水準而非簡單統計差價，衡量了本幣與外幣間的相對價格水準，因此真實匯率也可以解釋一國在國際貿易中競爭力的變動。

3. It holds that the devaluation will be successful if the price elasticities of demand for exports and imports are large enough so that the increase in exports sold to foreigners and the reduction in imports bought by domestic residents together more than offset the terms of trade loss caused by the devaluation

當一國出口和進口需求的彈性足夠大時，彈性分析法認為本幣貶值是有效的，因為本國出口的增長和進口的減少之和足以彌補貶值帶來的貿易損失。

Lesson 2　Real Exchange Rate and the Balassa-Samuelson Effect

Last lesson we have mentioned the concept of real exchange rate (RER). In fact, it is still a vague field in modern economic research that we still can not find its exact definition, sources and affection to economy.

Usually, we divide RER as external exchange rate and internal exchange rate. The external one refers to our natural understanding of RER, it reveals the relative price difference between countries. This index equals to the nominal exchange rate multiply by the ratio of foreign good's price level and native good's price level. Obviously, we could review the purchasing power parity (PPP) which means the price change in countries would be reflected on the nominal exchange rate and be offset immediately. Thus, the RER would identically remain constant over time if the PPP has empirical evidence.

However, there is hardly efficient test to support the PPP in short terms or most countries. Maybe it could be robust in long term tendency, but its volatility and path backwards to equilibrium is unacceptable even unpredictable, because the assumptions of this concept like the absence of friction in trade and a common goods basket in two countries are too idealized to realize. Then we consider PPP is not valid in reality, hence the RER would be variable.

Another way to examine the reason why PPP invalid is the sector effect in economy. The PPP might be effective in the tradable sector, but no doubt some industries are nontradable like services, retail and information technology. These sectors have remarkable influence on the nationwide price level, but its influence is not reflected in the nominal exchange rate which is only concerned with tradable goods.

We can assume a world with only two sectors, tradable sector and nontradable sector. And there are two small open economy countries in the world which use capital and labor to produce tradable goods priced in the world market and the nontradablespriced in the domestic market. Labor is immobile between countries and capital is perfectly mobile internationally. The economy is fully employed and the labor supply is unlimited.

The tradable sector is much easier to develop since it has better involved in the world economy with higher skill and technology level. So the productivity of tradable sector would grow faster than the nontradable sector, the tradable sector needs to pay more to workers if it wants to attract enough workers. We have hypothesized the labor mobility is perfect in domestic, then the nontradable sector has to raise the pay to avoid labor shortage, too. The increasing wage level would lead to the inflation in nontradable goods price despite its productivity is barely growing. Then we can express the RER as internal exchange rate which replaces the price part in external one as a weighted price of tradable goods and nontrable goods.

Accordingly, we can find out that the nontradable goods price raising will lead to the total inflationin one country, and the RER will appreciates immediately. Besides, we can source the inflation from the raise of productivity in tradable sector, then the price differential actually is the gap between the productivity of two sectors. This inference highlights that the alternation of relative productivity in tradable and nontradable sector is the reason of RER changes. And it is also the conclusion of Balassa-Samuelson Effect.

Perhaps it is the most far-reaching theory in RER and PPP, and there are large number of literatures to examine this effect in different period and countries, and the results are widely divergent. Nowadays, some criticize the effect concerns too much on the supply-side and ignores the attribution from demand-side, some of its assumptions like unlimited labor supply is far from real world, and researchers find that it is hard to distinguish the tradable and nontradable sector in economy for the boundary is confusing and their functions are mixing. Notwithstanding, the Balassa-Samuelson Effect still recasts the theory of RER even the international finance by the brief outline and clear propositions to explain the long-run trend deviation from PPP and wide variety of related economy phenomena, which has profound influence on RER estimation and evaluating terms of trade, both theoretical and empirical.

New Words, Phrases and Expressions
vague   adj. 模糊的；不清楚的
robust   adj. 穩健的；堅定的
sector   n. 部門；領域
mobile   adj. 可移動的；行動自如的；易變的；流動性的
inflation   n. 通貨膨脹
divergent   adj. 發散的；有分歧的
attribution   n. 歸因；歸屬
recast   vt. 重塑；再鑄
outline   n. 梗概，大綱，提綱，草稿，要點，主要原則
proposition   n. 命題；主張；建議
Notes

1. However, there is hardly efficient test to support the PPP in short terms or most countries. Maybe it could be robust in long term tendency, but its volatility and path backwards to equilibrium is unacceptable even unpredictable, because the assumptions of this concept like the absence of friction in trade and a common goods basket in two countries are too idealized to realize.

但是，在實證中很難得到購買力平價在短期或者在大多數國家成立的結果。或許在長期中購買力平價能夠得到證實，但是它的波動性和迴歸均衡的路徑對於我們來說是不可接受甚至難以預測的，因為對購買力平價的假設比如貿易環境無摩擦、各國一籃子商品組成一致，就現實而言很難實現。

2. Nowadays, some criticize the effect concerns too much on the supply-side and ignores the attribution from demand-side, some of its assumptions like unlimited labor supply is far from real world, and researchers find that it is hard to distinguish the tradable and nontradable sector in economy for the boundary is confusing and their functions are mixing.

如今，一些對巴拉薩-薩繆爾森效應的批評認為，這一理論過於關注供給面而忽略了需求面的因素，它的一些假設比如勞動力無限供給也與現實差異很大，研究者們在實證時也難以區分可貿易部門和不可貿易部門，這些部門的功能也存在交叉。

Section 3 A Brief Introduction to Financial Instruments and Assets

Lesson 1 Basic Financial Instruments and the Derivatives

Financial assets refer to the reasonable requirement on behalf of future income that as a symmetry concept of physical assets, also known as the financial instruments. Naturally, anyone who chooses to hold a financial instrument must contend with the risk associated with the instrument. For one thing, there is always a possibility that issuer of the instrument may default. Holding financial instruments, such as a loans or securities, issued by individuals, companies, or governments entails an extension of credit. The amount of credit via the purchase of a financial instrument is the principal amount of the loan or the security. Payments from the issuers that compensate the purchasers for the use of their funds constitute interest. The amount of interest as a percentage of the principal of a financial instrument is the interest rate. The issuer of a bond often sells the bond at a discount, meaning that the bond's selling price is less than its fact value. Therefore, the holder of the bond automatically earns a capital gain. This is an increase in the bond's market value, relative to its market value at the time of purchase, if he or she holds the bond until maturity.

Most bonds have a fixed term to maturity, or a finite period between the initial purchase of the instrument and the eventual receipt of principal and promised interest payments. It turns out that the term to maturity is a key factor influencing the degree of interest-rate risk associated with a financial instrument. This is the risk of variations in the market value of the financial instrument due to interest rate variations. Interest rate may arise from length of maturity and frequency of payments. In addition, equities are another kind of elementary instruments. They are ownership shares that might, or might not, pay the holder a dividend. Usually values of equities rise and fall with investors' expectations of the value of certain enterprise, but the change of interest rates would affect people's expectation profoundly.

Thus, allocating most funds to holding of short-term instruments is a potentially costly strategy for addressing interest-rate risk. An alternative strategy is to hedge interest-rate risk using other financial instruments in ways that reduce portfolio risks. A perfect hedge constitutes a strategy that fully eliminates such risks. Holders of financial instruments must meet two key requirements to hedge against portfolio risks. One is sufficient fluidity to adapt to various situations. Another is the capability to conduct that combine both flexibility and speed has led to the development of sophisticated financial instruments called derivatives.

There are several categories of derivative securities. Traders may use them as hedging instruments, or use them in speculative strategies. The first instrument is forward contracts, which are contracts guaranteeing the future sale of a financial instrument at a specified interest rate as of a specific date. But there are some factors may limit forward contracts are setting terms of the contract and default risk. Sometimes it is difficult for two parties to reach agreement, and the default risk might increase the risk and less the liquidity of contracts that restraining its trading volume.

Then the futures exchange is in need. It is, on contrast of forward contracts an agreement of specifying standardized quantities and terms of exchange, then parties do not have to spend time working out contracts terms. Futures contracts require daily cash-flow settlements so that holders of futures experience profits or losses in the contract time. Furthermore, the futures exchange is an organized market that simplifies the process of trading. This makes futures contracts highly liquid.

As another type of derivative instrument, European-type call (put) option is a security that gives its owner the right to buy (sell) a specified quantity of a financial or real asset at a specified price, the 「exercise price」, on a specified date, the 「expiration date」. An American-type option provides that its owner can exercise the option on or before the expiration date. If an option is not exercised on or before the expiration date, it expires and becomes worthless. Depending on these, call options are options that allow the holder to purchase a financial instrument at the exercise price, and put options are which allow the buyer to sell such instrument at a price. Options and forward or futures contracts are fundamentally different securities. Both provide for the purchase (or sale) of the underlying asset at a future date. A long position in a forward contract obliges its holder to make an unconditional purchase of the asset at the forward price. Moreover, the holder of a call option can choose whether or not to purchase the asset at the exercise price. Thus, a forward contract can have a negative value whereas an option contract never can.

The last contracts are swaps in which parties to the transactions exchange flows of payments. And the currency swap is one of them that come into common use, which is an exchange of payment flows denominated in different currencies. Firms always use it to lock in the domestic currency value of a debt payment or a future receipt or to arrange better loan terms.

In fact, these financial instruments' returns depend on the returns of other financial instruments. As a result, traders can use these securities to hedge against or speculate on interest rate risks.

And these instruments mentioned before are also tools for the typical risk, the foreign exchange rate risk, in international finance. Exchange rates vary over time may expose the members who engage in international transactions to potential risk. An individual or firm may be exposed to foreign exchange risk in any of three types. The first type is transaction exposure, which is the risk that the cost of a transaction in terms of the domestic currency, may change. It is created when a firm agrees to complete a foreign-currency-denominated transaction in the future. The second type is the translation exposure, which arises when translating the value of foreign-currency-denominated assets and liabilities into a single currency value. The last type of such risk is economic exposure, which is the effect that exchange-rate changes have on a firm's present value of future income screams, and it affects the ability of a firm to compete in a particular market over an extended period.

A change in the exchange rate may be positive or negative from the perspective of an individual or firm, the uncertainty it introduces is the individual has the incentive to reducing or eliminating the volatility, thus hedging with relevant financial instruments is an effective act of offsetting exposure to risk partially or completely, and once the risk is offset we may say the exposure is covered.

New Words, Phrases and Expressions
default　v, n. 未履行，拖欠；未參加或完成（例如，比賽）
entail　vt. 需要；牽涉；使必要
compensate　vt. 抵消；補償；賠償；報酬
alternative　adj. 替代的；另類的；備選的；其他的；n. 可供選擇的事物
maturity　n. 成熟；完備；（票據等的）到期
expiration　n. 截止；滿期；呼氣
elementary　adj. 初等；基本的；初級的
sophisticated　adj. 複雜的；精致的
denominated　adj. 計價的
exposure　n. 暴露；曝光；揭發；公開
perspective　n. 透鏡，望遠鏡；觀點，看法
incentive　n. 動機；誘因；刺激；鼓勵
contend with　與（某人）競爭；與（某事物）抗爭

Notes
1. Naturally, anyone who chooses to hold a financial instrument must contend with the risk associated with the instrument. For one thing, there is always a possibility that issuer of the instrument may default.

任何持有金融工具的投資者都必須面對金融工具帶來的風險，因為首先，任何金融工具的發行者都存在違約的可能。

2. Holders of financial instruments must meet two key requirements to hedge against portfolio risks. One is sufficient fluidity to adapt to various situations. Another is the capability to conduct that combine both flexibility and speed has led to the development of sophisticated financial instruments called derivatives.

金融工具的持有者要實現組合風險條件下的套期保值，必須滿足兩個關鍵要求：一是有充足的流動性以應對不同的情況；另一個則是有能力去滿足套期保值策略靈活性和流動速度，而這也導致了叫做金融衍生品的更為複雜的工具的產生。

3. Firms always use it to lock in the domestic currency value of a debt payment or a future receipt or to arrange better loan terms.

公司機構通常運用互換協議來鎖定債務償還、未來支付時的本幣價格或者實現更好的貸款期限配置。

Lesson 2 The Optimization of Portfolio

Why optimize? How to optimize? These questions are investors who holds bunch of assets and derivatives wants to acquaint. In general, investment is not a simple object process with unique asset or instrument. We all know the old proverb that don't put the eggs in the same basket, but it is not enough. The more important is to put eggs into suitable basket with appropriate arrangement. And this arrangement is so-called optimization.

There will be astronomical sum of portfolios to compound if we step into global financial markets. Stocks, bonds and other derivatives would consist a large pool to hedge risk. Optimization aims to narrow choices to the very best when there are virtually innumerable options and comparing them is difficult. Optimization is valuable in environments where complex decisions have to be made and where doing so is daunting because of massive amounts of data, a huge number of choices, and intricate relations.

Such optimization relies on advanced analytical methods and clean data to seize the sign from information and offersthe basis of decision. Consistently adding value through selecting stocks or other assets is an obvious challenge for every rational investor. As we mentioned in past lessons, a portfolio is considered efficient if it achieves maximum expected return under a given level of expected risk or it bears minimum risk at a given level of expected return. And optimization is a way to find this kind of portfolio in short time.

For the international fund allocation, optimization is much more difficult than domestics because the constraints and rules in markets are distinct from each other. For example, the transaction fee and tax would be much different in China and US market, some markets or products not allow fictitious transaction. These restrictions are obstacles to our estimation or modelling for improving the portfolio. To simplify the process of quantization, we often use factor model instead of direct calculating the expected return. The factor model uses a series factor of fundamental affairs, industries, macro-economics, or statistic method to reflect the sources of expected return, we might analyze the return into different area. Then we have a much clear path to attribute the reason of invest performance. Moreover, every asset has different exposure to these factors, we may take this advantage to hedge the risk, in other words, to optimize the portfolio.

The model to optimize is called optimizer. It is designed to find the very best solution and will always do so given enough time. Unfortunately, the timerequired to find the very best solution (and prove that it is the very best solution) may be more than is acceptable. As a compromise, optimizers produce a very good solution (although not necessarily the very best) in a short amount of time. This compromise is achieved in two ways. First, the search for the very best solution is terminated prematurely and the best solution found thus far is returned. (Often this is the very best solution, but the optimizer has not had enough time to prove that.) Second, certain goals and business rules are treated approximately rather than exactly.

Through optimizing we might find that whether we over or under weighted some instruments. But the actual aporia is the multinational transaction. You may know many about the local market, but the market oversea is a black box with hidden risks. The local funds would be a good choice. They are more professional than you in their markets and the fund has been a portfolio already. But fund is also kind of instrument itself. It will be a cross-nation portfolio if we consist different fund from different country, we call it fund of fund (FOF). This is a smart and active way to avoid extra risk from information shortage.

Without doubt, FOF is not the ultimate answer for global investment. Funds supply the liquidation of market and face the threaten of crisis at first. The global markets are much correlated as same as the invest strategies of funds, so FOF is efficient to solve the manage risk but is not compliant to deal with fat-tail risk, sometimes it might get the situation worse in market recession. Typically, optimization is a tool created to measure the quality of our investment, an evaluation of a portfolio management optimizer should not only focus on the decisions that have to be made now, but also on decisions that (may) have to be made in the future.

New Words, Phrases and Expressions
optimize   vt. 使最優化，使盡可能有效
acquaint   vt. 使熟悉；使認識
proverb   n. 諺語，格言；話柄，笑柄
appropriate   adj. 適當的；合適的；恰當的
daunting   adj. 令人畏懼的；使人氣餒的；令人卻步的
intricate adj. 錯綜複雜的；難理解的
compromise   n. 妥協；（名譽等的）損害
aporia   n. 難題，難點
recession   n. 經濟衰退，不景氣
fictitious transaction   買空賣空
fat-tail   肥尾分佈

Notes
1. Optimization aims to narrow choices to the very best when there are virtually innumerable options and comparing them is difficult.
優化的目的在於當我們有許多種選擇並且難以比較選擇間的優劣的時候，來幫助我們縮小選擇的範圍並找到最好的那一個。

2. Typically, optimization is a tool created to measure the quality of our investment, an evaluation of a portfolio management optimizer should not only focus on the decisions that have to be made now, but also on decisions that (may) have to be made in the future.

通常來說，優化是一個用來衡量我們投資質量的工具，但是對投資管理優化器的評估，不僅要立足於當下的決策質量，也要展望優化結果的未來。

Section 4　A General View of International Financial Institution

Lesson 1　Banks: From Domestic to Global

In the era of globalization, banks are no longer isolated storehouse for currency and deposits denominated in domestic currencies, nor do they look only at home for potential customers. Similarly, bank customers are becoming keener to the international opportunities to decrease their costs or improve their returns. And undoubtedly, regulators and central banks must pay more attention to the global interdependence of the world's financial institutions. And this section we will focus on the international dimensions of financial intermediaries and organizations.

Corporations and individuals have two main channels for finance. Under the situation of indirect finance, banks are the center intermediaries to transfer the deposits to the fund demanders. But for the direct way, the financing is a process that completely happens in capital markets as stock markets.

There are many reasons why a saver might wish to hold bonds issued by other country's company. One reason is to earn an anticipated higher return. Another reason might be to avoid country risk. More broadly, the saver might want to achieve overall risk reductions via international financial diversification. The role that banks has played in indirect finance is working as an effective way to avoid information asymmetry so that reduce the costs of matchmaking and regulatory for both sides. Moreover, by holding a world index fund, which portfolio is consisted of bonds issued in various nations, the saver can earn the average yield on securities of a number of nations while keeping overall risk to minimum. Holding shares like this makes the saver a part of process of international financial intermediation. It is kinds of indirect finance of capital investment across national border by financial intermediaries such as investment banks or pension fund companies.

The business of banking varies from nation to nation. In some countries, banks are the key vehicle for financial intermediation, whereas in others, banking is only one component of a high diversified financial system. The key attribution of this difference is whether the bank is the predominance of firm capital working. In Western Europe, Japan, and China, bank loans have dominated larger shares of their finance as compared with the business located in United States, where prefers to direct finance in capital markets. There is a boom for investment banks after the Great Recession, some says the investment banks characterize the history of Wall Street in 20th century.

These banks establish a bridge or a channel for corporations and capital markets by the services of IPOs, hedge funds, consultations and capital managements. Their growth is accompanied by the breakthrough of financial theory and the new design of derivatives. The investment banks claim themselves as the role of profession and modern, and release an impression of profitable and geek. The direct way improves the efficiency and the information transparency of modern finance on one side, for another the liquid of international capital fasten the linkage among global markets. Then an alternation of the economy indicator in a major country would affect the expectation of investors all over the world. It brings a chance and also a challenge to the future of investment banks. Failing to quantify the precise risk worldwide and guard against the black swan is the limitation of the modern financial model. The bankruptcy of Long Term Capital Management (LTCM) in 1990s and the breakdown of Lehman Brothers in 2008 are cases to warn investment banks.

Nowadays, the traditional gap between banks and investment bank is getting blur, universal banking, which offers full range of financial services and to own equity shares in corporations, is the general trend of finance. Those who favor universal banking point out that bank holds individually risky equity shares may reduce overall bank risk by the balance of different returns in portfolio.

Banks located in various countries are the foundation of finance globalization by using the funds around the world to loans to individuals or companies in other nations. Such banking institutions, often called megabanks, have profound influence in international finance. The reason of these large banks exist is the economies of scale. By increasing their asset portfolios through regional or worldwide expansion, megabanks may be able to reduce average operating costs, thereby gaining efficiency. Some economists believe that the expanding dimension of these banks is the requirement of evaluating and monitoring the creditworthiness of these multinational enterprises. This is also a better way to overcome the information asymmetry than just operating domestically.

New Words, Phrases and Expressions
storehouse   n. 倉庫；貨棧
keen to   熱衷於
intermediary   n. 媒介；中間人；調解人
anticipate   vt. 預見；預料；預感；先於……行動
matchmaking   n. 作媒；撮合
vehicle   n. 交通工具；媒介
yield   vi. 放棄，屈服；生利；退讓，退位；   n. 產量，產額；投資的收益

diversification   n. 變化，多樣化；多種經營
blur   n. 污跡，污斑；模糊不清的事物；   v. 塗污，弄臟；（使）變模糊
megabank   n. 大型銀行
profound   adj. 深厚的；意義深遠的
creditworthiness   n. 商譽
the Great Recession   大蕭條
black swan   黑天鵝，小概率事件

Notes

1. The role that banks has played in indirect finance is working as an effective way to avoid information asymmetry so that reduce the costs of matchmaking and regulatory for both sides.

在間接金融條件下，銀行的功能在於可以有效規避信息不對稱，這使得金融交易雙方撮合和監管的成本都顯著降低。

2. By increasing their asset portfolios through regional or worldwide expansion, megabanks may be able to reduce average operating costs, thereby gaining efficiency. Some economists believe that the expanding dimension of these banks is the requirement of evaluating and monitoring the creditworthiness of these multinational enterprises.

基於跨地區乃至全球範圍內的資產規模擴張，跨國銀行可以通過降低營運成本來提高效率。而許多經濟學家也相信，銀行擴張業務維度，也是出於評估和管理跨國公司商譽的需要。

Lesson 2   International Financial Institution

Though the coordination of economics is embedded, there is no any international center bank with legal force to restrain the global capital flow. But it does not mean that there are no corresponding regulators to monitor the megabanks and coordinate the global monetary policies.

In 1989 the regulators of banks in most advanced-market-based economies gathers at the Bank for International Settlements in Basel, Switzerland, to announce a system of risk-based capital requirements. The Basel Accords, since now it has updated to the third edition, the Basel Committee intends to account for varying asset risk characteristics into the calculation of required bank capital. The committee does not have the authority to enforce recommendations, although most members as well as some other countries and regions tend to implement the Committee's policies. Under this accords, banks must compute ratios of capital in relation to their risk-adjusted asset, and confront the rigid management requirement in overall. This global standard has a significant effect on international banking competition, and triggers a comprehensive adjustment in banking services.

International Monetary Fund (IMF) was founded in 1946, working to foster global monetary cooperation, secure financial stability, facilitate international trade, promote high employment and sustainable economic growth, and reduce poverty around the world. Generally speaking, IMF is treated as a monitor to coordinate global monetary policy by offering loans to its members when they face to the dilemma in balance of payments. Members have their own shares in fund which depend on the member's trade scale and the reserve they have paid as dues, and such shares are equal to corresponding rights to vote in operation of fund. Actually the regular dues are a part of the member's foreign exchange reserve, and yet for all that, IMF issues another asset which named Special Drawing Rights (SDR) as an addition global reserve, which is a composite currency to serve as the currency that provides extra liquidity needed so that central banks would use to settle transactions. IMF also supports the member who falls into financial crises, especially to developing countries, by offering loans. But such loans are conditional that require a promise of specific policies or indicators adjustments, so sometimes recipient member may upset about it and announces the so-called aids as political interferences. In conclusion, IMF is not so much a center bank as a consulting organization, for its weak constraint to major members but great influence to developing nations. The division of IMF's duties and services leads it to play an awkward role so far.

Another worldwide financial organization is World Bank Group. It is founder with IMF simultaneously, and the initial goals of World Bank are to help Europe recovers from the World War II. But now it has change the mission in global scale as ending extreme poverty within a generation and boosting shared prosperity. World Bank offers loan to members for industry, agriculture, education, transportation and so on, backing underprivileged countries and regions to improve living standards and infrastructure levels. Besides, it also provides consultations to transition countries in long term planning. But the growing gulf between rich and poor countries is a huge test to evaluate the work of World Bank, as same as the execution of its policies are questioned. Traditionally, based on a tacit understanding between the United States and Europe, the president of the World Bank has always been selected from candidates nominated by the United States. In 2012, for the first time, two non-US citizens, Jim Yong Kim, were nominated.

IMF and World Bank Group also contribute to overcome international information asymmetry by their scale advantages. These institutions are a microcosm of global governance, which reflects the opportunities and challenges in the international financial markets. The emerging markets countries have more demands on financial co-operation, the Asian Infrastructure Investment Bank (AIIB) and the New Development Bank attract the interests and curiosity of the world, we need to wait and see whether it is the time to turn the international structure around.

New Words, Phrases and Expressions
embedded   adj. 植入的，深入的，内含的

committee   n. 委員會
trigger   vt. 引發，觸發；扣⋯的扳機；n. 扳機
foster   v. 培養；撫育；促進
consulting   adj. 商議的，顧問資格的，諮詢的
awkward   adj. 笨拙的；令人尷尬的；難對付的；不方便的
interference   n. 干涉，干擾，衝突
prosperity   n. 繁榮；興旺，昌盛；成功
underprivileged   adj. 被剝奪基本權力的，窮困的，下層社會的
transition   n. 過渡，轉變，變遷
gulf   n. 海灣；分歧
microcosm   n. 縮影；微觀世界
scale advantage   規模優勢

Notes

1. The committee does not have the authority to enforce recommendations, although most members as well as some other countries and regions tend to implement the Committee's policies.

儘管所有的成員和其他國家及地區都直接採用了巴塞爾協議的內容，但實際上巴塞爾委員會的規定並沒有強制性。

2. Actually the regular dues are a part of nation's foreign exchange reserve, and yet for all that, IMF issues another asset which named Special Drawing Rights (SDR) as an addition global reserve, which is a composite currency to serve as the currency that provides extra liquidity needed so that central banks would use to settle transactions.

事實上除了普通的外匯儲備，國際貨幣基金組織還提供一種叫做特別提款權的儲備資產，這一資產由混合的一攬子貨幣組成，像貨幣一樣可以在需要時提供給央行來進行償付。

3. World Bank offers loan to members for industry, agriculture, education, transportation and so on, backing underprivileged countries and regions to improve living standards and infrastructure levels.

世界銀行向會員提供工業、農業、教育和交通等一系列貸款，並且支持不發達國家和地區提高生活水準和基礎設施水準。

Section 5　International Financial Coordination and Crisis

Lesson 1　Coordination of International Monetary Policies

Any monetary authorities have the motivation and obligation to launch series policies for expected international financial outcomes, and the first step of such policies is to set an exchange rate regime.

In a simple partition, exchange rate regime could beclassified as the fixed and floating rate system. Under the extreme and traditional explanation, the monetary authority will promise a permanent exchange rate in fixed rate system, the typical example is the gold standard that official can deal the current price ratio by selling or buying gold to ensure currency stability. Meanwhile, the floating rate system means that the authority maintains neutrality to the exchange market at any time and never announces or maintains any exchange prices, thus there is only indirect influence from government.

Actually, the choice of exchange rate regime is constrained by the domestic and international economics, so the completely fixed or floating rate systems are rare and aborted in practice now. Between these two extremely arrangement, there are some feasible rules such as crawling peg and dirty float, which are compromises of fixed and floating rate system.

The arrangement of exchange rate system is the center of world monetary system. Indeed, the infrastructure of world exchange framework is decided by the major countries, especially where issue international reserve currencies. For example, almost all the other currencies' exchange rates have pegged to the U. S. dollar, any fluctuations in policy of Federal Reserve would affect global exchange markets.

There is a basic macroeconomic policy trilemma for open economies. Of three goals that most countries share—independence in monetary policy, stability in the exchange rate and the free movement of capital—only two can be reached simultaneously. Such trilemma is established for every country, but unlike developed countries, developing countries have less ability to influence their terms of trade so that their exchange policies are also dependent. While a major economy like the United States changes policies, a smaller economy often finds the costs of such volatility hard to sustain.

Therefore, countries usually seek for cooperation mechanisms to maximum their financial purposes while minimize the costs of spillover effect from elsewhere. However, the defect of an international financial system is the absence of a worldwide central bank with absolute power and obligation, and the international monetary policy has been the balance the decentralize decision making among countries.

In the period of fixed exchange rate such as the Bretton Woods System, countries work together under the agreements of rate. Once the regime turns to be flexible, the excess volatility of exchange rate increases the burden to macroeconomic management, notably its transmission mechanism to domestic price level, even though the floating rate system offers a greater autonomy to make independent policy. The divergent monetary target in each country may conduct a selfish policy decision that worsen the global economic, thus coordination is still an ultimate goal of globalization anticipants, not only an indirect result of making exchange rate rules.

Except the simple exchange rate agreements, there are some other theories and practices for internationalfinancial coordination. Monetary Union is a relatively fledged framework which has been carried out in Europe as European Union (EU). Since 1999, members of EU, excluding the UK, have adopted a common currency named the euro. In fact, this union is a regional fixed exchange rate zone which members have similar economic structures and requirement of integration to reduce the cost of trade and settlement. It is the experiment of the optimum currency area theory, which set a fixed exchange rate internal and a floating one external. For Europe, the euro has much impressive political influence than economic so far, partly because a standard optimum currency area has severe criterions for the mobility of productive factors, also for the process of European integration is too soon to consummate corresponding fiscal system. But it is a meaningful trial in human history due to its complex and interesting cost-benefit analysis.

In the context of treating U. S. dollar as the major reserve currency, another radical monetary policy is dollarization. There is no specific definition about it, generally dollarization is not only about completely choosing U. S. dollar as the legal tender instead of domestic currency, but also including the announcement of a permit that U. S. dollar could circulate as the domestic one simultaneously. As the macroeconomic policy trilemma we mentioned earlier, under the condition of capital unrestricted flow globally, the exchange rate stability and the independent monetary policy making are alternative. Then ignoring the national sentiments, dollarization might be a reasonable choice as a restriction to incompetent monetary authority and malignant inflation. But the loss of seignorage and the explicit fiscal deficit are the other side of dollarization which is too costly to stand in economic cycle. Dollarization is a unilateral demisability of monetary sovereignty from certain country to USA, as the case stands, but what the role Fed should play in this process is ambiguous.

In a few words, the priority of between the domestic price level reasonableness and the exchange rate stability is inconclusive. Like most theory, coordination of international monetary policy is a procedure of game with multi-solution, but unstable in dynamic states.

New Words, Phrases and Expressions
obligation   n. 債務；義務，責任；證券，契約
regime   n. 政治制度，政權，政體；管理，方法
permanent   adj. 永久（性）的，永恆的，不變的，耐久的，穩定的
neutrality   n. 中立，中立地位
aborted   adj. 流產的，失敗的
feasible   adj. 可行的；可用的；可實行的
establish   vt. 建立，創建
decentralize   vt. 權力分散
transmission   n. 傳送；播送
autonomy   n. 自主權；自治，自治權
ultimate   adj. 極限的；最後的；最大的；首要的；n. 終極；頂點；基本原理

criterions   n. 規範；(批評、判斷等的) 標準，準則
optimum   adj. 最適宜的；n. 最佳效果
unilateral   adj. 片面的；單邊的，一方的
demisability   n. 讓渡，遺贈
sovereignty   n. 主權國家；國家的主權
ambiguous   adj. 模棱兩可的；含糊的，不明確的；引起歧義的

Notes

1. Of three goals that most countries share—independence in monetary policy, stability in the exchange rate and the free movement of capital—only two can be reached simultaneously.

一國貨幣政策的三個目標——貨幣政策獨立性、匯率穩定和資本自由流動只有兩個可以同時實現。

2. The divergent monetary target in each country may conduct a selfish policy decision that worsen the global economic, thus coordination is still an ultimate goal of globalization anticipants, not only an indirect result of making exchange rate rules.

各國由於目標不同而獨立制定的貨幣政策往往是自私的，並可能會惡化國際經濟形勢，因此合作是全球化參與者們的共同目標，而不僅僅是制定匯率規則的間接產物。

3. But the loss of seignorage and the explicit fiscal deficit are the other side of dollarization which is too costly to stand in economic cycle.

但是鑄幣稅的損失及財政赤字是美元化帶來的另一方面的影響，這一影響對於一個經濟體來說通常成本太高而難以承受。

Lesson 2 International Financial Markets and the Financial Crisis

In the past decades, the world trade has expanded significantly because of reductions in trade barriers, lower transportation costs and improvement in telecommunications. International financial markets also experienced a rapid growth in the same period, due to relaxation of capital controls, innovation in financial instruments and integration of markets. It is difficult to distinguish the causality between these two processes in an empirical world, but the financial markets were growing remarkably quicker than goods transactions.

Generally, we divide financial markets as capital market and money market. International capital markets are the markets for cross-border exchange of financial instruments that have a maturity of a year or more, and sometimes these instruments have no distinct maturity. And the international money markets, as compared, are the markets exchange instruments which expired within a year.

For capital markets, the most important instruments are bonds and equities. They are the foundation of almost all the other derivatives, as we discussed earlier. Meanwhile, traders exchange a number of different types of instruments in money markets, foreign exchange instruments are most actively traded. The foreign exchange market is a system of private banks, foreign exchange dealers and central banks buy and sell foreign exchange publicly. The media publish the exchange rates on TVs or newspapers everyday are spot exchange rates, and the spot exchange market is a market for immediate purchase and delivery of foreign exchange. A large portion of the transactions in this market consists of exchanges of foreign-currency-denominated deposits among traders, in fact, rather than the actual flow of a currency from one nation to another.

And the forward exchange market is another part of money markets. It is a market for contracts ensuring the future delivery of a currency at specified exchange rate. Because a forward contract guarantees a rate of exchange at a future date, it can cover the exposure of exchange risk. For most currencies the forward exchange rates differ from the spot rate, and the forward rates with different maturity periods are different, too. If there are no currency restrictions or government interventions, the supply and demand determines forward exchange rates. Therefore, the forward rate reflects the supply and demand of a currency for future delivery, and the rate provides information of the future spot exchange rate. An efficient market is one where market prices adjust quickly to relevant information. Thus, an efficient foreign exchange market means the forward exchange rate is a good predictor of the future spot rate, and it should adjust duly to eliminate the opportunities of arbitrage profits.

The most well-known international financial market is the Eurodollar market, and it must be said that it is a debt market instead of a money market. Other than the traditional financial instruments, Eurodollar is a bankdeposit denominated in a currency other than that of the nation in which the bank deposit is located. At first it refers in particular to the deposit denominated in U. S. dollar, but now it is a general designation of similar assets, and its geographical concept is spread to all over the world. This market is not the extension of domestic currency system because of there is no any regulator to supervise or support its operation, its practicability lies on the solvency of borrowers, banks in this market, to attract individuals, corporations, monetary institutions and even governments to save denominated currency in maintaining the liquidity. But the absence of regulators is a shadow over the Eurodollar market all along, the obscure credit terms of lenders has deteriorated the precondition of international coordination, especially when the lenders are low credit sovereignties.

Even though the expansion and integration of global financial markets has reduced the cost of investment and fundraising, the financialliberalization is also an incentive to global financial crisis. The capital flow which concentrates on the cross-border securities investments, also known as hot money, is free to enter and depart from domestic market so that causes uncertainty to the market especially in developing countries. Furthermore, the crisis would be highly contagious from one nation to another in an interdependent markets system. Thus a domestic financial problem would evolve in a regional and even global turbulence, which is a threat to the world financial system.

There are three basic theories to distinguish reasons in financial crisis. The first focuses on the potential element in economic. The authority's monetary and fiscal policies might be mismatch to its exchange rate target, and then the center bank will exhaust the reserves to withstand international speculation, finally beaten in the depression by currency depreciation. The second model aims to the effect of self-fulfilling expectation. It brings that, even the center bank has enough reserves, the trader will take measures to arbitrage once the real exchange is deviated from the expected rate, and such actions would lead others to follow the trend that aggravate the departure, and the crisis may just occur by an expected successful speculation. The last theory considers that the defects of nation's monetary system must hold responsible for the crisis. The twist of corporations and government is kind of moral hazard to market, which is the reason of non-performing loans and financial collapse.

Economists expect to set an early warning system to financial crisis, despite there are still plenty explanations to the same economic phenomena, and the last three theories are thought insufficient to analysis a crisis. The financial markets globalization is irreversible, however, so perhaps reconsidering about the system structure and supportive measures of international financial markets is much more urgent today.

New Words, Phrases and Expressions
innovation   n. 改革, 創新; 新觀念
distinguish   vt. 區分, 辨別, 分清; 辨別出, 識別
integration   n. 結合; 混合; 一體化
designation   n. 指定; 名稱; 選派
obscure   adj. 昏暗的, 朦朧的; 晦澀的, 不清楚的; 隱蔽的
deteriorate   v. 惡化, 變壞
turbulence   n. 騷動, 騷亂
exhaust   vt. 排出; 用盡, 耗盡
withstand   vt. 經受, 承受, 禁得起; 反抗
aggravate   vt. 激怒; 加重, 使惡化
duly to   適當地
deviated from   偏離
self-fulfilling   自我滿足的
evolve in   進化

Notes

1. Thus, an efficient foreign exchange market means the forward exchange rate is a good predictor of the future spot rate, and it should adjust duly to eliminate the opportunities of arbitrage profits.

所以一個有效的外匯市場意味著其遠期匯率是未來即期匯率良好的預測器, 而

且遠期匯率還會為消除套利機會而適當變動。

2. But the absence of regulators is a shadow over the Eurodollar market all along, the obscure credit terms of lenders has deteriorated the precondition of international coordination, especially when the lenders are low credit sovereignties.

但是缺乏監管者一直是歐洲美元市場的問題所在，借款者不清晰的信用狀況危害了國際合作的前提，尤其是在這些借款者是低信用主權國家的時候。

3. The financial markets globalization is irreversible, however, so perhaps reconsidering about the system structure and supportive measures of international financial markets is much more urgent today.

但是金融市場的全球化趨勢已經不可逆轉，所以對金融系統結構和國際金融市場支持手段的重新思考，在今天尤為緊迫。

Exercises

Ⅰ. Multiple Choice

1. International economic integration refers to：

A. the expansion of world governance and society.

B. the increased mobility of peoples and information.

C. the strengthening of existing international linkages of commerce and the addition of new linkages.

D. the strengthening of existing international linkages of commerce but not the addition of new linkages.

2. A market that involves the immediate sale or purchase of an asset is known as a (n)：

A. currency market.

B. bond market.

C. spot Market.

D. free market.

3. The act of offsetting exposure to risk is known as：

A. dodging.

B. hedging.

C. speculating.

D. avoidance.

4. The existence of unexploited profit opportunities is referred to as：

A. the parity condition.

B. equilibrium.

C. interest rate parity.

D. an arbitrage opportunity.

5. Which of the following instruments provides that its owner can exercise it on or before the expiration date?

A. Bonds.
B. Equities.
C. European-type option.
D. American-type option.
6. Which of the following exchange rates is adjusted for price changes?
A. Nominal exchange rate.
B. Real exchange rate.
C. Effective exchange rate.
D. Forward exchange rate.
7. Which of the following options is not included in the current account of payments?
A. Goods.
B. Services.
C. Foreign direct investments.
D. Current transfer.

Ⅱ. Define the following concepts
1. efficient market
2. foreign exchange risk
3. purchasing power parity
4. absorption approach
5. basic macroeconomic policy trilemma
6. forward contracts financial crisis
7. the Special Drawing Rights?

Ⅲ. Short Answer Questions
1. Collectingthe relative information, trying to evaluate the works of IMF and World Bank Group.
2. Based on the macroeconomic policy trilemma theory, discussing about the policy choice of developing and developed countries.
3. Discuss the prospect of international monetarypolicies coordination.
4. Why did some economists announce that the Europe's debt crisis is unavoidable?
5. Discuss the future of RMB globalization.

Further Reading Comprehension

Trading the yuan: Yuawn

Buzz about the rise of China's currency has run far ahead of sedate reality

IF HEADLINES translated into trading volumes, the yuan would be well on its way to dominating the world's currency markets. It once again graced front pages this week after moves to lift its status in London, the world's biggest foreign-exchange market. This was the latest instalment of a five-year-long public-relations campaign. Since 2009, when China first declared its intention to promote the yuan internationally, a string of announcements and milestones has cast the Chinese currency as a putative rival to the dollar.

The hype rests on several seemingly impressive numbers. Yuan deposits beyond China's borders have increased tenfold in the past five years. The「dim sum」bond market for yuan-denominated debt issued outside China has gone from non-existence to a dozen issuances a month. And the yuan is the second-most-used currency in the world for trade finance.

Adding to the impression that something big is afoot is the competition between cities around the world to establish themselves as yuan-trading hubs. London puffed up its chest this week after the Chinese government designated China Construction Bank as the official clearing bank for yuan-denominated transactions in Britain and agreed to launch direct trading between the pound and the yuan in China. These announcements were made to coincide with a trip to London by Li Keqiang, China's prime minister.

The designation of a clearing bank creates a channel for yuan held in Britain to flow into Chinese capital markets, boosting London's appeal as a trading centre for the currency. Other cities such as Frankfurt and Singapore have also been awarded clearing banks, but London already controls nearly 60% of yuan-denominated trade payments between Asia and Europe, and this week's agreement will shore up its position.

London's currency traders, however, will not be hyperventilating. The rapid growth in the use of the yuan outside China, whether for trade settlement or investment, has been from a minuscule base. The yuan is the seventh-most-used currency in international payments, according to SWIFT, a global transfer system. That is up from 20th place at the start of 2012. However, the Chinese currency still accounts for a mere 1.4% of global payments, compared with the dollar's 42.5%. Given that many of those deals just shuffle cash between Chinese companies and their subsidiaries in Hong Kong, there is much less than meets the eye to the yuan's stature as a trade-settlement currency.

Even more telling is the yuan's standing as an investment currency. The dollar's biggest selling point as a global reserve currency is the deep, liquid pool of American assets open to international buyers. Despite the barrage of reports in recent years about the dim-sum bond market, China's offerings are much sparser. Jonathan Anderson of Emerging Advisors Group calculates that global investors have access to $56 trillion of American assets, including bonds and stocks. They can also get their hands on $29 trillion of euro-denominated assets and $17 trillion of Japanese ones. But when it comes to Chinese assets, just $0.3 trillion or so are open to foreign investors. This puts the yuan on a par with the Philippine peso and a bit above the Peruvian nuevo sol, Mr. Anderson notes.

What is holding the yuan back? The answer is China itself—both by circumstance and, more importantly, by design. For a currency to go global, there has to be a path for it to leave its country of origin. The easiest route is via a trade deficit. For example, since the United States imports more than it exports, it in effect adds to global holdings of dollars on a daily basis. That does not work for China, which almost always runs a large trade surplus. It has tried to solve this problem by offering to pay for imports in yuan, while still accepting dollars for its exports.

Yet this approach can go only so far, because of the design of the Chinese system. Foreigners paid in yuan cannot do much with the currency and thus look askance at it. China could change this at astroke by flinging open its capital account. There is speculation that it might do just that as debate about financial reform intensifies in Beijing. But Yu Yongding, a former adviser to the central bank, predicts that caution will prevail, with the government slowly lowering its wall of capital controls rather than demolishing it. That would be far better for China's financial stability. But it also means that the chasm between the hype about the yuan and the mundane reality is likely to widen.

From: http://www.economist.com/news/finance-and-economics/21604579-buzz-about-rise-chinas-currency-has-run-far-ahead-sedate-reality-yuawn? zid=306&ah=1b164dbd43b0cb27ba0d4c3b12a5e227.

Chapter 6
International Marketing

Chapter checklist

When you have completed your study of this chapter, you will be able to

1. Know the core concepts on marketing and identify the three principles of marketing.
2. Understand the basic theories and methods of marketing.
3. Describe the environmental forces that affect the company's ability to serve its customers.
4. Identify the components of the marketing mix in the marketing of a product or service.
5. Know the new development of international marketing.

In this chapter, we will introduce you something about international marketing. There are five sections in this chapter, they are: an introduction to marketing, basic theories and methods of marketing, the marketing environment, marketing mix and the new development of international marketing. The more important and difficult parts include the basic theories and methods of marketing, marketing environment and the marketing mix.

Section 1　An Introduction to Marketing

If you want to found a successful marketing program, you must have a sound understanding of the marketing discipline. Marketing is the process of focusing the resources and objectives of an organization on environmental opportunities. The first and most fundamental fact about marketing is that it is a universal discipline. It is as applicable in New York of United States as it is in Tokyo of Japan. Marketing is a set of concepts, tools, theories, practices and procedures, and experience. In this section, we will first introduce you some core concepts on marketing; and then describes the basic principles of marketing.

Lesson 1　The Core Concepts on Marketing

When comes to marketing, many people think that marketing merely involves selling something or advertising for something. Indeed, marketing can be found in everywhere of human life. As Philip Kotler, one of the world leading authorities of marketing says, 「…every day, we are bombarded with television commercials, newspaper ads, direct mail campaigns and sales calls.」However, marketing is more than just selling and advertising which only constitute the tip of the marketing iceberg. Though quite important, they are only two of many marketing functions, and could not be regarded as the most important ones.

The Original Marketing Concept

During the long history of the development of marketing thought and practice, the concept of marketing has changed dramatically. The original concept of marketing focused marketing on the product, on making a 「better」 product for the objective of profits, and the means to achieving the objective was selling, or persuading the potential customers to exchange their money for the company's products. In fact, marketing practice has passed different stages and is guided by difficult concepts. Table 6.1 shows the evolution of the various marketing concepts.

Table 6.1 the Evolution of Marketing Concept

	Concept		
	Old	New	Strategic
Era	Pre-1960	1960-1990	1990-now
Focus	Product	Customer	Way of doing business
Means	Telling and selling	Integrated marketing mix	Knowledge and relationships
End	Profit	Value	Mutually beneficial relationship
Marketing is	Selling	A function	Everything

Source: Warren Keegan Associates, Inc., 1994.

The New Concept of Marketing

It is believed that the new concept of marketing appeared about 1960, since then, the focus of marketing changed from the product to the customer. Marketing people found that marketing can no longer be understood in the old sense— 「telling and selling」 —but in the new sense of satisfying customers. If the marketer does well in understanding customer needs, wants, and desires, and promotes them effectively, these products will find right market and sell easily. The objective was still profits, but the means to achieve it has been expanded to include the entire 「marketing mix」 or the 「Four Ps」, which includes a set of marketing tools that work together to affect the marketplace, including product, price, place (distribution) and promotion.

The Strategic Concept of Marketing

We can see that by the 1990s the 「new」 concept of marketing was again out of date and the new era required a strategic concept. The strategic concept of marketing, a major revolution in history of marketing thought, shifted the focus of marketing from the customer or the product to the customer in the context of broader external environment. By this concept, marketers must know the customer in a context consisting of the competition, government policy and regulation, and the macro economic, social, cultural, and political forces that affect markets.

Meanwhile, the objectiveof marketing also experienced a revolutionary change in strategic marketing concept, shifting from profits to stakeholder benefits. Stakeholders are individuals or groups who have an interest in the activity of a company. They include the employees and management, customers, society, and government, to mention only the most prominent. More and more people recognize that profits are a reward for performance (defined as satisfying customers in a socially responsible or acceptable way). To survive in today's competitive market, it is necessary to build up a team which is devoted to continuing innovation and to producing superior products. This means marketing must focus on the customers, the stakeholders' benefits and deliver value to them.

In a word, Marketing is about finding out who the best customers are, what they want, how much they will pay, how to promote the product and how to sell it, it is an attitude of mind which places the customer at the very center of a business activity, and automatically orientates a company towards its markets rather than its operations. To explain this definition, we first examine the following important terms: needs, wants, demands and exchange.

Basic Terms on Marketing

The most important and basicconcept on marketing is that of human needs. Human needs are states of felt deprivation. They describe basic human requirements including physical needs for food, clothing, shelter, and safety, social needs for belonging and affection, and individual needs for knowledge and self-expression. These needs are not invented by marketers; they are a basic part of human makeup.

Needs become wants when they are directed to specific objects that might satisfy the needs. Wants can be defined as the form taken by human needs as they are shaped by culture and individual personality. A hungry American may want a hamburger, French fries, and a Coke. A hungry Chinese may want a bowl of rice with sweet and sour fish and a cup of tea after the meal. Wants are described in terms of objects that satisfy needs.

People have almost unlimited wants, but have limited resources. Thus, they want to choose products that provide the most value and satisfaction for their money. When backed by buying power, wants become demands. Thus, a Mercedes means comfort, luxury, and status. Many people want them, but not all of them are able and willing to buy one. Companies must measure not only how many people want their product but also how many people would actually be willing to buy and can afford it.

People can obtain products in one of the following four ways. The first way is self-production. In this case, there is no market and no marketing actions. The second way is coercion. Hungry people can wrest or steal from others. No benefits can be offered to the others except that they will not be harmed. The third way is begging. Hungry people can asks others or beg them for food. They have nothing tangible to thank them except their gratitude. The fourth way is exchange. Hungry people can offer a resource in return for food such as money, some goods, or a service. Marketing emerges when people decide to satisfy their needs and wants by exchanging.

Exchange is the act of obtaining a desired product from someone by offering something in return. For exchange potential to exist, the following conditions must be satisfied. There are at least two parties. (1) Each party has something that might be of value to the other party. (2) Each party is capable of communication and delivery. (3) Each party is free to accept or reject the exchange offer. (4) Each party believes it is appropriate or desirable to deal with the other party.

New Words, Phrases and Expressions
context   n. 環境；上下文；來龍去脈
coercion   n. 強制；強迫；高壓政治；威壓
external   n. 外部；外觀；外面 adj. 外部的；表面的
iceberg   n. (地理) 冰山；顯露部分
interest   n. 興趣，愛好；利息；趣味；同行
           vt. 使……感興趣；引起……的關心；使……參與
prominent adj. 突出的，顯著的；傑出的；卓越的
revolutionary n. 革命者；adj. 革命的；旋轉的；大變革的
stakeholder n. 利益相關者；賭金保管者
sour   n. 酸味；苦事 adj. 酸的；發酵的；v. 使變酸；使失望，發酵
underlying adj. 潛在的；根本的；在下面的；
            v. 放在……的下面；為……的基礎；優先於
wrest   vt. 用力擰；搶奪；歪曲 n. 扭，擰
bombard with   連續提出，用……轟擊
backed by   依靠，在……支持/掩護/配合下
take form   成型，形成

Notes

1. The original concept of marketing focused marketing on the product, on making a 「better」 product for the objective of profits, and the means to achieving the objective was selling, or persuading the potential customer to exchange his or her money for the company's product.

市場行銷的原始概念認為市場行銷重點研究的是產品，是製造出「更好」的產

品來獲利，而實現獲利的途徑是銷售產品或說服潛在客戶購買公司產品。

2.「Marketing mix」or the「Four Ps」, which includes a set of marketing tools that work together to affect the marketplace, including product, price, place (distribution) and promotion.

市場行銷組合（4Ps）：一組結合起來使用以影響市場的行銷工具，包括產品、價格、地點（分銷）和促銷四部分。

3. The strategic concept of marketing, a major revolution in history of marketing thought, shifted the focus of marketing from the customer or the product to the customer in the context of broader external environment.

市場行銷的戰略概念：市場行銷理論發展史上的重大革命，行銷的焦點由此從消費者或產品轉移至更廣外部背景下的消費者。

4. Stakeholders are individuals or groups who have an interest in the activity of a company. They include the employees and management, customers, society, and government, to mention only the most prominent.

利益相關者：在公司行為中有利益關係的個人或組織，最主要的利益相關者有員工、管理層、客戶、社會和政府等。

5. Profits are a reward for performance (defined as satisfying customers in a socially responsible or acceptable way).

利潤是對公司業績（以對社會負責的或得到社會認可的方式滿足客戶需求）的回報。

6. Marketing can be defined as a social and managerial process by which individuals and groups obtain what they need and want through creating and exchanging products and value with others.

所謂市場行銷，是指個人和組織通過創造並同他人交換產品和價值換取可以滿足其需要和慾望的一種社會性管理過程。

7. Wants can be defined as the form taken by human needs as they are shaped by culture and individual personality.

需求是由文化和個體的個性共同形成的人類需要的表現形式。

Lesson 2　Three Principles of Marketing

We can summarize the essence of marketing in three important principles. The first principle emphasizes the purpose and task of marketing, the second the competitive reality of marketing, and the third the principal means for achieving the first two.

Principle One: Customer Value and Satisfaction

The product or offering will be successful if it delivers value and satisfaction to the targetcustomers. The buyer chooses among different offerings on the basis of which is perceived to deliver the most value. We define customer value as the customer's assessment of the product's overall capacity to satisfy his or her needs. It is the difference between the value that customer gains from owning and using a product and the cost of obtaining the product. For example, UPS customers gain a number of benefits. The most obvious are fast and reliable package delivery. However, when using UPS, customers may also receive some status and image values. When deciding whether to send a package via UPS, customers will weigh these values against the money, effort, and psychic costs of using the service. Moreover, they will compare the value of using UPS against the value of using other shippers—Federal Express, the U. S. Postal Service—and select the one that gives them the greatest delivered value.

Very often, customers do not make accurate or objective judgment on product values and costs. They tend to act on the value perceived. For example, does UPS really provide faster, more reliable delivery? If so, is this better service worth the prices that the firm charges? The FedEx argues that its express service is comparable. However, in terms of market share, most consumers prefer UPS, which dominates relatively larger share of the global express-delivery market, compared with other American shippers. The FedEx and the Postal Service's challenges are to change these customer value perception on express delivery service.

One of the important objectives of marketing activities is to make customers satisfied so that they can become repeat or loyal customers. But how can marketers keep their customers pleased and how to measure the degree of satisfaction? In fact, customer satisfaction depends on a product's perceived performance in delivering value relative to a buyer's expectation. If the product's performance falls short of the customer's expectations, the buyer is dissatisfied. If the performance matches expectations, the buyer is delighted. Smart companies aim to delight customers by promising only what they can deliver, then delivering more than they promise.

Principle Two: Competitive or Differential Advantage

Another key principle of marketing is competitive advantage, which is a total offer vis-à-vis relevant competition that is more attractive to customers. This strength exists in any element of the company's offering: the product, the price, the advertising and point-of-sale promotion, the distribution channels, and the brand image.

As is known by marketing people, one of the most powerful strategies forpenetrating a new market is to offer a superior product at a lower price. The price advantage may get immediate customer attention from price-sensitive consumers, while for those who are more concerned with life quality, the superior performance and brand reputation will come first.

Principle Three: Focus

The third marketing principle is focus, or the concentration of attention. Focus is required to succeed in the task of creating customer value at a competitive advantage. Any companies, large or small, achieve success because they have understood and applied this great principle. IBM, having long been regarded as a great company in the emerging data processing industry, found itself in crisis in the early 1990s. The major reason was that its competitors had become much more focused on customer needs and wants. Dell and Compaq, for example, focused on giving customers computing power at low prices, whereas IBM was offering the same computing power with higher prices.

Therefore, marketing managers need to have a clear focus on customer needs and wants, and on the competitive offering so as to mobilize the effort necessary to maintain a differential advantage. This can be accomplished only by focusing company's resources and efforts on the needs and wants of its target customers, and on the way to deliver a product that meets those needs and wants.

New Words, Phrases and Expressions
assessment   n. 評定；估價
comparable   adj. 可比較的；比得上的
dominate   vt. 控制；支配；占優勢；在……中占主要地位；
             vi. 占優勢；處於支配地位
delight   n. 高興；vt. 使高興
essence   n. 本質，實質；精華；香精
mobilize   vt. 動員，調動；集合，組織；使……流通；使……鬆動
perceive   vt. 察覺，感覺；理解；認知
penetrate   vt. 滲透；穿透；洞察；   vi. 滲透；刺入；看透
psychic   n. 靈媒；巫師；   adj. 精神的；心靈的；靈魂的；超自然的
perception   n. 知覺；感覺；看法；洞察力；獲取
repeat   n. 重複；副本
relevant   adj. 有關的；中肯的；有重大作用的
relative to   相對於；涉及
vis-à-vis   面對面（的），對面的人；對應的事物或人；在……對面，與……相比，關於

Notes

1. Customer value as the customer's assessment of the product's overall capacity to satisfy his or her needs. It is the difference between the value that customer gains from owning and using a product and the cost of obtaining the product.

客戶價值指的是客戶對產品滿足其需要的總體能力的評價。客戶價值是客戶從擁有和使用某種產品中得到的價值與其獲得該產品的成本之間的差額。

2. Customer satisfaction depends on a product's perceived performance in delivering value relative to a buyer's expectation.

顧客滿意度取決於產品在傳遞價值的過程中相對於買主預期的可感知性能。

3. Competitive advantage, which is a total offer vis-à-vis relevant competition that is more attractive to customers.

競爭優勢意味著企業為客戶提供的、比同類產品更具有吸引力的產品和服務的總和。

4. Focus is required to succeed in the task of creating customer value at a competitive advantage.

要成功實現在競爭優勢的基礎上為客戶創造價值的目標，重點戰略不可或缺。

Section 2　Basic Theories and Methods of Marketing

Now we have known the core concepts of marketing, and learned something important in the process of marketing, but we also need to know some basic theories in international marketing. The following content will tell us the five forces competition model which is the basic theory in marketing and then introduce two basic methods that is useful in the process of marketing.

Lesson 1　The Five Forces Competition Model

The five forces competition model (see Figure 6.1) was put forward by Michael Porter in early 1980s, which made a far-reaching global impact in the development of the company's strategy. It is mainly used for analyzing the competitive strategy, as well as the competitive environment of the customers.

```
                    ┌──────────────────────┐
                    │ threat of new entrants│
                    └──────────┬───────────┘
                               ↓
┌──────────────────┐    ┌──────────────────────────┐    ┌──────────────────────┐
│bargaining of buyers│──→│competition of competitors│←───│bargaining of supplier│
└──────────────────┘    └──────────┬───────────────┘    └──────────────────────┘
                                   ↓
                    ┌──────────────────────┐
                    │ threat of substitutes │
                    └──────────────────────┘
```

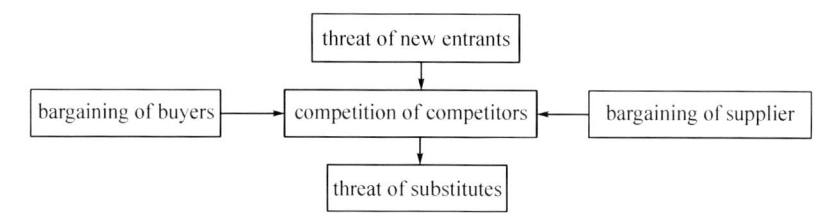

Figure 6.1　The five forces competition model

In this model, the five forces are:

(1) Threat of New Entrants

When the new entrants bring the new productive capacity and new resource to the industry, they would like to win the market divided by those existing companies. It will bring the competition about raw materials and market shares with existing companies, and eventually reduce the level of the current corporate earning, moreover, perhaps endanger the survival of these companies.

The possibility of the new company entering an industry depends on the proportion among the potential interest, the expense and the hazard estimated by the new entrant.

(2) Threat of Substitutes

Two companies in different fields could compete with each other, possibly because their products become the substitute of each other. This competition of substitute affects the competitive strategy of existing companies in different ways.

The lower price, the better quality and the lower transferred cost of the substitutes are, the stronger the competitive stress would be. The strength of this competitive stress could be described by the sales growth rate of the substitutes, the productive ability of substitute companies and the profitable expansion through investigation.

(3) The Bargaining of Purchasers

The purchaser affects theprofitability of the existing companies mainly through the ability of forcing prices down and demanding better quality of products and services.

(4) The Bargaining of Suppliers

The supplier affects the profitability and competitive power of products mainly through the ability of increasing the input prices and decreasing the quality of single value.

The power of the supplier depends on the input factors provided for the purchases. When the value of the input factor seizes a large scale in the total cost, and plays an important role or effect in the quality of the purchaser's products, the bargaining of the supplier would enhance a lot.

(5) The Competition of Competitors

The interest of most companies in the same trade often goes hand in hand, however, as an individual company, the purpose of each competitive strategy lies in achieving the advantage to other competitors. Therefore, in the process of implementation, there would appear conflicts and confrontations, which develop into competition among these existing companies.

According to the discussion of the above five competitive forces, the companies could take many measures to deal with the five forces competition on strengthen the market status and competitive power, such as dividing the management and competition, affecting the competitive rule from self-interest, occupying the favorable market status first and then starting aggressive competition, and so on.

New Words, Phrases and Expressions
bargain   n. 討價還價
conflict   n. 衝突，抵觸
confrontation   n. 對抗，衝突
entrant   n. 加入者，進入者
endanger   v. 危及
enhance   v. 提高，增強
hazard   n. 風險，危險
potential adj. 潛在的，可能的
proportion   n. 協調，比例

substitution   n. 代替，替換
survival   n. 幸存，生存
hand in hand   密切聯繫
lie in   在於
put forward   提出
take measures   著手，採取措施

Notes

1. When the new entrants bring the new productive capacity and new resource to the industry, they would like to win the market divided by these existing companies.

新進入者在給行業帶來新生產能力、新資源的同時，也希望在已被現有企業瓜分完畢的市場中贏得一席之地。

2. The lower price, the better quality and the lower transferred cost of the substitutes are, the stronger the competitive stress would be.

替代品價格越低、質量越好、用戶轉換成本越低，其所能產生的競爭壓力就越強。

3. The purchasers affect the profitability of the existing companies mainly through the ability of forcing prices down and demanding better quality of products and services.

購買者主要通過其壓價與要求提供較高的產品或服務質量的能力來影響行業中現有企業的盈利能力。

4. The interest of most companies in the same trade often goes hand in hand, however, as an individual company, the purpose of each competitive strategy lies in achieving the advantage to other competitors.

同行業中的大部分企業相互之間的利益都是緊密聯繫在一起的。作為企業整體戰略一部分的各企業競爭戰略，其目標都在於使得自己的企業獲得相對於競爭對手的優勢。

5. Therefore, in the process of implementation, there would appear conflicts and confrontations, which develop into competition among these existing companies.

所以，在實施中就必然會產生衝突與對抗現象，這些衝突與對抗就構成了現有企業之間的競爭。

Lesson 2　SWOT Analysis

The SWOT analysis is also called situation analysis, which was put forward by a professor of management in University of San Francisco in the early period of 1980s. The SWOT analysis (the situation analysis) lists various major internal advantages, disadvantages, opportunities and threats which are closely related to the researching object and arranges them according to a matrix form. Then these elements are matched and analyzed through the idea of systematic analysis, from which a series of policy-making conclusions are obtained. By using this method, we could study completely, systematically and accurately about the condition of the researching object, and formulate the corresponding development strategy, plans and measures according to the results.

The SWOT analysis is often used to formulate the development strategy of group and analyze the condition of competitors. It is one of the most common methods in strategy analysis. The approach includes. (1) Analysis of environment elements; (2) SWOT matrix structure; (3) Action plan.

Analyzing various environment elements about the company uses all kinds of researching methods, namely, macro and micro environmental elements, external and internal environmental elements.

External environmental elements consist of opportunities and threats factors, which are also the favorable and unfavorable factors affecting directly the development of the company from the external environment, belonging to objective factors, and classified as the factor of the economy, politics, society, demography, product and service, technology, market, competition and other different aspects.

Internal environment elements consist of advantage and disadvantage factors, which are also the positive and negative factors existing in the development of the company, belonging to active factors. It is mainly in comparison with the competitors (or industry average) to know the own strength and weakness of the business unit, in order to highlight its advantages and features in formulating the business strategy and to avoid defeat in the competition.

The SWOT matrix is constructed according to the importance or effect of the investigated elements. In this process, the direct, important, substantial, immediate and profound factors contributed to the development of the company shall be prioritized, and the indirect, secondary, little, delayed and temporary ones are located in the back.

After completing the analysis of environmental factors and SWOT matrix structure, we can lay down an appropriate action plan. The basic idea of the plan is: playing dominant factors; overcoming weaknesses; taking advantage of opportunities and resolving threats. Matching and mixing the considered factors with each other by using the comprehensive analysis, we could draw up a series of alternative measures for the growth of the company in the future.

New Words, Phrases and Expressions
corresponding   adj. 相當的，相應的；一致的
demographic   adj. 人口統計的
external   adj. 外部的，表面的
favorable   adj. 有利的，合適的
formulate   v. 規劃，制定
highlight   v. 強調，突出
internal   adj. 內部的，本身的

matrix   n. 矩陣
preferential   adj. 優先的，優惠的
systematically   adv. 有系統地，有組織地
temporary   adj. 臨時的，暫時的
as follows   如下
classify as   把……分類
in order to   為了
lay down   制定，規定
take advantage of   利用

Notes

1. The SWOT analysis (the situation analysis) lists various major internal advantages, disadvantages, opportunities and threats which are closely related to the researching object and arranges them according to a matrix form.

SWOT 分析，即態勢分析，就是將與研究對象密切相關的各種主要內部優勢、劣勢、機會和威脅等，通過調查列舉出來，並依照矩陣形式排列。

2. Then these elements are matched and analyzed through the idea of systematic analysis, from which a series of policy-making conclusions are obtained.

然後用系統分析的思想，把各種因素相互匹配起來加以分析，從中得出一系列具有決策性的結論。

3. The SWOT analysis is often used to formulate the development strategy of group and analyze the condition of competitors.

SWOT 分析法常常被用於制定集團發展戰略和分析競爭對手情況。

4. Analyzing various environment elements about the company uses all kinds of researching methods.

運用各種調查研究方法，分析出公司所處的各種環境因素。

5. It is mainly in comparison with the competitors (or industry average) to know the own strength and weakness of the business unit.

它主要是通過同競爭對手（或行業平均水準）的比較，瞭解業務單位自身的優勢和劣勢。

Lesson 3   BCG Matrix Analysis

BCG Matrix is a product mix planning method, firstly introduced by a largecommercial consulting firm in the United States—Boston Consulting Group. BCG Matrix considers that there are two basic determinants in the product structure: the marketing attraction and the company strength. Through the interaction of these two factors, there would appear four different types of products to form the following different development perspectives: (1) High sales growth, high market share product group (stars) (2) Low sales growth, low market share product group (thin dogs) (3) High sales growth, low market share product group (question marks) (4) Low sales growth, high market share product group (cash cow).

Basic principles of BCG Matrix are as follows: BCG Matrix reorders all the products from the aspect of sales growth and market share. In Figure 6.2, it shows sales growth on the longitudinal axis and market share on the horizontal axis, each with 10% and 20% for distinguishing high and low mid-point. There will be 「question marks (?)」「stars (★)」「cash cow ($)」 and 「thin dogs (x)」 if we divide the graph into four quadrants. The company could arrange the products into different quadrants according to their sales growth and market share. It could make clear about the existing product portfolio of the company; meanwhile, it is easier to form different development decisions for the products in different quadrants to achieve the virtuous circle of the allocation structure of the products and resources.

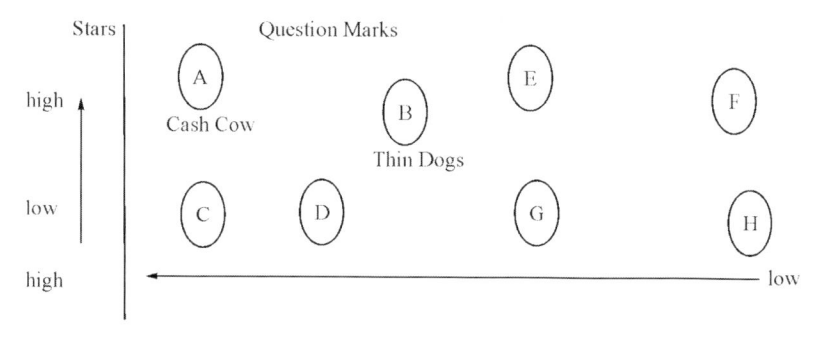

Figure 6.2 BCG Matrix Analysis

Basic steps of BCG Matrix includes the following: first, account sales growth and market share of various products. Second, draw four-quadrant diagram. Third, formulate the following strategic response of each quadrant.

「Stars」 means the product group with high sales growth and high market share; these products may become the cash cow products. It requires increased investment to support its rapid development. The products within this quadrant should be applied with development strategy.

「Cash Cow」, also called beneficial product, refers to the product group with low sales growth but high market share, having entered the mature period. The decline in market share of most products has become an unstoppable momentum, so it can be applied with harvest strategy: the invested resources are limited once achieving maximum short-term gains. For the products in this quadrant with increased sales growth should be further broken down to maintain the existing market growth rate or delay the decreased rate.

「Question Marks」 is the product group in the quadrant with high sales growth but low market share. These question marks should be applied to selective investment strategy, that is, at first identify those which may be improved to become star products, and then invest to improve the market share to transfer to 「stars」, while the other products that will become the stars in the future should take affirmative response in a period of time.

「Thin Dogs」 is also called recession product. It is the product group in the quadrant with low growth and low market share. The retreating strategy should be taken on this kind of products in order to deliver the resources to the more potential field.

New Words, Phrases and Expressions
axis　n. 軸，中心線
allocation　n. 配給，分配
affirmative　adj. 肯定的，樂觀的
diagram　n. 圖表，圖解
determinant　n. 決定因素
perspective　n. 觀點，希望，態度
horizontal　adj. 水準的，橫的
harvest　n. 收穫；產量；結果
longitudinal　adj. 經度的，縱向的
momentum　n. 動力，勢頭
quadrant　n. 象限
recession　n. 經濟衰退，不景氣
retreat　v. 撤退，退出
unstoppable　adj. 無法停止的，無法阻礙的
break down　拆散，粉碎，損壞
divide into　分成，分為
deliver to　轉交，傳達
make clear　解釋，說明，弄清楚

Notes

1. BCG Matrix considers that there are two basic determinants in the product structure: the marketing attraction and the company strength.

波士頓矩陣認為，一般決定產品結構的基本因素有兩個，即市場引力和企業實力。

2. It means the product group with high sales growth and high market share; these products may become the cash cow products.

這是指處於高增長率、高市場佔有率象限內的產品群，這類產品可能成為企業的現金牛產品，需要加大投資以支持其迅速發展。

3. The decline in market share of most products has become an unstoppable momentum, so it can be applied with harvest strategy: the invested resources are limited once achieving maximum short-term gains.

對於這一象限內的大多數產品，市場佔有率的下跌已成不可阻擋之勢，因此可

採用收穫戰略：所投入資源以達到短期收益最大化為限。

4. It is the product group in the quadrant with high sales growth but low market share. 這是指處於高增長率、低市場佔有率象限內的產品群。

Section 3　The Marketing Environment

　　Before companies decide to enterthe market, they must scan the environment in the whole market. The environmental scanning should consists of the microenvironment and the macro environment, so that the company can implement their market strategy more successfully. This section we will learn something about the marketing environment, including the microenvironment and the macro environment and know the important factors that influences the company's marketing decisions.

Lesson 1　What Is Marketing Environment?

　　This section addresses the key forces in the firm's marketing environment and how they affect its ability to maintain satisfying relationships with target customers. A company's marketing environment consists of the factors and forces outside marketing that affect marketing management's ability to develop and maintain successful transactions with its target customers. The marketing environment offers both opportunities and threats. Successful companies know the vital importance of using their marketing research and intelligence systems constantly to watch and adapt to the changing environment. Too many other companies, unfortunately, fail to think of change as opportunity.

　　The micro-environment consists of the forces close to the company that affect its ability to serve its customers—the company, suppliers, marketing channel firms, customer markets, competitors and publics. The macro-environment consists of the larger societal forces that affect the whole micro-environment—demographic, economic, natural, technological, political and cultural forces.

　　What is the company's micro-environment? The micro-environment consists of the forces close to the company that affect its ability to serve customers—the company, suppliers, marketing channel firms, customer markets, competitors and publics.

The Company

　　In designing marketing plans, marketing management should take other company groups, such as top management, finance, research and development, purchasing, manufacturing and accounting, into consideration. All these interrelated groups form the internal environment. Top management sets the company's mission, objectives, broad strategies and policies. Marketing managers must make decisions consistent with the plans made by top management, and marketing plans must be approved by management before they can be implemented.

Marketing managers must also work closely with other company departments. Finance is concerned with finding and using funds to carry out the marketing plan. The R&D department focuses on the problems of designing safe and attractive products. Purchasing worries about getting supplies and materials, whereas manufacturing is responsible for producing the desired quality and quantity of products. Accounting has to measure revenues and costs to help marketing know how well it is achieving its objectives. Therefore, all of these departments have an impact on the marketing department's plans and actions. Under the marketing concept, all of these functions must 「 think customer 」 and they should work together to provide superior customer value and satisfaction.

Suppliers

Suppliers are an important link in the company's overall customer 「 value delivery system 」. They provide the resources needed by the company to produce its goods and services. Supplier developments can seriously affect marketing. Marketing managers must watch supply availability—supply shortages or delays, labor strikes and other events can cost sales in the short run and damage customer satisfaction in the long run. Marketing managers must also monitor the price trends of their key inputs. Rising supply costs price increases that can harm the company's sales volume.

Marketing Intermediaries

Marketing Intermediaries are firms that help the company to promote, sell and distribute its goods to final buyers. They include resellers, physical distribution firms, marketing services agencies and financial intermediaries.

Resellers are distribution channel firms that help the company find customers or make sales to them. These include wholesalers and retailers which buy and resell merchandise. Selecting and working with resellers is not easy. No longer do manufacturers have many small, independent resellers from which to choose. They now face large and growing reseller organizations. These organizations frequently have enough power to dictate terms or even shut the manufacturer out of large markets.

Physical distribution firms help the company to stock and move goods from their points of origin to their destinations. Working with warehouse and transportation firms, a company must determine the best ways to store and ship goods, balancing such factors as cost, delivery, speed and safety.

Marketing services agencies are the marketing research firms, advertising agencies, media firms and marketing consultancies that help the company target and promote its products to the right markets. When the company decided to use one of these agencies, it must choose carefully because the firms vary in creativity, quality, service and price. The company has to review the performance of these firms regularly and consider replacing those that no longer perform well.

Financial intermediaries include banks, credit companies, insurance companies and other businesses that help finance transactions or insure against the risks associated with the buying and selling of goods. Most firms and customers depend on financial intermediaries to finance their transactions. The company's marketing performance can be seriously affected by rising credit costs and limited credit. For example, small and medium-sized businesses in China have often found difficulty in obtaining finance for market and product development activities.

Like suppliers, marketing intermediaries form an important component of the company's overall value delivery system. In its quest to create satisfying customer relationships, the company must do more than just optimize its own performance. It must partner effectively with suppliers and marketing intermediaries to optimize the performance of the entire system.

Customers

Customer is king. The company must study its customer markets closely. There are six types of customer markets. Consumer markets consist of individuals and households that buy goods and services for personal consumption. Business markets buy goods and service for further processing or for use in their production process. Reseller markets buy goods and services to resell at a profit. Institutional markets are made up of schools, hospitals, nursing homes, prisons and other institutions that provide goods and services to people in their care. Government markets are made up of government agencies that buy goods and services in order to produce public services or transfer the goods and services to others who need them. International markets consist of buyers in other countries. including consumers, producers, resellers and governments.

Competitors

To be successful, a company must provide greater customer value and satisfaction than its competitors do. Thus, marketers must do more than simply adapt to the needs of target consumers. They must also gain strategic advantage by positioning their offerings strongly against competitors' offerings in the minds of consumers.

No single competitive marketing strategy is best for all companies. Each firm should consider its own size and industry position compared to those of its competitors. Large firms with dominant positions in an industry can use certain strategies that smaller firms cannot afford. But being large is not enough. There are winning strategies for large firms. But there are also losing ones. And small firms can develop strategies that give them better rates or return than large firms enjoy.

Publics

A public is any group that has an actual or potential interest in or impact on an organization's ability to achieve its objectives. There are seven types of publics: (1) Media publics are those that carry news, features and editorial opinions. They include newspapers, magazines and radio and television stations. (2) Financial publics influence the company's ability to obtain funds. Banks, investment houses and stockholders are the principal financial publics. (3) Government publics Marketers must often consult the company's lawyers on issues of product safety, truth-in-advertising and other matters. (4) Citizen action publics A company's marketing decisions may be questioned by consumer organizations, environmental groups. (5) General public A company needs to be concerned about the general public's attitude towards its products and activities. The public's image of the company affects its buying. Thus, many large corporations invest huge sums of money to promote and build a healthy corporate image. (6) Local publics Every company has local publics, such as neighborhood residents and community organizations. (7) Internal publics A company's internal publics include its workers, managers, volunteers and the board of directors. Large companies use newsletters and other means to inform and motivate their internal publics.

New Words, Phrases and Expressions
constantly　adv. 不斷地；時常地
channel　vt. 引導，開導；形成河道；　n. 通道；頻道；海峽
demographic　adj. 人口統計學的；人口學的
Micro-environment　微觀環境
Macro-environment　總體環境
transaction　n. 交易；事務；辦理；會報，學報
vital　adj. 至關重要的；生死攸關的；有活力的
dictate　v. 命令；口述；使聽寫
household　n. 家庭；一家人
monitor　n. 監視器；監聽器；監控器；顯示屏；班長
merchandise　n. 商品；貨物
intermediary　adj. 中間的；媒介的；中途的；n. 中間人；仲裁者；調解者；媒介物
optimize　v. 使最優化，使完善
reseller　n. 經銷商；代理商
be concerned with　參與，干預；關心；相干；涉及
consistent with　符合；與……一致
physical distribution　[貿易] 物流，物資調運
take into consideration　顧及；考慮到……

Notes
1. No longer do manufacturers have many small, independent resellers from which to choose.
廠商不再有很多可供選擇的小而獨立的分銷商。

2. In its quest to create satisfying customer relationships, the company must do more than just optimize its own performance.

公司在尋求創造滿意的客戶關係的時候，要做的不僅僅只是優化自己的表現。

3. They must also gain strategic advantage by positioning their offerings strongly against competitors' offerings in the minds of consumers.

他們還必須通過在消費者的心目中定位自己的產品來強烈地對抗競爭者所提供的產品來得到戰略優勢。

4. A public is any group that has an actual or potential interest in or impact on an organization's ability to achieve its objectives.

公共是指任何一個有實際或潛在的利益或者影響一個組織實現自己目標的能力的團體。

Lesson 3　Macro-environment

A PEST analysis of the company's macro-environmental factors will be made in this section. PEST analysis involves identifying the political, economic, social and technological influences on an organization. P means political environment, E means economic environment; S means sociocultural environment; T means technological environment.

Political/Legal Environment

An organization doesn't function strictly by its own set of rules. It has to serve its customers as well as special interest groups; Together, these comprise the political-legal environment. This environment influences marketing strategies through laws, regulations, and political pressures.

Laws and regulations cover many marketing activities, including product testing, packaging, pricing, advertising, and sales to minors. Laws and regulations can limit marketing activities or be a source of opportunity for organizations that provide goods and services. Abiding by laws not only helps organizations avoid fines and lawsuits, but promotes confidence among customers as well. Marketers must be familiar with the foreign, state, and local laws and regulations in areas where they seek to do business.

Global marketers are affected by agreements between countries and by the laws in countries in which they operate. Some of the more vital political and legal factors in the global environment are international trade agreements. For example, in America, the North American Free Trade Agreement (NAFTA) is designed to drop trade barriers among Canada, the United States, and Mexico through means such as the elimination of tariffs. Similarly, there are European Union (EU), Association of South East Asian Nations (ASEAN), Asia Pacific Economic Cooperation (APEC) and Organization of Petroleum Exporting Countries (OPEC).

Economic Environment

The economic environment for marketing comprises the overall economy, including business cycles and spending patterns, as well as consumer income issues.

Learning about the economic environment helps marketers determine whether customers will be willing and able to spend money on products and services. Spending patterns are linked to the business cycle, or the pattern in the level of business activity that moves from prosperity to recession, to recovery.

Prosperity During times of prosperity, production and employment are high. Many consumers demand more goods and services, and they spend freely not only on basics but on luxuries such as vacations, designer clothing, and entertainment. In addition, they may upgrade high-ticket items such as houses and cars. Many consumers in prosperous times want the 「best」 of everything and are willing to pay for it, so marketers introduce new luxury versions of products.

Recession During a recession, consumers fold their wallets and snap their purses closed; production decreases and unemployment generally rises. Decreased consumer demand leads organizational buyers to reduce their spending as well. Both types of buyers may stick to purchasing basics and look for the best deal.

Marketers of private-label and generic goods may find they have an edge in creating value by offering good quality at less than brand-name prices. Some companies even offer 「price rollbacks」—price decreases that echo previous, lower prices for individual products—or they advertise that they haven't raised a price in a certain number of years. In such ways, resourceful and creative marketers can proper during a recession.

Recovery While the economy is in the recovery stage, progressing from recession to prosperity, the level of production and employment increases. Consumers and organizational buyers may have more money to spend, but may still be reluctant to increase their purchases. They may recall the recent recession and be wary of another slump. Many consumers try harder to save money and buy few items on credit. As the economy becomes stronger, buyers begin to relax and spend more freely and prosperity can return.

Although business cycles reflect the overall health of the economy. The income of individual households influences whether or not consumers buy products. Disposable income is the money an individual or household has left after paying taxes. This is the money consumers can save or spend on rent or a mortgage, groceries, clothing, and any other essentials or luxuries.

The society that people grow up in shapes their basic beliefs, values, and norms. People absorb, almost unconsciously, a world view that defines their relationship to themselves, to others, to nature, and to the universe. Here are some of the main cultural characteristics and trends of interest to marketers.

The people living in a particular society hold many core beliefs and values that tend to persist. Core beliefs and values are passed on from parents to children and are reinforced by major social institutions—schools, religious groups, business, and government.

Brand's Essence of Chicken found that its TV commercials in Shanghai showing a chicken in a pot of broth too simplistic for its audience. A new version depicting a child doing homework with a computer proved more successful because it hinged on the high regard for education by the Chinese.

To encourage more Singaporean Chinese to speak mandarin and preserve their cultural values, the government has regular「Speak Mandarin」campaigns. These increased the use of Mandarin from 26% to 65% within 12 years.

Each society contains subcultures, that is, various groups with shared values emerging from their special life experiences or circumstances. Buddhists, teenagers, South Korea's Orange youths, and Japan's Harajuku kids all represent subcultures whose members share common beliefs, preferences, and behaviors. According to the extent that subcultural groups exhibit different wants and consumption behavior, marketers can choose subcultures as their target markets.

In Guangzhou, Chinese youths seem to be placing personal interests above all else. Some 64% feel that money works wonders; 63% believe that nothing great can be accomplished without telling lies; and 57% have a「I use you, you use me」mentality. Further, modernization has resulted in more Chinese men engaging in bigamy and taking on concubines.

Scientific knowledge, research, inventions and innovations that result in new or improved goods and services all make up the technological environment of marketing. Technological developments provide important opportunities to improve customer value. Also new technologies create new markets and opportunities.

When organizations fail to keep up with technological change, technology becomes a threat. IBM's dominance of the market for large mainframe computers did not spare the company from posting a huge loss when customers found they could get all the computing power they needed from personal computers. To succeed in this volatile industry, computer makers innovate continually.

Keeping up with technological developments is especially important for marketers who serve business customers. organizational buyers may rely on technological innovations for their very survival in competitive markets.

Technology life cycles are getting shorter. For example, the average life of some computer software products is now well under one year or even half a year. Firms must track technological trends and determine whether or not these changes will affect their products' continued ability to fulfill customers' needs. Technologies arising in unrelated industries can also affect the firm's fortunes. Businesses must assiduously monitor their technological environment to avoid missing new product and market opportunities.

The Internet is potentially useful throughout the marketing process. It provides access to a wealth of demographic data and other information about the external environment. Web sites also are a means to communicate with existing and potential customers. Organizations can make available information about themselves and their products, cultivate long-term business relationships, and offer products for sale.

New Words, Phrases and Expressions
assiduously   adv. 刻苦地；勤勉地
bigamy   n. 重婚罪，重婚
broth   n. 肉湯；液體培養基
concubine   n. 妾；情婦；姘婦
echo   v. 反射；重複；隨聲附和；發出回聲
generic   adj. 類的；一般的；屬的；非商標的
minor   adj. 未成年的；次要的；較小的
mainframe   n. [計] 主機；大型機
resourceful   adj. 資源豐富的；足智多謀的；機智的
snap   vt. 突然折斷，拉斷；猛咬；「啪」地關上
simplistic   adj. 過分簡單化的；過分單純化的
volatile   adj. [化學] 揮發性的；不穩定的；爆炸性的；反覆無常的
be wary of   提防，當心
designer clothing   品牌服裝
disposable income   可支配收入
hing on   堅持下去；不掛斷；握住不放
spare from   使某人免受（不願意之事）

Notes

1. Learning about the economic environment helps marketers determine whether customers will be willing and able to spend money on products and services.

學習經濟環境有助於行銷者確定消費者是否願意並且能夠把錢花在商品和服務上。

2. Marketers of private-label and generic goods may find they have an edge in creating value by offering good quality at less than brand-name prices.

自有品牌和一般商品的行銷者會發現他們在以低於品牌商品的價格提供優質產品來創造價值上的優勢。

3. Some companies even offer「price rollbacks」—price decreases that echo previous, lower prices for individual products—or they advertise that they haven't raised a price in a certain number of years.

有些公司甚至實行「價格回落」，即對於個人的消費品實行降價以呼應先前較低的價格，或者他們做廣告說自己很多年都沒有漲過價。

4. Scientific knowledge, research, inventions and innovations that result in new or improved goods and services all make up the technological environment of marketing.

所有由科學知識、研究、發明和創新所得來的新的或者改進的產品和服務共同組成了市場行銷的技術環境。

5. Organizations can make available information about themselves and their products, cultivate long-term business relationships, and offer products for sale.

組織可以得到關於自身和自己的產品的信息，培養長期的業務關係和提供用於出售的產品。

Section 4　Marketing Mix

Marketing mix combines various marketing strategies of companies by focusing on the integrated marketing goal to reduce the marketing cost and achieve marketing objectives.

Marketing mix mainly includes four major aspects: product, price, place and promotion. The marketing strategy of companies forms four different types of strategy mix centering on these four aspects: product strategy, pricing strategy, placing strategy and promotion strategy. These four marketing strategies are often called「4P」because of the first letter「P」of these strategies.

Lesson 1　Product Strategies

The concept of product as a whole in modem marketing includes three levels of the core product. the tangible product and the extended product. The core of product mainly refers to the basic utility or basic function of the product; the product form mainly refers to the product appearance and major features, which is the main reason for consumers to identify and select products. Its general performances are the product's quality, style, feature, packaging and brand. The product form determines differences in characteristics of the product. The extended product is not only the physical product that companies provide to consumers with a variety of functions, but also the excellent pre-sale, sale and after-sale service.

Product Strategies mainly refer to the companies providing a variety of tangible and intangible products to market for consumers' demand to achieve their marketing goals. After identifying the product marketing strategies, the companies should take a series of specific marketing strategies of the product itself, including specific implementing strategies about trademarks, brands, packaging, product mix, product life cycle, and so on. The product strategy plays an important role in its marketing mix strategy.

Product life cycle theory was put forward firstly by Harvard University Professor Vemon in 1966 in International Trade and International Investment in Product Cycle. Product life cycle generally is divided into four stages: introduction, growth, mature and decline.

The characteristics in introduction is low sales, but high promotion and manufacturing cost, low or even negative profits. In the period of introduction, different marketing strategies have been combined by four basic elements of product, distribution, price and promotion: (1) rapid skimming strategy; (2) slow skimming strategy; (3) rapid penetration strategy; (4) slow penetration strategy.

After the introduction of new products, the consumers have already been familiar with the product, and their habits have been formed. When the sales grow rapidly, the new product enters the growth stage. Several strategies can be taken as follows: (1) to improve product quality; (2) to search for new market segments; (3) to change the focus of advertising; (4) to reduce price timely.

After entering mature period, sales of product grow slowly, and gradually reach a peak and then fall slowly, meanwhile the sales profit begins to decline from the highest growth period, and the market competition is fierce, similar products with various brands and styles appear continually. The mature product should adopt proactive strategy to extend the maturity, or recycling the product life cycle. Three strategies are: (1) market adjustment; (2) product adjustment; (3) marketing mix adjustment.

After entering the decline period, product sales decline sharply. The profit of such products is very low or even zero. A lot of competitors drop out of the market and the habits of consumers have changed. To deal with the product in this period, there are usually several strategies to choose from: (1) to continue strategy; (2) to focus strategy; (3) to shrink strategy; (4) to give up strategy.

New Words, Phrases and Expressions
extend　v. 延長；擴展
penetration　n. 滲透；侵入
proactive　adj. 主動的
skimming　n. 撇脂
shrink　v. 收縮
segment　n. 部分，片段
tangible　adj. 實際的，有形的
trademark　n. (註冊) 商標
utility　n. 功用，效用
drop out of　不參與，退出
give up　放棄，認輸
product life cycle　產品生命週期
Notes

1. The core of product mainly refers to the basic utility or basic function of the product;

the product form mainly refers to the product appearance and major features, which is the main basis to identify and select by consumers.

產品核心主要是指產品的基本效用或基本功能；產品形態主要是指產品外觀形態及其主要特徵，是消費者得以識別和選擇產品的主要依據。

2. In the period of introduction, different marketing strategies have been combined by four basic elements of product, distribution, price and promotion.

在產品的介紹期，由產品、分銷、價格、促銷4個基本要素組合成各種不同的市場行銷策略。

3. When the sales grow rapidly, the new product enters the growth stage.

銷售量迅速增長時，新產品就進入成長期。

4. To deal with the product in this period, there are usually several strategies to choose from.

面對處於衰退期的產品，通常有幾種策略可供選擇。

Lesson 2　Pricing Strategies

As an important component of marketing mix, the pricing strategy could adjust and guide the market demand. It is also an important means in marketing competition to influence the realization of company marketing objectives, at the same time it is restricted by company marketing environment and conditions. Price generally includes four elements: production costs, distribution costs, taxes to the state and profit. Reasonable price is in relation to the benefit of producers and operators and their market image and it has an important position in marketing activities.

Choosing pricing methods and pricing strategies by companies are based on pricing targets. Different companies have a varietyof pricing objectives. For example: to pursue profit, to increase the sales; to advance the market share, to improve the amage, to compete, and so on. The main factors that affect price are product cost, market supply and demand, competition, policies and regulations.

After determining the objective of pricing, the company should predict sales and competitive reaction according to the general procedure developed in commodity prices. The program of pricing generally includes the following steps: (1) to determine the objectives of pricing; (2) to estimate sales; (3) to analyze the competitors' response; (4) to predict market share; (5) to consider the relevant situations of business activities.

There are many pricing methods, whose specific application should not be directly restricted by pricing objectives. The pricing method applied by different companies, companies with different market competitiveness, and companies in different market environments varies. Pricing methods are generally divided into the following three categories: (1) Cost-oriented pricing: two forms of cost-plus pricing, target profit pricing. (2) Demand-oriented pricing: two forms of habit pricing, perceived-value pricing. (3) Competition-driven pricing. It has been widely adopted by companies. There are three forms of going-rate pricing, competitive pricing and sealed bid pricing.

Price strategy focuses on the specific circumstances of the market from the pricing objective. In order to realize the marketing objectives of companies, it uses the price measures to adapt to different market conditions. Price strategies are as follows: There are four forms in different price strategies: geographical different strategy, time different strategy, usage different strategy, and quality different strategy. Discount price strategy mainly includes three forms: cash discount, quantity discount and business discount. Psychological price strategy includes four forms: mantissa price strategy, integer price strategy, prestige price strategy, and attraction price strategy.

In short, the concrete market prices are changing all the time, so the company should pay attention to the application of the price means. Take care of coordination of the other non-price competition means in pricing.

New Words, Phrases and Expressions
category   n. 種類, 類別
circumstance   n. 環境, 條件, 情況
distribution   n. 經銷, 分銷
discount   n. 折扣
integer   n. 整數
mantissa   n. 尾數
predict   v. 預言, 預測
prestige   n. 威信, 威望, 聲望
relevant   adj. 有關的, 切題的
target   n. 對象, 目標
at the same time   同時; 一起
for example   例如, 譬如
going-rate   通行價
in relation to   與……有關
perceived-value   認知價值
pay attention to   注意
sealed bid   密封競標
take care of   照顧, 對付, 處理

Notes

1. The main factors that affect price are product cost, market supply and demand, competition, policies and regulations.
影響企業定價的主要因素有產品成本、市場供求、競爭情況和政策法規。

2. There are many pricing methods, whose specific application should not be directly

restricted by pricing objectives.

企業的定價方法很多，定價方法的具體運用不受定價目標的直接制約。

3. In short, the concrete market prices are changing all the time, so the company should pay attention to the application of the price means.

簡而言之，市場上具體的行銷價格是變化多端的，故企業要重視價格手段的應用。

Lesson 3　Distribution Strategies

Distribution is to make products and services appropriate to the number and geographical distribution of the target market in a timely manner to meet customer needs. Distribution channel strategy is involved with the distribution channels and structure, selection and management of strategies, wholesalers and retailers and the physical distribution, etc.

Distribution channels are the ownership of goods and services in transferring from producers to consumers or all companies and individuals helping to transfer such ownership. They include businessmen, brokers, agent brokers, as well as the producers and consumers at the start and the end of channels.

In accordance with the belonging of ownership, the distribution sector will be generally divided into three categories: distributors, agents and support organizations. Common distributors include general wholesalers and retailers. The common agents include company agents, sales agents, purchasing agents, commission agents and brokers. Support organizations include distribution centers, transportation companies. independent warehouses, banks and advertising agents. According to the circulation number of distribution channels, it can be divided into direct channels and indirect channels. The difference between these two is the broker.

The main function of distribution channels is to collect, study and formulate plans and information, to design and disseminate information related to commodities, to formulate program of promotion, to encourage consumers to buy, to expand channels, to coordinate the market and production, to adjust product according to the demand; to negotiate on behalf of the buyer or the seller to participate in the price and other terms of trade, to store and transport, to finance, to share some costs or all of work, to share the risks.

Channel design refers to strategy activities of establishing marketing channels that have never existed before, or changing the existing channel. When choosing the specific mode of distribution channel, the administrators should follow the principles of smoothness and efficiency, the principles of moderation, stability control, the coordination and equilibrium, and the principle of advantages for whatever consideration, or wherever to start. There are many factors affecting the design, such as market factors, product factors, company factors, brokers factors, environmental factors and behavioral factors.

With the rapid development of the Internet, the development of Internet technology makes a tremendous impact on all aspects of social and economic life. As an important part in marketing system, distribution channels and their structures have also made such a profound change under this influence. Since the emergence of Internet technology, the traditional models of distribution channels are facing intense shock. The Internet has an incomparable advantage to the traditional communication. It makes companies to grasp the accurate market information, decrease transaction costs, improve product competitiveness and minimize company's inventories; it also helps companies to provide personalized products. Either as a new distribution channel, or as a strategic means of reshaping business processes, Internet determines the transformation of traditional distribution channels. Today's distribution channel is no longer an external value-added service independent outside of the company, but the company's core assets deciding success or failure. Companies should pay attention to the significance and function of network distribution channels, adjust and configure the distribution channels reasonably.

New Words, Phrases and Expressions
broker   n. 中間人，代理商
belonging   n. 附屬品，附件，屬性
commission   n. 授權，委託
circulation   n. 流通，循環
distributor   n. 經銷商，分銷商
inventory   n. 存貨，庫存
ownership   n. 物主的身分，所有（權），所有制
retailer   n. 零售商，零售店
transfer   v. 轉移，遷移
transaction   n. 處理，辦理，執行
wholesaler   n. 批發商
warehouse   n. 倉庫，貨棧
no longer   不再，已不
on behalf of   為了……的利益，代表……

Notes
1. Distribution is to make products and services appropriate to the number and geographical distribution of the target market in a timely manner to meet customer needs.
分銷就是使產品和服務以適當的數量和地域分佈來適時地滿足目標市場的顧客需要。
2. When choosing the specific mode of distribution channel, the administrators should follow the principles of smoothness and efficiency, the principles of moderation, stability control, the coordination and equilibrium, and the principle of advantages for whatever

consideration, or wherever to start.

分銷渠道管理人員在選擇具體的分銷渠道模式時，無論出於何種考慮，從何處著手，一般都要遵循暢通高效的原則、覆蓋適度的原則、穩定可控的原則、協調平衡的原則和發揮優勢的原則。

3. Since the emergence of Internet technology, the traditional models of distribution channels are facing intense shock.

由於互聯網技術的出現，傳統分銷渠道模式正在受到強烈的衝擊。

4. Either as a new distribution channel, or as a strategic means of reshaping business processes, Internet determines the transformation of traditional distribution channels.

互聯網無論是作為一種新的分銷渠道，還是作為重塑業務流程的戰略手段，都決定了傳統企業分銷渠道的轉型。

Lesson 4　Promotion Strategies

The promotion strategy is one of the basic strategies in marketing mix, which refers to all kinds of promotion ways of marketing by staff, advertising, public relations and sales promotion and so on. In order to achieve the purpose of expanding sales, promotion strategy conveys product information to consumers or users to stimulate their desire to buy and purchase by drawing their attention and interest.

In modem marketing, promotion strategies include advertising, sales promotion, personal selling, public relations. Promotion mix refers to the communication methods used to provide information for the customers and the market. The company shall consider the characteristics and marketing goals of products and all other factors to choose, allocate and apply a variety of promotions, so that their cooperation could develop maximally to implement promotional objectives successfully.

Advertising is a very important part in marketing mix, which is also the most widely used and most effective marketing tool. Generally, advertising refers to the economic activities that the company pays for various means of communication to the target market to transfer the information of the product or service for expanding sales and obtaining the profit.

Sales promotion is a kind of strategy that the company applies special methods to implement a strong stimulus to consumers to promote a rapid growth of sales in a certain period of time.

The methods commonly used in sales promotion are giving samples, issuing coupons, prize-winning sales, old for new service, organizing competitions and live demonstrations and so on. Sometimes it is also used for promotion to intermediaries, such as the transferring rebate, paying promotional allowances, organizing sales contests, etc. Exhibition and exposition are also means frequently used in promotion.

Personal marketing refers that the company sends out salesmen to talk to one or more potential buyers, as an oral presentation, to promote products and expand sales. It is a process of salesman helping and persuading the buyers to purchase some products or services. Personal marketing possesses the advantages of flexibility, operating elasticity, holding other functions of marketing. Being lack of personal selling is when the market is vast and dispersed, marketing costs would be higher and the employee would be difficult to manage. Personal selling strategy generally includes three types as marketing strategy, transaction strategy and satisfied strategy.

The public relation is another important strategy in business promotion. The public relation is an activity in which the company establishes a good social image and marketing environment by using various ways of communication, including all aspects of the public, customers, brokers, community residents, government agencies and media. The public relation is mainly to improve the marketing environment through establishing the public image. There are many ways of communication in public relations, all kinds of media or various forms of direct communication. The application of the media in public relations is usually in the form of news report. The role of public relations is played in all aspects of the company. The strategy of public relations can be divided into three levels: public relations campaign, public relations activities and public relations consciousness.

New Words, Phrases and Expressions
allocation　n. 配給，分配，調撥
allowance　n. 津貼，補助，補貼
coupon　n. 優惠券
campaign　n. 宣傳活動
consciousness　n. 意識，觀念，覺悟，看法
demonstration　n. 表明，證明，示範
disperse　v.（使）散開，驅散
exhibition　n. 展覽，展覽會
exposition　n. 展覽會，博覽會
elasticity　n. 彈性，彈力
maximize　v. 使（某事物）增至最大限度
intermediary　n. 中間人
rebate　n. 可減免的款額，折扣
stimulus　n. 刺激物，促進因素
transaction　n. 交易，協議，協定
public relation　公共關係
send out　發送，散布，派遣

Notes
1. Promotion mix refers to the communication methods used to provide information for the customers and the market. The company shall consider the characteristics and

marketing goals of products and all other factors to choose, allocate and apply a variety of promotions, so that their cooperation could develop maximally to implement promotional objectives successfully.

促銷組合就是企業根據產品的特點和行銷目標，綜合各種影響因素，對各種促銷方式進行選擇、編配和運用，使企業的全部促銷活動互相配合協調，最大限度地發揮整體效果，以順利實現促銷目標。

2. Advertising is a very important part in marketing mix, which is also the most widely used and most effective marketing tool.

廣告是企業促銷組合中十分重要的組成部分，也是運用最為廣泛和最為有效的促銷手段。

3. Generally, advertising refers to the economic activities that the company pays for various means of communication to the target market to transfer the information of the product or service for expanding sales and obtaining the profit.

通常情況下，廣告是指企業為擴大銷售獲得盈利，以付酬的方式利用各種傳播手段向目標市場的廣大公眾傳播商品或服務信息的經濟活動。

4. Sales promotion is a kind of strategy that the company applies special methods to implement a strong stimulus to consumers to promote a rapid growth of sales in a certain period of time.

營業推廣（銷售促進）是企業在某一段時期內採用特殊的手段對消費者進行強烈的刺激，以促進企業銷售迅速增長的一種策略。

5. Personal marketing refers that the company sends out salesmen to talk to one or more potential buyers, as an oral presentation, to promote products and expand sales.

人員推銷是指企業通過派出銷售人員與一個或一個以上可能成為購買者的人交談，做口頭陳述，以推銷商品，促進和擴大銷售。

6. The public relation is mainly to improve the marketing environment through establishing the public image.

公共關係主要是通過樹立企業的良好形象來改善企業的經營環境。

Section 5　the New Development of International Marketing

With the development of economy, international marketing has also developed to some new stages. More and more kinds of international marketing has come to our eyes, in this section, we will introduce you some new forms of international marketing, including on-line marketing, relation marketing, green marketing and service marketing.

On the basis of Internet technology, the on-line marketing is a kind of modern marketing mode of satisfying the demand of customers efficiently. As a product of high-level development of modern science and technology, the on-line marketing has many benefits:

improving efficiency and interactivity and reducing costs; getting rid of the space and time limit and the dependence on documents and providing more choices for customers; enhancing the value of products and service information and decreasing the drawback in expanding the international market, etc.

The common methods of on-line marketing include: searching engine, searching key words, on-line advertisement, e-shop, exchanging link, announcing information. marketing forum, mail list, permissive e-mail marketing, individual marketing, member marketing, virus marketing, and so on.

From the beginning of the 1990s, more and more attention is paid to the theory about the relation marketing. The relation marketing regards the marketing activities as a process of interactivities between the company, customer, supplier, distributor, competitor, governmental agency and other public institutions, in which the heart is to build and develop a good relationship among these public institutions.

From the definition, the relation marketing emphasizes the traditional conception, the correlation between every element in the personal nets. Therefore, the relation marketing has the characteristics of enhancing the two-way communication, collaboration strategy, reciprocity marketing and feedback information.

Green Marketing

The conception of the green marketing, which was put forward in the 1980s, refers to rethinking, designing, selling and producing about the products and services centered on the benefit of the company, customer and environment in the process of production and management. The function of the green marketing is mainly to promote the awareness of environment protection in the process of production and management to ensure the eco-balance. The practice of green marketing increases the production safety, the company's competitiveness and defends the benefit of customers as well.

Topractice the green marketing should start with formulating the programme and strategy of the green marketing. The green marketing programme includes the plans about green investment, green product development, green image-building, green education and green production. The green marketing strategy includes developing green products, pricing green products and structuring green distribution channels.

Service Marketing

With the economic development and structural promotion, the importance of service has been gradually improved. The service marketing refers to the marketing activities as the researching object of service behavior and service in products management. The essence of the service marketing is to win the customer's satisfaction and loyalty, through which the exchange of mutual benefit and the long-term development will be realized.

Compared with other forms of marketing, the service marketing has the features of dispersed supply and demand, single marketing form, flexible demanding of service and higher demanding of the service personnel.

New Words, Phrases and Expressions
collaboration   n. 合作，協作
channel   n. 通道，渠道
drawback   n. 缺點，障礙
distributor   n. 批發商
dispersed   adj. 散布的，被分散的
eco-balance   生態平衡
feedback   n. 反饋
flexible   adj. 靈活的，可變通的
institution   n. 慣例，習俗，機構
reciprocity   n. 相互性，互惠
supplier   n. 供應商
compare with   與……相比
get rid of   除掉，擺脫
image-building   形象塑造
personal net   關係網
regard... as   把……看做，認為
searching engine   搜索引擎
start with   以……開始
service personnel   服務人員

Notes

1. Relation marketing regards the marketing activities as a process of interactivities between the company, customer, supplier, distributor, competitor, governmental agency and other public institutions, in which the heart is building and developing a good relationship among these public institutions.

關係行銷，就是把行銷活動看成是企業與消費者、供應商、分銷商、競爭者、政府機構及其他公眾發生互動作用的一個過程，其核心是建立和發展與這些公眾的良好關係。

2. The conception of the green marketing. which was put forward in the 1980s, refers to rethinking, designing, selling and producing about the products and services centered on the benefit of the company, customer and environment in the process of production and management.

綠色行銷的行銷觀念於 20 世紀 80 年代提出，是指企業在生產經營過程中，將企業自身利益、消費者利益和環境保護利益三者統一起來，並以此為中心，對產品和服務進行再構、設計、銷售和製造。

3. The function of the green marketing is mainly to promote the awareness of environ-

ment protection in the process of production and management to ensure the eco-balance.

綠色行銷的作用主要是促使企業在生產經營活動過程中加強環保意識，保證生態平衡。

4. To practice the green marketing should start with formulating the programme and strategy of the green marketing.

實施綠色行銷需要從制定綠色行銷戰略規劃、綠色行銷策略等著手。

5. The essence of the service marketing is to win the customer's satisfaction and loyalty, through which the exchange of mutual benefit and the long-term development will be realized.

服務行銷的核心理念是顧客滿意和顧客忠誠，通過取得顧客的滿意和忠誠來促進相互有利的交換，最終實現企業的長期發展。

6. Compared with other forms of marketing, the service marketing has the features of dispersed supply and demand, single marketing form, flexible demanding of service and higher demanding of the service personnel.

與其他行銷方式相比，服務行銷具有供需分散、行銷方式單一、對服務的需求彈性大、對服務人員的要求較高等特點。

Exercises

Ⅰ. Define the following concepts

1. market
2. demands
3. relation marketing
4. macroenvironment
5. customer value
6. distribution channel

Ⅱ. True or False

1. Product life cycle generally divides into four stages: introduction, growth, mature and death. (　　)

2. Three strategies in mature period are as follows: (1) market adjustment; (2) product adjustment; (3) marketing mix adjustment. (　　)

3. There are four forms in different price strategy: geographical different strategy, time different strategy, usage different strategy, and quality different strategy. (　　)

4. Channel design refers to strategy activities of establishing marketing channels that never existed before. (　　)

5. Advertising is a very important part in marketing mix, which is also the most widely used and most effective marketing tool. (　　)

Ⅲ. Questions for Discussion

1. What's marketing and what environmental factors are affecting an organization?
2. Explain the five forces in the five forces competitive model.

3. What's marketing mix and discuss different strategies in the process of marketing.

4. Can you give any examples of the new development of marketing?

Further Reading Comprehension

Quantitative Easing: Turning over a New Leaf?

Mar 26th 2014, 13:33 by C. R. | LONDON

NORDIC and Germanic opposition to unconventional monetary policy is crumbling, so it seems. On March 25th, in interviews with Market News and the Wall Street Journal, two of the more hawkish members of the European Central Bank's governing council hinted that further monetary easing may be on the cards for the euro area. The euro obligingly sank on foreign-exchange markets.

First, Jens Weidmann, president of the German Bundesbank, told Market News that quantitative easing was no longer「out of the question」, having previously ruled it out as a legitimate policy tool for the ECB. Then, Erkki Liikanen, Governor of the Bank of Finland, also seemed to open the door to this type of policy. He told the Wall Street Journal that, even with interest rates at 0.5%「we haven't exhausted our manoeuvering room」on monetary policy, and that the question of negative deposit rates, in my mind, isn't any longer a controversial issue.

Other members of the ECB's governing council alsoseem to be more open to using a more unconventional range of policy tools. Jozef Makuch, the governor of the National Bank of Slovakia, said yesterday that quantitative easing was one option being considered, and said that there was growing support for reflationary policies:「Several policy makers are ready to adopt non-standard measures to prevent slipping into a deflationary environment.」

Such a change in attitudes is probably due to three factors. First, the consensus views among Nordic and Germans economists have shifted over the last year or so. The mood at academic conferences has become less hostile towards unconventional monetary policies, and the fear of unintended consequences from quantitative easing is much reduced (although a few economists are growing more concerned about the effect of monetary stimulus on inequality).

But the shift also reflects rising worries in two other areas. More economists are fretting about the risks of getting stuck in a deflationary trap. Inflation in the euro areahas been stuck at less than 1% since October; in February it fell to 0.7%.

Even worse, business confidence has been hit in central Europe by the Crimea crisis. On March 25th the latest results for the Ifo Institute for Economic Research's business-climate index suggested a sharp fall in economic confidence in Germany. The fear of a drop in demand in Europe's economic powerhouse has increased support for policies such as negative interest rates or quantitative easing being used to counteract the risk of a damaging slowdown.

However, launching unconventional monetary policies could still prove problematic.

The German constitutional court ruled last month that the ECB's bond rescue plan, Outright Monetary Transactions (OMT), is contrary to European law. And there has been little sign, as of yet, that German public opinion, or the attitudes of the government they recently elected, are becoming more favorable towards such a policy agenda. In short, there's still a long way to go before we see the ECB set QE free.

From: The Economic, http://www.economist.com/Fblogs/freeexchange/2014/03/quantitative-easing 2014/3/27.

Chapter 7
International Investment

Chapter checklist

When you have completed your study of this chapter, you will be able to
1. Define international investment and identify its relevant noun.
2. Know about environment of international investment.
3. Understand the types of international investment and its theories.
4. Identify the risks of international investments.
5. Know the process of portfolio management.

In this chapter, we will mainly introduce relevant knowledge for international investment. There are five sections in the chapter. They are overview of international investment, international investment environment, international direct investment, international securities investment, and management of international investments. The more important and difficult parts include international arrangements on investment environment, theories of FDI, currency risk, political risk and their management.

Section 1 An Overview of International Investment

Lesson 1 Relevant Definition for International Investment

Investment can be defined as any use of resources with the hope to increase future production output or income. This said, college education is an investment. Students pay tuition and spend four years at school hoping to make more money in the future than people without a college degree.

In economics, investment refers to the purchase of goods that are not consumed today but are used in the future to create wealth. For example, building a factory is an investment in the economic sense. In Finance, investment refers to the purchase of marketable securities with the hope that they will generate income in the future or appreciate and be sold at a higher price. For example, buying publicly traded stocks is an investment in the financial sense.

Investments are the study of the investment process. Although the field of investments encompasses many aspects, it can be categorized by two primary functions: analysis and management.

Traditionally, investment takes place within the investor's home country. It is called domestic investment. However, nowadays, with the integration of the world economy, globalized too. Investment across country borders is called foreign investment.

Some researchers use international investment to actually mean foreign investment, but use the term globalinvestment to include both domestic investment and foreign investment. In this book, we use foreign investment, international investment and global investment interchangeably. The reason is that basic investment principles such as the trade-off between expected return and risk are applicable to any investment, no matter it is within a country's border or beyond it. In the case of securities investment, foreign securities or funds are considered as part of a portfolio to mitigate risk. No investors will invest in foreign stocks only.

International investment falls into two major categories: foreign direct investment (FDI) and foreign portfolio investment (FPI). Foreign direct investment (FDI) refers to investment abroad in which the investor obtains alasting interest in an enterprise in another country. It may take the form of buying or constructing a factory in a foreign country or adding improvements to such a facility, in the form of property, plants, or equipment.

Through FDI, an investor supply capital, know-how, technology and management techniques to the foreign recipient. As a return, the investor collects profits in the form of dividends, retained earnings, management fees, and royalty payments. FDI implies a controlling stake in a business and often connotes ownership of physical assets such as equipment, buildings, and real estate. FDI is more difficult to pull out or sell off. Consequently, direct investors may be more committed to managing their international investments and less likely to pull out at the first sign of trouble.

Foreign portfolio investment (FPI) refers to the purchase of foreign securities, such as bonds and stocks. Investors can pick foreign securities and construct their own portfolios. They can also purchase shares of international mutual funds. The returns that an investor acquires on FPI take the form of interest payments, dividends and appreciation in values of the securities they hold. FPI is investing in foreign securities but investors do not seek to control the companies issuing the securities. Hence, investors do not actively participate in the management of the companies. Since the securities are easily sold in the securities market. FPI is more volatile than FDI. When the country's economy is rising, the large inflow of money by FPI can bring about rapid development. However, when the economy takes a downturn, sometimes just by failing to meet the expectations of international investors, FPI flees the country quickly. This affects the exchange rate of the nation's currency and finally produces widespread economic crisis as evidenced by the financial crisis in Southeast Asia in 1997-1998.

Sometimes FDI and FPI may overlap, especially in regard to equity investment. Ordinarily, the threshold for FDI is ownership of「10 percent or more of the ordinary shares of voting power」of a business entity (IMF Balance of Payments Manual, 1993).

New Words, Phrases and Expressions
investment　n. 投資
investments　n. 投資學（也用作 investment 的復數形式，指多種投資）
relevant　adj. 相關的；切題的；中肯的；有重大關係的；有意義的，目的明確的
defined　vt. 定義；使明確；規定
economics　n. 經濟學；國家的經濟狀況
in the economic sense　從經濟意義上來說
encompass　vt. 包含；包圍，環繞；完成
categorize　vt. 分類，把……分類（或分等級）；把……歸入一類（或同一等級）
home country　祖國，母國
integration　n. 集成，綜合；一體化
interchangeably　adv. 可交換地；可互換地；可交替地
trade-off　n. 交換，交易；權衡；協定
in the case of　至於，在……的情況下
mitigate　vt. 使緩和，使減輕
facility　n. 設施；設備；容易；靈巧
recipient　n. 容器，接受者；容納者
dividend　n. 紅利；股息；[數] 被除數；獎金
connote　vt. 意味著；含言外之意
pull out　離開，撤離；拔出；渡過難關；恢復健康；折疊的大張插頁；飛機進場重新飛起
sell off　廉價出清（存貨）
be committed to　致力於；委身於；以……為己任
volatile　adj. 揮發性的；不穩定的；爆炸性的；反覆無常的
downturn　n. 衰退（經濟方面）；低迷時期
overlap　vi. 部分重疊；部分的同時發生　vt. 與……重疊；與……同時發生
in regard to　關於；對於；至於；就……而言
equity　n. 公平，公正；衡平法；普通股；抵押資產的淨值
threshold　n. 入口；門檻；開始；極限；臨界值

Notes

1. In economics, investment refers to the purchase of goods that are not consumed today but are used in the future to create wealth.

在經濟學裡，投資是指購買物品，購買這些物品的目的不是用於今天的消費而

是為了將來的增值。

2. In Finance, investment refers to the purchase of marketable securities with the hope that they will generate income in the future or appreciate and be sold at a higher price. For example, buying publicly traded stocks is an investment in the financial sense.

在金融學裡，投資是指有價證券的購買，並希望其在將來能夠產生收入或者增值並且以更高的價格賣出。例如，購買公開交易的股票是金融意義上的投資。

3. When the economy takes a downturn, sometimes just by failing to meet the expectations of international investors, FPI flees the country quickly. This affects the exchange rate of the nation's currency and finally produces widespread economic crisis as evidenced by the financial crisis in Southeast Asia in 1997–1998.

當經濟出現衰退，或有時只是不符合國際投資者的期望，外國證券投資者便迅速逃離該國。這會影響該國的貨幣匯率，最終產生廣泛的經濟危機，正如1997—1998年東南亞金融危機。

Lesson 2　Markets of International Investment

When considering international investments, there are three major types of markets from which to choose: developed markets, emerging markets and frontier markets.

Developed markets include the largest, most industrialized economies. Their economic systems are well developed, they are politically stable, and the rule of law is well entrenched. Developed markets are considered the safest investment destinations. However, the growth rate is modest.

The following countries are often classified as developed markets: Australia, Canada, France, Germany, Japan, Singapore, Switzerland, the United Kingdom and the United States.

The term「emerging markets」was created by Antoine van Agtmael in the 1980s when he was a World Bank economist. It refers to countries that are in a transitional phase between developing and developed status. They are experiencing rapid industrialization and often demonstrate extremely high levels of economic growth as measured by GDP. Investment in emerging markets may bring about superior returns to those that are available in developed markets. However, investors are bearing much greater risks.

Emerging markets are generally less efficient than developed markets. Political instability, currency volatility, and lack of strict standards in accounting and securities regulations often cause concerns in investors. Also, local stock exchanges may notoffer sufficient equity opportunities because many large companies are still「state-run」or because it is not open to outside investors.

Examples of emerging markets include Argentina, Brazil, Chile, China, India, Indonesia, Malaysia, Mexico, Russia, Thailand, Turkey. FTSE Group, a provider of economic and financial data, distinguishes between advanced and secondary emerging markets on the basis of their national income and the development of their market infrastructure. Brazil and Mexico belong to the advanced emerging markets group, and other countries mentioned above belong to the secondary group.

The term 「frontier markets」 was coined by IFC's Farida Khambata in 1992. They are countries with investable stock markets that are less established than those in the emerging markets. They are also known as 「pre-emerging markets」. According to a Merrill Lynch report on frontier markets, there is no strict definition of a frontier market. They are simply developing economies with undeveloped equity markets. 「Frontier markets tend to be young. Thinly-traded equity markets with weak regulatory frameworks, low levels of transparency and low levels of foreign ownership.」 says the report. Examples of frontier markets include Jamaica, Jordan, Kenya, Kuwait, Romania, Sri Lanka and Vietnam.

Frontier markets offer investors potentially high returns but they are exposed to higher risks. These risks include political instability, poor liquidity, inadequate regulation, substandard financial reporting and large currency fluctuations. In addition, many markets are overly dependent on volatile commodities. The good news is that frontier markets have a low correlation with other markets hence provide additional diversification benefits when held in a well-round investment portfolio.

Notice that different agencies may classify a country into different markets and the list for the same agent may change over time. For example, FTSE Group categorized Luxembourg and South Korea into developed markets but they are not included in MSCI Barra's developed markets list. As of September 2009, Argentina was not classified by PTSE Group as frontier markets but it was included in the September 2010 indices as a frontier market.

New Words, Phrases and Expressions
politically   adv. 政治上
entrench   vt. 確立, 牢固；用壕溝圍住；挖掘
be classified as   被歸類為……；分類為……
transitional   adj. 變遷的；過渡期的
industrialization   n. 工業化
demonstrate   vt. 證明；展示；論證
extremely   adv. 非常, 極其；極端地
available   adj. 有效的, 可得的；可利用的；空閒的
instability   n. 不穩定（性）；基礎薄弱；不安定
volatility   n. [化學] 揮發性；易變；活潑
distinguish   vt. 區分；辨別；使傑出, 使表現突出
distinguish between... and...   把……和……區別
infrastructure   n. 基礎設施；公共建設；下部構造
regulatory   adj. 管理的；控制的；調整的

transparency   n. 透明，透明度；幻燈片
liquidity   n. 流動性；流動資產；[金融] 償債能力
fluctuation   n. 起伏，波動
in addition   另外，此外
classify   vt. 分類；分等

Notes

1. They are experiencing rapid industrialization and often demonstrate extremely high levels of economic growth as measured by GDP. Investment in emerging markets may bring about superior returns to those that are available in developed markets. However, investors are bearing much greater risks.

它們正在經歷快速的工業化和常表現出以 GDP 衡量極高水準的經濟增長。投資於新興市場對那些可進入發達市場的投資者來說可能帶來更高的回報。然而，投資者將會承擔更大的風險。

2. Frontier markets tend to be young. Thinly-traded equity markets with weak regulatory frameworks, low levels of transparency and low levels of foreign ownership.

前沿市場往往是年輕的：清淡交易的股票市場與薄弱的監管框架、透明度低、外資持股水準低。

3. The good news is that frontier markets have a low correlation with other markets hence provide additional diversification benefits when held in a well-round investment portfolio.

好消息是，前沿市場與其他市場的相關性低，因此當持有良好週期的投資組合時可以提供更多的多元化的好處。

Lesson 3   The Motivations of International Investment

Motivations for FDI

FDI is often undertaken by multinational corporations. They invest internationally for similar reasons as they expand their operations within their home country. The economist John Dunning has identified four primary reasons for corporate foreign investments: Market seeking: Firms may have saturated sales in their home market and go overseas to find new buyers; Resource seeking: Some companies in the developed countries are attracted by the cheap labor in less developed countries like China and India; Efficiency seeking: Multinational companies may shift the allocation of its resources to overseas in response to broader economic changes such as lower tariff rates and favorable fluctuation in exchange rates; Strategic asset seeking: Firms may seek to invest in other companies abroad to help build strategic assets, such as distribution networks or new technology.

Motivations for FPI

FPI comes from individuals through pension funds or mutual funds held by individuals. They invest internationally for two reasons: the lure of higher return than domestic investment and the benefit of diversification of portfolio risk. In the 1980s and early 1990s, a number of countries and areas in the East Asia (Indonesia, Japan, South Korea, Malaysia, Singapore, Thailand, etc.) began to experience enormous economic growth rates. Investors from around the world were attracted by the much higher potential the East Asia than that at home. In 2004, international stocks returned 19% while the S&P 500 (a broad measure of US stocks) posted a gain of only 11%.

The other motives for FPI is to diversify away some of the portfolio risk. Markowitz's model and modern portfolio theory argue that when two securities are not perfectly positively correlated, combining them will lower the risk of your investment. The correlation between foreign stocks and domestic stocks are thought to be lower than the correlation among domestic stocks. Therefore, international investment should be pursued by any investor who wants to manage the portfolio risk wisely. The economies of the world's different regions have tended to boom and bust in cycles and often offset each other in the past three decades. One region collapsed; another made up for it. A dramatic example came in the East Asian financial crisis when Japan's Nikkei index lost almost 40% between June 1997 and October 1998. Meanwhile, European markets blossomed as the continent came closer to true economic union and a rash of US-style corporate restructurings began to pay dividends.

New Words, Phrases and Expressions
motivation   n. 動機；積極性；推動
multinational   adj. 跨國公司的；多國的
saturate   vt. 浸透，使濕透；使飽和，使充滿   adj. 浸透的，飽和的；深顏色的
allocation   n. 分配，配置；安置
diversification   n. 多樣化；變化
portfolio   n. 公文包；文件夾；證券投資組合；部長職務
offset   vt. 抵消；彌補；用平版印刷術印刷
blossom   vi. 開花；興旺；發展成

Notes

1. Multinational companies may shift the allocation of its resources to overseas in response to broader economic changes such as lower tariff rates and favorable fluctuation in exchange rates.

跨國公司在應對更廣泛的經濟變革時，可將資源配置轉移到海外，這些變革如降低關稅稅率和匯率有利波動。

2. The economies of the world's different regions have tended to boom and bust in cycles and often offset each other in the past three decades.

世界上不同地區的經濟已經趨於繁榮和蕭條的週期，並且在過去的30年裡它們通常互相抵消。

Section 2　International Investment Environment

Lesson 1　An Introduction to International Investment Environment

International investment environment may be simply defined as the general conditions of the host country which affect the decision making of foreign investment. Some authors prefer to define it narrowly as the general economic conditions of the host county. In China, the international investment environment is usually divided into soft environment and hard environment. The hard environment for investment means the material conditions of the host country, including natural resources, communication and transportation, infrastructure, etc. In contrast, the soft environment for investment is referred to as the immaterial conditions, such as laws and policies, manners and customs. In western countries, the term investment climate is used commonly, with the similar meaning to the term investment environment.

When assessing the investment environment of a country, foreign investors usually take the following economic factors, political factors, legal factors, natural factors, social and cultural factors into consideration.

Economic Factors

Simply speaking, a country with a rapid and stable economic growth has a special attraction for foreign investors. There are some important indicators for investors to measure the economic development of a country, such as GDP, the economic growth rate, price indexes, rate of inflation, balance of payments, etc. Every investor hopes the host country has a sound and perfect market system. This required that the market for factors of production, commodity market, labor market, technology market all have matured considerably. Such a market system allows foreign investors to compete fairly with domestic firms. However, in many host countries, all markets are not at the same level of maturation. For example, in China, labor market lags behind in development in comparison with commodity market.

Good infrastructure service is also significant for the host country to attract foreign investment. Infrastructure service involves energy supply, transportation and communication, financial and information service, living facilities, education and health service, etc.

The economic policies of host country are sure toaffect the decision making of foreign investors. Especially, the trade policies, foreign exchange regulations, tax policies, industrial policies are paid much attention to. In general, a host country which exercise free trade policies and have loose regulations on foreign exchange and taxation, is more attractive for foreign investment.

Political factors include (1) Stability of politics, (2) Governance of government

(3) Political system and (4) External relations of government.

Stability of politics has two aspects: stability of government and continuity of policies. The government of a country is stable if it will be in power in the long rum. A stable government will help to keep the continuity of policies. Undoubtedly, a country which has its government and policies changed frequently or unpredictably will hold back foreign investment.

Governance of government involves the governing capability and efficiency of the government, corruption problems, etc. Of course, international investors want to invest in a host country governed by a government which has great capability and efficiency, and corruption-free officers.

Normally, investors like to invest in those countries with the similar political systems to their home countries. However, with the deepening of economic globalization, the differences of political systems between countries are making less sense for international investments. For instance, as a socialist state, China is increasingly attracting foreign investment from many capitalist countries because of its reforming and opening policies and rapid economic growth in the past decades.

Good and stable external relations with foreign governments help the government of host country to keep smooth international trade and cooperation. In contrast, bad and unstable external relations of host country bring more external risks for foreign investment. Such external risks may be trade banns or even wars.

Legal factors include (1) Completeness of laws; (2) Stability of laws; (3) Justice of laws.

The host country shall have a full and complete legal system to encourage and protect foreign investment. To build up a legal system, a wide range of laws in almost all sectors and industries of complete the society shall be enacted.

The laws in host country shallbe stable in order that the foreign investors may plan their investments and predict their profits in the long run.

Justice is the basic principle of laws in today's world. This principle requires both legislation and enforcement of laws to be just.

Natural factorsinclude natural resources、population and climate. Natural resources include mineral resources, water resources, forest resources, land resources and etc. If a host country is rich in one sort of resource, foreign investor may prefer to invest in this industry. For example, many investors from developed countries make investments in oil production in Middle-east countries because of their abundant oil resource.

Population is significant in determining both the market size and the work force of the host country. Generally, a large population means a large market and a large number of labors in the host country. However, this will not be the truth in those poor countries.

Temperature, sunlight, rainfall and wind may affect the production of certain industry.

Climate may also have influences on transportation and sale of product. So climate will be taken into account when foreign investors choose a country to invest.

Social and Cultural factors

Social and cultural factors may be the most complicated factors affecting the investment environment. Among these factors, language, education level, prevailing values, customs, religion and nationalist sentiment may be most important for foreign investors to consider.

New Words, Phrases and Expressions
be defined as   被定義為；被界定為
immaterial   adj. 非物質的；無形的；不重要的；非實質的
take into consideration   顧及；考慮到
maturation   n. 成熟
lag behind   落後；拖欠
significant   adj. 重大的；有效的；有意義的；值得注意的；意味深長的
transportation   n. 運輸；運輸系統；運輸工具；流放
pay attention to   注意；重視
stability   n. 穩定性；堅定，恒心
undoubtedly   adv. 確實地，毋庸置疑地
capability   n. 才能，能力；性能，容量
efficiency   n. 效率；效能；功效
in contrast   與此相反；比較起來
external   adj. 外部的；表面的；[藥] 外用的；外國的；外面的
completeness   n. 完整；完全；完成；圓滿；結束
justice   n. 司法，法律制裁；正義；法官，審判員
legislation   n. 立法；法律
enforcement   n. 執行，實施；強制
prevailing   adj. 流行的；一般的，最普通的；占優勢的；盛行很廣的
sentiment   n. 感情，情緒；情操；觀點；多愁善感

Notes

1. International investment environment may be simply defined as the general conditions of the host country which affect the decision making of foreign investment. Some authors prefer to define it narrowly as the general economic conditions of the host county.

國際投資環境，可以簡單地定義為東道國影響外來投資做出決策的東道國總體環境。一些學者寧願狹義地定義它為東道國的整體經濟狀況。

2. When assessing the investment environment of a country, foreign investors usually take the following economic factors, political factors, legal factors, natural factors, social and cultural factors into consideration.

當評估一個國家的投資環境時，外國投資者通常會將經濟因素、政治因素、法

律因素、自然因素、社會文化因素考慮在內。

3. China is increasingly attracting foreign investment from many capitalist countries because of its reforming and opening policies and rapid economic growth in the past decades.

中國吸引了越來越多的來自資本主義國家的投資，因為在過去幾十年它實行了改革開放政策，並且經濟快速增長。

Lesson 2 International Arrangements on Investment Environment

After evolution of 60 years, the volume and range of international rules and standards dealing with investment environment issues has shot up dramatically. There are now more than 2,857 bilateral investment treaties, 200 regional cooperation arrangements, and some 500 multilateral conventions and instruments. These arrangement cover most areas of the investment environment—from property fights protection, taxation, and corruption, to regulation in areas as diverse as banking, shipping, telecommunications, labor, and natural environment. In this section, several most important international arrangements on investment environment will be briefed.

The Agreement on Trade Related Investment Measures (TRIMs) are rules that apply to the domestic regulations a country applies to foreign investors, often as part of an industrial policy. The agreement was agreed upon by all members of the World Trade Organization. The agreement was concluded in 1994 and came into force in 1995. The WTO was not established at that time, it was its predecessor, the GATT (General Agreement on Trade and Tariffs. The WTO came about in 1994-1995.).

The agreement recognizes that certain investment measures restrict and distort trade. It provides that no contracting party shall apply any trade related investment measures inconsistent. The agreement requires mandatory notification of all non-conforming TRIMs and their elimination within two years for developed countries, within five years for developing countries and within seven years for least-developed countries. It establishes a Committee on TRIMs which will, among other things, monitor the implementation of these commitments. The agreement also provides for consideration. At a later date, of whether it should be complemented with provisions on investment and competition policy more broadly.

The agreement only deals with those investment measures relating to trade, and there may be some other domestic regulations and restrictions on foreign investors in a host country. Therefore, the TRIMs agreement is not enough for the purpose of improving international investment environment.

In May 1995, the OECD began negotiations on a Multilateral Agreement on Investment (MAI). The US, eager to avoid interference from poor countries, considered the OECD Council a「safe」body since only rich countries are members of the organization. Secret negotiations took place from 1995 until 1997 when an OECD source leaked a copy of the draft agreement to a Canadian citizen group. The leak revealed that the MAI sought to establish a new body of universal investment laws that would guarantee corporations unconditional rights to buy, sell and do financial operations all over the world, without any regard for national laws and citizens' rights. The draft gave corporations a right to sue governments if national health, labor or environment legislation threatened their interests.

However, the negotiations failed in 1998 when first France, and then other countries, successively withdrew after pressure from a global movement of NGOs, citizens groups and governments of poor countries. MAI opponents saw the agreement as a threat to national sovereignty and democracy and argued that it would lead to a 「race to the bottom」 in environmental and labor standards. Many critics also argued that the OECD was more susceptible to direct influence by transnational corporate forces than alternative fora with more universal membership such as the UNCTAD and the WTO. This view was finally accepted by governments and turned out to be a crucial factor in the MAI's defeat.

The failure of the MAI negotiations proved a success for the global movement against the MAI. However, rich governments continue to push for similar investment provisions in regional trade agreements and the World Trade Organization (WTO). At the WTO Ministerial in Cancun in September 2003, the richer WTO members tried to introduce a Multilateral Investment Agreement (MIA) through the 「Singapore」 issues. These efforts failed again, however, as a group of more than twenty poor countries united in demanding a fairer trade deal. This page follows the MAI and other related initiatives on international investment.

In addition, from May 2006, the OECD has promoted a non-binding set of 「good practices」 for attracting investment, known as the 「Policy Framework for Investment」 (PFI). The PFI is the most comprehensive and systematic approach for improving investment conditions ever developed. It covers ten policy areas and addresses some questions to governments to help them design and implement policy reform to create a truly healthy, robust, attractive and competitive environment for domestic and foreign investment. The PFI is neither prescriptive nor binding. It emphasizes the fundamental principles of rule of law, transparency, non-discrimination and the protection of property rights but leaves for the country concerned the choice of policies, based on its economic circumstances and institutional capabilities. One size does not fit all.

New Words, Phrases and Expressions
shoot up　射出；發芽；暴漲；迅速成長
dramatically　adv. 戲劇地；引人注目地；顯著地，引人注目地
bilateral　adj. 雙邊的；有兩邊的；相互的；互惠的；對等的
multilateral　adj. 多邊的；多國的，多國參加的
telecommunication　n. 電訊；[通信] 遠程通信；無線電通信
illustrative　adj. 說明的；作例證的；解說的

be inconsistent with　與……不一致
mandatory　adj. 強制的；託管的；命令的
implementation　n. 實現；履行；安裝啟用
commitments　n. 承諾，保證；委託；承擔義務；獻身
complement　vt. 補足，補助
sovereignty　n. 主權；主權國家；君主；獨立國
systematic　adj. 系統的；體系的；有系統的；分類的；一貫的，慣常的
transparency　n. 透明，透明度；幻燈片；有圖案的玻璃
susceptible　adj. 易受影響的；易感動的；容許……的
transnational　adj. 跨國的；超越國界的
fora　n. 論壇；討論會；法庭；專題講話節目（forum 的複數形式）
prescriptive　adj. 規定的；指示的；命令的；規範的；約定俗成的；慣例的
leave for　動身去；把……留給（某人）做；把……留在（另一時間做）

Notes

1. These arrangement cover most areas of the investment environment—from property fights protection, taxation, and corruption, to regulation in areas as diverse as banking, shipping, telecommunications, labor, and natural environment.

這些協議覆蓋了大部分地區的投資環境，從物權保護、稅收、腐敗、監管到不同領域的金融、航運、通信、勞動、環境和自然環境。

2. The agreement requires mandatory notification of all non-conforming TRIMs and their elimination within two years for developed countries, within five years for developing countries and within seven years for least-developed countries.

該協議規定了所有不符合要求的投資措施，並要求兩年內在發達國家清除，五年內在發展中國家的清除，七年內在最不發達國家清除。

3. Many critics argued that the OECD was more susceptible to direct influence by transnational corporate forces than alternative fora with more universal membership such as the UNCTAD and the WTO.

許多評論家認為，經合組織比更普遍成員替代論壇更容易受到跨國企業的直接影響，如聯合國貿發會議和世界貿易組織。

4. It covers ten policy areas and addresses some questions to governments to help them design and implement policy reform to create a truly healthy, robust, attractive and competitive environment for domestic and foreign investment.

它覆蓋了 10 個政策領域，並解決了一些問題，以幫助政府設計和實施政策改革，從而創造一個真正健康、穩健、有競爭力、有吸引力的國內國外投資環境。

Section 3　International Direct Investment

Lesson 1　An Overview of International Direct Investments

International direct investment, or foreign direct investment (FDI), plays an extraordinary and growing role in global business. It can provide a firm with new markets and marketing channels, cheaper production facilities, access to new technology, products, skills and financing. For a host country or the firm which receives the investment, it can provide a source of new technologies, capital, processes, product, organizational technologies and management skills, and as such can provide a strong impetus to economic development. Foreign direct investment is classically defined as a company from one country making a physical investment into building a factory in another country. It is the establishment of an enterprise by a foreigner. Nowadays, its definition can be extended to include investments made to acquire lasting interest in enterprises operating outside of the economy of the investor. The FDI relationship consists of a parent enterprise and a foreign affiliate which together form a multinational corporation (MNC).

In order to qualify as FDI, the investment must afford the parent enterprise control over its foreign affiliate. The IMF defines control in this case as owning 10% or more of the ordinary shares or voting power of an incorporated firm or its equivalent for all unincorporated firm. By direction FDI can be categorized into inward and outward. Inward foreign direct investment is when foreign capital is invested in local resources; and outward foreign direct investment, sometimes called 「direct investment abroad」, is when local capital is invested in foreign resources.

Modes of international direct investment have many forms. Mainly, there are greenfield investment, mergers and acquisitions, etc.

Greenfield investment

Greenfield investment refers to the direct investment in new facilities or the expansion of existing facilities in the host country.

Greenfield investments are the primary target of a host country's promotional efforts because they create new productioncapacity and jobs, transfer technology and know-how, and can lead to linkages to the global marketplace.

However, there are also some criticisms of greenfield investment that domestic firms may lose market shares for the competition from foreign firms, and that profits are perceived to bypass local economies and instead flow back to the foreign investor's home economy.

Mergers and acquisitions

Mergers and acquisitions (M&A) means the transfers of existing assets from local firms to foreign firms which paythe price. Cross border mergers and acquisitions are involved in international direct investment.

Cross-border mergers occur when the assets and operation of firms from different countries are combined to establish a new legal entity. Cross-border acquisitions occur when the control of assets and operations is transferred from a local to a foreign firm, with the local firm becoming an affiliate of the foreign firm.

Unlike greenfield investment, M&A provide no long term benefits to the local economy, even in most deals the owners of the local firm are paid in stock from the acquiring firm, meaning that the money from the sale could never reach the local economy. Nevertheless, M&A is a significant mode of FDI and the most common way for multinationals to do FDI in recent years.

Principal determinants of FDI mode choice

For international investors, it is of great significance to choose the FDI mode which meets its need best. The following are principal determinants may be taken into consideration by the investors when making decision of FDI modes.

Market competition. Generally speaking, when fewer firms control most or the entire market share of an industry, they can individually and collectively profit more by cooperation than by competition. Under such conditions, if the foreign investor creates new capacity in this market, it will usually provoke aggressive responses from established domestic firms, especially from market leaders. Therefore, it may be easier for foreign investors to acquire the established firms in the host country.

Market transaction situation. When the host equity market is imperfect, a foreign investor's acquisition activity may be delayed while it searches for suitable targets, analyzes their economic viability, conducts relevant negotiations, and endures inefficient bureaucracy. This may be especially true in host markets in the developing world. The firm's acquisition activity may thus be inhibited by high market transaction costs owing to a lack of resources and bargaining power in the host market. However, greenfield entry can avoid such market transaction costs by reducing exposure to the host equity market.

Complementary assets. When a firm invests in a foreign market, it may attempt to acquire complementary assets (such as brand image, real estate, market power, marketing channels, network relationships, and raw materials) through the host market.

However, it may encounter hurdles with high transaction costs in its quest to acquire the complementary assets it needs in the host market. For instance, transactions necessary to acquire complementary assets may be inefficient, or complementary assets may be inseparable from the host firm, or the host firm may be not be willing to sell complementary assets unbundled from its operations. The investing firm may thus find itself compelled to acquire the host firm's equity through acquisition in order to obtain the complementary assets. It needs to reduce transaction costs. Hence, it will be more likely that a firm will choose acquisition over greenfield when investments in imperfect foreign markets include complementary assets.

Relocation cost. A firm transfers its resources and facilities to operate in the foreign market. The cost will be increased. For instance, it must train and remunerate expatriates, its assets have to be adapted to host cultures and standards, and it has to rebuild network relationships in the host market.

In general, greenfield investment may incur high relocation costs because it often has engaged in cross-border resource transfer on a large scale. Hence, in order to reduce such relocation costs, a firm may prefer acquisition instead of greenfield. A firm with structural weaknesses may find it difficult to bear such costs. So it will be more likely to enter through acquisition.

Cultural distance. When a firm invests in foreign countries, it must face both national and corporate culture. Hence, cultural differences between home and host country may affect how a firm operates in the international market. Such cultural differences may be one of the most important decision, making determinants for the firm's FDI mode choice.

Entering a foreign market through acquisition, the acquired firm is often troubled by 「double-layered acculturation」 problems resulting from cultural distance between the acquirer and the acquired firm. Hence, in order to reduce the risks of cultural distance between home and host market, the investor is more likely to enter through greenfield.

Laws and policies in the host country. Some countries have strict M&A laws that make international acquisition more difficult than greenfield investment. Moreover, in many developing countries, there are favorable policies to attract greenfield investments because the host governments want more production capacity and jobs to be created. In addition, different accounting systems between the home and host country will also bring troubles to the acquiring firm.

New Words, Phrases and Expressions
extraordinary   adj. 非凡的；特別的；離奇的；臨時的；特派的
impetus   n. 動力；促進；衝力
classically   adv. 擬古地；古典主義地
affiliate   n. 聯號；隸屬的機構等
acquisitions   n. 兼併，購並；獲得，所獲之物
promotional   adj. 促銷的；增進的；獎勵的
know-how   n. 訣竅；實際知識；專門技能
linkage   n. 連接；結合；聯接；聯動裝置
flow back   逆回；回流
nevertheless   adv. 然而，不過；雖然如此   conj. 然而，不過
generally speaking   adv. 一般而言
provoke   vt. 驅使；激怒；煽動；惹起

aggressive   adj. 侵略性的；好鬥的；有進取心的；有闖勁的
viability   n. 生存能力，發育能力；可行性
inhibit   vt. 抑制；禁止
hurdle   n. 障礙；跨欄；跨欄跑；障礙賽跑
inseparable   adj. 不可分割的；不能分離的
complementary   adj. 補足的，補充的
unbundled   adj. 非捆綁的；分類定價的 v. 分類定價；把……分開
compel   vt. 強迫，迫使；強使發生
relocation   n. 重新安置；再布置，變換布
remunerate   vt. 酬勞；給予報酬；賠償
expatriates   n. 被流放者；移居國外者   vt. 使移居國外；流放，放逐；使放棄國籍
favorable   adj. 有利的；良好的；讚成的，讚許的；討人喜歡的

Notes

1. For a host country or the firm which receives the investment, it can provide a source of new technologies, capital, processes, product, organizational technologies and management skills, and as such can provide a strong impetus to economic development.

對於東道國或接收投資的公司，它可以提供新的技術、資金、工藝、產品、組織技術和管理技能等資源，因此能提供強大的動力來發展經濟。

2. Greenfield investments are the primary target of a host country's promotional efforts because they create new production capacity and jobs, transfer technology and know-how, and can lead to linkages to the global marketplace.

綠地投資是東道國推廣工作的主要目標，因為它們能創造新的生產能力和就業機會，轉讓技術和訣竅，並可能導致聯繫到全球市場。

3. There are also some criticisms of greenfield investment that domestic firms may lose market shares for the competition from foreign firms, and that profits are perceived to bypass local economies and instead flow back to the foreign investor's home economy.

也存在一些關於綠地投資的指責，（他們認為）在與外國企業競爭的過程中，國內企業可能會失去市場份額，其利潤被認為繞過了當地經濟而回流到外國投資者國內。

4. When the host equity market is imperfect, a foreign investor's acquisition activity may be delayed while it searches for suitable targets, analyzes their economic viability, conducts relevant negotiations, and enduresinefficient bureaucracy.

當東道主股權市場不完善時，外國投資者的併購活動可能會被延遲，同時它會搜索合適的目標，分析其經濟上的可行性，進行相關的談判，並忍受效率低下的官僚機構。

Lesson 2   Theories of FDI

Foreign Direct Investment（FDI）acquired an importantn role in the international economy after the Second World War. Theoretical studies on FDI have led to a better understand of the economic mechanism and the behavior of economic agents, both at micro and macro level allowing the opening of new areas of study in economic theory.

To understand foreign direct investment must first understand the basic motivations that cause a firm to invest abroad rather than export or outsource production to national firms. the empirical results reveal that for FDI there is not a unified theoretical explanation, and it seems at this point very unlikely that such a unified theory will emerge.

Theories of FDI may be classified under the following headings:

Monopolistic Advantage Theory

In 1960, Stephen Hymer put forward monopolistic advantage theory to explain FDI. He saw the role of firm-specific advantages as a way of marrying the study of direct foreign investment with classic models of imperfect competition in product markets. He argued that foreign direct investment took place because of the product and factor market imperfections. The direct foreign investor is a monopolist or, more often, an oligopoly in product or factor markets and possesses some kind of proprietary or monopolistic advantage unavailable to local firms which make it competent in an unfamiliar market. These advantages may be economies of scale, superior technology, or superior knowledge in management or finance, famous brand name, easy access to raw materials, etc. Monopolistic advantage theory is regarded as the first theory on FDI but it has its own limitations. It does not dealing with the location-decision of FDI and is unable to explain the choice between export, license agreement and FDI.

Production Cycle Theory

Production cycle theory developed by Raymond Vernon in 1966 was used to explain certain types of foreign direct investment made by U. S. companies in Western Europe after the Second World War in the manufacturing industry.

Vernon believes that there are four stages of production cycle: innovation, growth, maturity and decline. According to Vernon, in the first stage the U. S. transnational companies create new innovative products for local consumption and export the surplus in order to serve also the foreign markets. According to the theory of the production cycle, after the Second World War in Europe has increased demand for manufactured products like those produced in USA. Thus, American firms began to export, having the advantage of technology on international competitors.

If in the first stage of the production cycle, manufacturers have an advantage by possessing new technologies, as the product develops also the technology becomes known. Manufacturers will standardize the product, but there will be companies that you will copy it. Thereby, European firms have started imitating American products that U. S. firms were exporting to these countries. US companies were forced to perform production facilities on the local markets to maintain their market shares in those areas.

This theory managed to explain certain types of investments in Europe Western made by U. S. companies during 1950-1970. However, it does not explain why it is profitable to undertake FDI rather than continue to export from the home production base, or license a foreign firm to produce the product. What PLC theory does is simply to say that when a foreign market is large enough to support local production, FDI will occur.

Theory of Internalization

The internalization theory is an extension of the market imperfection theory. A firm operating in an imperfect market wants to maximize the profit. When external markets for supplies, production, or distribution fails to provide efficiency, the firm can invest FDI to create their own supply, production, or distribution streams. Certain transactions may work much more efficiently within the framework of a company than when externally organized by the external market, such as: production, services, marketing, research and development, training of human resources and etc.

By investing directly in a foreign market, the company is able to send the knowledge across borders while maintaining it within the firm, where it presumably yields a better return on the investment made to produce it. By internalizing across national boundaries, a firm becomes multinational.

Internalization helps a company to avoid search and negotiating costs, avoid cost of violated contracts and litigation, capture economies of interdependent activities, avoid government intervention, control supplies and market outlets, apply cross-subsidization and transfer pricing.

However, the internalization theory failed to explain the location decision of FDI either.

The Eclectic Theory of International Production

The eclectic theory of international production is developed by John Dunning. He attempted to integrate a variety of strands of thinking about FDI and tried to offer a general framework for determining the extent and pattern of both foreign-owned production undertaken by a country's own enterprises and also that of domestic production owned by foreign enterprises.

If a company wants to service a local or foreign market from a foreign localization, it must have access to firm-specific advantages or be able to acquire these at lower cost. This is called ownership-specific advantages or O-advantages. Such advantages include technologies, reputation, brand names, human capital, etc. This kind of advantages call be replicated in different countries without losing its value, and easily transferred within the firm without high transaction costs.

Given that ownership-specific advantages are present, it must be in the best interest for the firm to use these itself, rather than sell them or license them to other firms. These are Internalization-specific advantages or I-advantages. Why don't firms just sign a contract with a subcontractor (external agent) in a foreign country? Because contracting out is risky: it implies transferring the ownership-specific advantages outside the firm and revealing the proprietary information (e. g. how to use the technology or the patent).

In addition to ownership specific advantages as well as internalization advantages are necessary, it must be in the firm's interest to use these in combination with a least some factor inputs located abroad — so called location-specific advantages or L-advantages. These location-specific advantages may be producing close to final consumers or downstream customers, saving transport costs, obtaining cheap inputs, circumventing trade barriers, etc.

By combining O-advantages, L-advantages and I-advantages, we get the 「eclectic」 approach to FDI — the so called OLI paradigm of international production.

The eclectic, or OLI paradigm, suggests that the greater the ownership-specific and internalization specific advantages possessed by firms, and the more the location advantages provided in a foreign country, the more FDI will be undertaken.

Where firms possess substantial ownership-specific and internalization-specific advantages but the location-advantages favor the home country, then domestic investment will be preferred to FDIand foreign markets will be supplied by exports.

When firms possess ownership-specific advantages which are not best acquired, augmented and exploited from a foreign market, but by way of inter-firm alliances or by the open market (without internalization-specific advantages), then FDI will be replaced by a transfer of at least some assets normally associated with FDI and a transfer of these assets or the right to their use.

New Words, Phrases and Expressions

monopolistic adj. 壟斷的；獨占性的；專利的
put forward 提出；拿出；放出；推舉出
firm-specific adj. 公司特有的
proprietary adj. 所有的；專利的；私人擁有的
competent adj. 勝任的；有能力的；能幹的；足夠的
presumably adv. 大概；推測起來；可假定
violate vt. 違反；侵犯，妨礙；褻瀆
litigation n. 訴訟；起訴
cross-subsidization n. 交叉補貼
critique n. 批評；評論文章 vt. 批判；評論
neoclassical adj. 新古典主義的
innovate vt. 改變；創立；創始；引入； vi. 創新；改革；革新
alternatively adv. 非此即彼；二者擇一地；作為一種選擇
rationale n. 基本原理；原理的闡述

eclectic   adj. 折中的；選擇的；折中學派的   n. 折中派的人；折中主義者
replicate   vt. 複製；折疊   vi. 重複；折轉
hierarchy   n. 層級；等級制度
reveal   vt. 顯示；透露；揭露；洩露
downstream   adj. 下游的；順流的；   adv. 下游地；順流而下
circumvent   vt. 用計謀取勝，智勝；（巧妙地）避開，規避；包圍，陷害；繞行
paradigm   n. 範例；示例；模範；規範
augment   vt.（在數量、尺寸、強度、效力等方面）擴大，擴張；增大，加強；提高；補充，增補
theoretical   adj. 理論的；理論上的；假設的；推理的

Notes

1. The direct foreign investor is a monopolist or, more often, an oligopoly in product or factor markets and possesses some kind of proprietary or monopolistic advantage unavailable to local firms which make it competent in an unfamiliar market.

外商直接投資者是一個壟斷者，或者更多的時候，在產品或生產要素市場的寡頭壟斷，它具備某種本土公司沒有的專利或壟斷的優勢，這使得它在一個陌生的市場具有競爭力。

2. By investing directly in a foreign market, the company is able to send the knowledge across borders while maintaining it within the firm, where it presumably yields a better return on the investment made to produce it.

通過直接在國外市場投資，該公司能夠在維持自身發展的同時跨越國界傳送知識，通過這種方式投資生產通常能夠獲得更好的回報。

3. He attempted to integrate a variety of strands of thinking about FDI and tried to offer a general framework for determining the extent and pattern of both foreign-owned production undertaken by a country's own enterprises and also that of domestic production owned by foreign enterprises.

他試圖整合各種考慮外商直接投資的設想，並試圖提供一個總體框架來確定範圍以及由一國自己的企業承擔的國外生產，也是外國企業所擁有的國內生產的模式。

4. Where firms possess substantial ownership-specific and internalization-specific advantages but the location-advantages favor the home country, then domestic investment will be preferred to FDI and foreign markets will be supplied by exports.

如果公司持有具體的所有權和內部特有的優勢，但位置優勢有利於本國，那麼國內的投資將是首選外國直接投資和外國市場由出口供應。

5. When firms possess ownership-specific advantages which are not best acquired, augmented and exploited from a foreign market, but by way of inter-firm alliances or by the open market (without internalization-specific advantages), then FDI will be replaced by a transfer of at least some assets normally associated with FDI and a transfer of these assets or the right to their use.

當企業擁有的所有權優勢不是被最好的收購、擴張並從國外市場開發，而是通過企業間的聯盟的方式，或通過公開市場（不含內部化特定優勢）獲得的，那麼至少部分外國直接投資就會被一些與國際直接投資相聯繫的資產轉換取代，或者獲得對它們的使用權利。

Section 4　International Securities Investment

Lesson 1　An Overview of International Securities Investment

In contrast to foreign direct investment (FDI) introduced in section 3, international securities investment is considered indirect investment. According to the US government's definition, FDI is 「ownership or control of 10 percent or more of all enterprises voting securities, or the equivalent interest in an unincorporated business」. The purpose of FDI is to control the invested company. However, international securities investors passively hold the securities that they purchase. They do not actively manage or control the securities' issuer.

International securities investment is also called foreign portfolio investment (FPI) where investors hold a collection of foreign securities. Narrowly speaking, it refers to the purchase of foreign bonds and stocks (the two common securities investors hold).

The official definition of security, from the Securities Exchange Act of 1934, is: 「Any note, stock, treasury stock, bond, debenture, certificate of interest or participation in any profit-sharing agreement or in any oil, gas, or other mineral royalty or lease, any collateral trust certificate, reorganization certificate or subscription, transferable share, investment contract, voting-trust certificate, certificate of deposit. For a security, any put, call, straddle, option, or privilege on any security, certificate of deposit or group or index of securities (including any interest therein or based on the value thereof), or any put, call, straddle, option, or privilege entered into on a national securities exchange relating to foreign currency, or in general, any instrument commonly known as a 『security』; or any certificate of interest or participation in temporary or interim certificate for, receipt for, or warrant or right to subscribe to or purchase, any of the foregoing; but shall not include currency or any note, draft, bill of exchange, or banker's acceptance which has a maturity at the time of issuance of not exceeding nine months, exclusive of days of grace. or any renewal there of the maturity of which is likewise limited.」

In essence, a security can be regarded as a contract, which documents the holder's right to claim financial benefits in the future. For example, bond holders receive periodical interest payments and receive the face value when the bond matures. Stock holders receive dividends and/or realize return from the appreciation of stock price.

Depending on the length of investment, securities can be categorized into two major types: money market securities and capital market securities. Money market securities are short-term securities with a maturity of less than one year. They are liquid and of low risk. A good example is US Treasury bills. Capital market securities are long-term securities with a maturity of longer than one year. Examples include bonds and stocks. A bond matures eventually but a stock does not.

In addition to these categories, securities include a broad range of financial assets. Many people also refer to derivatives (futures, options, swaps, etc.) as securities. Another example is mutual fund shares. The third one is the mortgage backed securities (MBS) that is often quoted when we discuss the financial crisis that the US is still struggling with. Derivatives and MBS are innovative and complicated investment instruments.

New Words, Phrases and Expressions
in contrast to　與……形成對照
royalty　n.（國王授予個人或公司對礦山等的）開採特許權；（專利權的）使用費；（著作的）版稅，稿酬
lease　n. 租約；租期；租賃物；租賃權
collateral　adj. 側面的，旁邊的；伴隨的；輔助的，附屬的；附加的
interim　adj. 臨時的，暫時的；中間的；間歇的；　n. 過渡時期，中間時期；暫定
renewal　n. 更新，恢復；復興；補充；革新；續借；重申
in essence　本質上；其實；大體上
periodical　adj.［數］週期的；定期的
derivative　n.［化學］衍生物，派生物；導數；　adj. 派生的；引出的

Notes
1. A security can be regarded as a contract, which documents the holder's right to claim financial benefits in the future. For example, bond holders receive periodical interest payments and receive the face value when the bond matures. Stock holders receive dividends and/or realize return from the appreciation of stock price.

有價證券可以被視為一個合同，它記錄了持有人在未來索取經濟利益的權利要求。例如，債券持有者接受定期利息支付和當債券到期接收面值。股票持有人收取股息及/或實現股票價格的上升的回報。

Lesson 2　International Securities Markets and Their Characteristics

The primary market is where investors have the first opportunity to buy a newly issued security. The issuer receives cash proceeds from the buyers of these securities, who in turn receive financial claims that previously did not exist. If the issuer is a publicly traded company, then the sale of new securities is called seasoned new issues. The first time for a private company to sell stocks to the public is called initial public offering (IPO). IPOs are often issued by smaller, younger companies seeking the capital to expand, but can also be done by large privately owned companies looking to become publicly traded. In an IPO, the issuer obtains the assistance of an underwriting investment bank, which helps it determine what type of security to issue (common or preferred), the best offering price and the time to bring it to market. Investing in IPOs can be very risky because of the lack of historical data about the issuing company.

Once new securities have been sold in the primary market, subsequent trades occur in the secondary markets. It provides an efficient mechanism for investors to sell and buy previously-issued securities without the involvement of the issuing companies. When we talk about stock market normally we are referring to the secondary market because it is where most investors are active.

There are some common characteristics such as: liquidity (marketability), expected return and risk.

Securities are traded on securities markets. However, liquidity refers to the ease with which you can turn your investment into cash without losing much of its value. Obviously, T-bills are more liquid than bonds and stocks.

Investors purchase securities to make money. Expected return is the overall profit that you might expect to receive from your investment, in the form of interest or dividend (income yield) or from changes in the market value of the security (capital gains or losses).

In 2008, US stock holders on average lost almost 40 percent of their investments. It shows that investors' expected return might not be realized. The degree of uncertainty about your expected return from an investment is the risk of a security. The realized return might be positive and even higher than the expected return, but investors more often view the risk as the possibility of losing money.

New Words, Phrases and Expressions
previously　adv. 以前；預先；倉促地
seasoned　adj. 經驗豐富的；老練的；調過味的
underwriting　n. 保險業；[金融] 證券包銷；　v. 認購；寫在……下面；經營保險業
historical　adj. 歷史的；史學的；基於史實的
subsequent　adj. 後來的，隨後的；連續的，接下去的
involvement　n. 牽連；包含；混亂；財政困難
marketability　n. 流動性；可銷售，市場性；適銷性
uncertainty　n. 不確定；無法預測；模糊不清；不明確

Notes
1. IPOs are often issued by smaller, younger companies seeking the capital to expand, but can also be done by large privately owned companies looking to become publicly traded.
新股上市往往是由更小的、更年輕的公司尋求擴大資本發行的，但也可以是大

型民營企業希望實現公開交易而進行的。

2. Once new securities have been sold in the primary market, subsequent trades occur in the secondary markets. It provides an efficient mechanism for investors to sell and buy previously-issued securities without the involvement of the issuing companies.

一旦新的證券已在一級市場售出，後續的交易發生在二級市場。在沒有發行公司的參與時，它提供了一個有效的機制，方便投資者出售和購買以前發行的證券。

3. Expected return is the overall profit that you might expect to receive from your investment, in the form of interest or dividend (income yield) or from changes in the market value of the security (capital gains or losses).

預期收益是，你可能期望從你的投資獲得的以利息或股息（收益率）的形式，或從證券的市場價值的變化（資本利得或損失）獲得的整體利潤。

Lesson 3　Stock Investments

Whether we say shares, equity, or stock, we mean the same thing —a security that represents ownership of a company. The ownership stake in the company is proportionate to the number of shares you have. For example, Coca-Cola Co. is the most famous company of nonalcoholic beverages. Their Coca—Cola, Diet Coke, Fanta, and Sprite brand name drinks are popular around the world. If you own 1,000 shares of Coca-Cola Co., then you own 0.000,000,04 percent of Coca-Cola (Coca-Cola has roughly 2.30 billion shares outstanding). As a partial owner, you have claim on the company's assets and earnings. Though not guaranteed, you may receive cash dividend. If Coca-Cola were to be sold or liquidated, you would receive your share of whatever was left over after all of Coca-Cola's debts and other obligations were paid. As an owner, you also have the right to vote on important matters regarding Coca-Cola.

There are two main types of stocks: common stock and preferred stock. When we talk about stocks, we are usually referring to common stocks. Common stocks represent ownership in a company. Investors are entitled to anything paid out by the company. They get one vote per share to elect the board of directors at annual meetings.

Preferred stock represents some degree of ownership in a companybut usually doesn't come with the same voting rights. Preferred stocks are「preferred」to common stocks in two aspects. First, preferred stockholders are usually guaranteed a fixed dividend forever. For example, if a corporation issues 9% preferred stock with a par value of ＄100, the preferred stockholder will receive a dividend of ＄9 per share per year. If the corporation issues preferred stock with a par value of ＄25 and paying a quarterly 2% dividend. This would translate into a 50 cents dividend paid each quarter. Second, in the event of liquidation, preferred stockholders are paid off before the common stockholders. Preferred stocks are considered hybrid securities. The fixed dividend rate of a preferred stock is similar to a bond's coupon rate, but like a stock, preferred stock does not mature. For accounting and tax purposes, preferred stocks are treated as equity.

There are many different classifications for stocks. Here we list the terms you often hear that include Chinese stocks, blue chips and red chips.

Chinese stocks come in different classes. Shanghai and Shenzhen stock exchanges list A shares and B shares. A shares are traded in Chinese Yuan and are primarily available to domestic Chinese investors, with some exceptions for large, qualified foreign institutional investors.

B shares are traded in US dollars in Shanghai and Hong Kong dollars in Shenzhen. B shares were created for foreign investors but Chinese investors can invest in B shares too. Eventually, this separation of A shares and B shares might be eliminated.

H shares are shares of a company incorporated in China's mainland that is listed on the Hong Kong Stock Exchange. In the eyes of foreign investors, Hong Kong's market environment is transparent and free of most restrictions that exist in China's mainland. Most international investors participate in China via H-shares. Investors in the mainland of China were not allowed to invest in H shares until 2007. This change greatly increased the demand for H shares. China's prestigious companies such as Petro China, China Mobile, and the Aluminum Corporation of China, all list in Hong Kong.

Chinese companies also list on other foreign exchanges, like the London Stock Exchange (L shares), the Tokyo Stock Exchange (J shares), the New York StockExchange (N shares), and the Singapore exchange (S shares).

Blue chips are the stocks of nationally recognized, well-established and financially sound companies. The term derives from casinos, where blue chips stand for counters of the highest value. Oliver Gingold of Dow Jones sometime in 1923 or 1924 used the term in reference to the high-priced stocks. It has been used ever since then. Today, it commonly refers to high-quality stocks. Most blue chips pay regular dividends even when the economy moves adversely and the stock price closely follows the S&P 500. A few examples of blue chips are Wal-Mart, Coca-Cola, Gillette, Berkshire Hathaway and Exxon-Mobile.

Red chips are the stocks of companies in the mainland of China that are incorporated outside China's mainland and listed on the Hong Kong Stock Exchange. The actual business is based in China's mainland and usually controlled by shareholders in the mainland of China. The word「red」comes from the color of the Chinese national flag and hence represents the Peoples Republic of China. The term was coined by Hong Kong economist Alex Tang in 1992. Red chip stocks are expected to maintain the filing and reporting requirements of the Hong Kong Stock Exchange, which makes them a main outlet for foreign investors who wish to participate in the rapid growth of the Chinese economy.

New Words, Phrases and Expressions

proportionate   adj. 成比例的；相稱的；適當的；   vt. 使成比例；使相稱

nonalcoholic　adj. 不含酒精的

beverage　n. 飲料

roughly　adv. 粗糙地；概略地

guarantee　n. 保證；擔保；保證人；保證書；抵押品　vt. 保證；擔保

be entitled to　有權；有……的資格

come with　與……一起供給；伴隨……發生；從……開始

hybrid　n. 雜種，混血兒；混合物；　adj. 混合的；雜種的

coupon　n. 息票；贈券；聯票；[經] 配給券

restriction　n. 限制；約束；束縛

prestigious　adj. 有名望的；享有聲望的

casinos　n. 賭場；娛樂場（casino 的復數）；（意）鄉間住宅

outlet　n. 出口，排放孔；[電] 電源插座；發泄的方法；批發商店；（商品的）銷路，市場

Notes

1. If Coca-Cola were to be sold or liquidated, you would receive your share of whatever was left over after all of Coca-Cola's debts and other obligations were paid. As an owner, you also have the right to vote on important matters regarding Coca-Cola.

如果可口可樂公司要被出售或清算，你會得到你的那份——可能是可口可樂公司的債務或其他支付義務遺留下來的。作為業主，你也擁有投票決定關於可口可樂公司的重大事項的權利。

2. Blue chips are the stocks of nationally recognized, well-established and financially sound companies.

藍籌股是國家認可的、成熟的和財務健全的公司的股票。

3. Red chips are the stocks of companies in the mainland of China that are incorporated outside China's mainland and listed on the Hong Kong Stock Exchange.

紅籌公司是在中國大陸的公司，但在中國大陸以外的地方上市，如香港聯合交易所。

Lesson 4　Bond Investments

A bond is a long-term debt instrument in which the issuer promises to make paymentsof interests and principal, on specific dates, to the holders of the bond. A typical bond is an interest-only loan, meaning that the borrower will pay the interest every period, but none of the principal will be repaid until the end of the loan. Bond is commonly referred to as a fixed-income security because the amount and timing of future cash inflows are fixed, known to the investors when they purchase the bond.

Terminology and conventions vary a lot across the global bond markets. This section presents the key features of a typical US bond. The par value (face value): It is the principal amount of a bond that is repaid at the end of the term. It is usually $1,000 for corporate bonds. Government bonds frequently have much larger par values. The bond could be sold at a price the same as or different from the face value. If the selling price equals me par, the bond is called a par value bond. If the selling price is greater than the par, the bond is called a premium bond. If the selling price is lower than the par, the bond is called a discount bond.

Coupon and coupon rate: Coupon is the stated interest payment made on a bond. The annual coupon divided by the face value gives the coupon rate. If a corporate bond has a 12 percent coupon rate, it means that the issuing corporation will make an annual payment of $ 120 to the bond holders as interests. In the US corporate bonds normally make semi-annual coupon payments, which in our case, means $ 60 every half year.

Maturity: It is the specified date on which the principal amount of a bond is paid. A corporate bond frequently has a maturity of 30 years when it is originally issued. Obviously, once the bond has been issued, the number of years to maturity declines as time goes by.

The global bonds market is divided into three broad groups: domestic bonds, foreign bonds, and Eurobonds. Domestic bonds are issued locally by a domestic borrower and are usually denominated in the local currency. Foreign bonds are issued on a local market by a foreign borrower and are usually denominated in the local currency. Foreign bond issues and trading are under the supervision of local market authorities. Eurobonds are underwritten by a multinational syndicate of banks and are placed mainly in countries other than the one in whose currency the bond is denominated.

Domestic bonds make up the bulk of a national bond market. Different issuers belong to different market segments. For example, in the US, bonds are issued by federal government (Treasury bonds), state and local governments (municipals bonds or just munis), and corporations.

Although stocks occupy the headlines of all financial news, the US Treasury market is the largest securities market in the world. In 2008, the total debt of the US government was $ 9.4 trillion and it is growing. The US Treasury issues have no default risk because (we hope) the Treasury can always come up with the money to make the payments. The coupons received on a Treasury issue are exempt from state income taxes. They are taxed only at the federal level.

Unlike Treasury issues, municipal bonds have varying degrees of default risk. The coupons are exempt from federal income taxes. This is the most intriguing feature of munis, especially for investors in the high-tax bracket. Because of this tax benefit, munis have much lower returns as compared corporate bonds on the pre-tax basis. Investors need to compare the after-tax returns when they compare two bonds. For example, a muni that pays 4.5% is equivalent to a corporate bond that pays 6.4% if the tax rate is 30% (0.045/0.70=0.064 or 6.4%). A company can issue bonds just as it can issue stocks. Large corporations have a lot of flexibility as to how much debt they can issue: the limit is whatever the market will bear. Generally, a short-term corporate bond is less than 5 years; intermediate is 5 to 12 years, and long term is over 12 years. Generally speaking, corporate bonds offer higher return than government bonds because of there is a higher risk of a company defaulting than a government.

Foreign bonds issued on local markets have existed for a long time. They often have colorful names referring to the domestic market in which they are being offered.

A Yankee bond is a bond issued in the US by foreign banks and corporations and is denominated in US dollars. Accordingto the Securities Act of 1933, these bonds must first be registered with the Securities and Exchange Commission (SEC) before they can be sold. Yankee bonds are often issued in tranches and each offering can be as large as $ 1 billion. Foreign issuers tend to prefer issuing Yankee bonds when US interest rates are low, because this means lower interest payments for the foreign issuer to pay out.

A Bulldog bond is issued in London by a non-British company. It is denominated in sterlings. Bulldog is a national symbol of England. A Samurai bond is a yen-denominated bond issued in Tokyo by a non-Japanese company. Other examples of foreign bonds include Maple bonds (in Canada), Matador bonds (in Spain), Matilda or Kangaroo bonds (in Australia).

Investors in foreign bonds are usually residents of the domestic country. Since foreign bonds are issued by foreign entities. They add foreign content into an investor's portfolio. The foreign bonds are denominated in the domestic currency. Therefore, investors do not need to worry about the effects of currency exchange fluctuations.

The name Eurobond comes from the historical fact that the banks placing the Eurobond are located in Europe. Usually, a Eurobond is issued by an international syndicate and categorized according to thecurrency in which it is denominated. A Eurodollar bond that is denominated in US dollars and issued in Japan by an Australian company would be an example of a Eurobond. The Australian company in this example could issue the Eurodollar bond in any country other than the US. The Eurobond issued by a Japanese company is also called Sushi Bond.

In 1999, all bonds denominated in one of the former currencies of Euro-land were translated into Euros. For example, all French government bonds denominated in French francs became bonds denominated in Euros. They are bonds issued in Euros or euro bonds but not Eurobonds. It is likely that the terminology will evolve to clear the confusion. Eurobonds are attractive financing tools as they give issuers the flexibility to choose the country in which to offer their bond according to the country's regulatory constraints. They may also denominate their Eurobond in their preferred currency. Eurobonds are attractive to investors as they have small par values and high liquidity.

Many bonds have fancy features. This creates many different types and names of bonds. Here are a few examples such as zero coupon bonds (or just zeroes) do not make coupon payments. They are initially priced at a deep discount. Floating-rate bonds (or floaters) make adjustable coupon payments. A particularly interesting type of floating-rate bond is an inflation—linked bond. Such bonds have coupons adjusted according to the rate of inflation. In the US, the Treasury began issuing such bonds in 1997. The issues are sometimes called Treasury Inflation-Protected Securities or「Tips」. The income bonds make coupon payments to bondholders only if the company's income is sufficient. A callable bond could be repurchased or called by the issuer before the bond matures. A put bond allows the holder to force the issuer to repurchase the security at specified dates before maturity. A convertible bond can be converted into a predetermined amount of the company's stock at certain times during its life, usually at the discretion of the bondholder.

New Words, Phrases and Expressions
principal   adj. 主要的；資本的   n. 首長；校長；資本；當事人
terminology   n. 術語，術語學；用詞
semi-annual   n. 半年度
maturity   n. 成熟；到期；完備；（票據的）到期；到期日；期限
denominate   vt. 為……命名；把……稱作……；   adj. 有特定名稱的
supervision   n. 監督，管理
syndicate   n. 辛迪加；企業聯合；財團   v. 聯合成辛迪加；在多家報刊上同時發表
municipal   adj. 市政的，市的；地方自治的   n. 市政府債券；市公債
munis   n. 市政債券；市政公債
headline   n. 大標題；內容提要；欄外標題；頭版頭條新聞
treasury   n. 國庫，金庫；財政部；寶庫
federal   adj. 聯邦的；同盟的
exempt   vt. 免除；豁免；   adj. 被免除的；被豁免的
intriguing   adj. 有趣的；迷人的；   v. 引起……的興趣；策劃陰謀；私通（intrigue 的 ing 形式）
flexibility   n. 靈活性；彈性；適應性
defaulting   n. 違約；不付欠款   v. [會計] 拖欠（default 的 ing 形式）；不履行義務
tranche   n. （國際貨幣基金組織貸款劃分的）部分；一期款項
sterling   adj. 純正的；英幣的；純銀制的   n. 英國貨幣；標準純銀
constraint   n. [數] 約束；局促，態度不自然；強制
initially   adv. 最初，首先；開頭
sufficient   adj. 足夠的；充分的
convertible   adj. 可改變的；同意義的；可交換的
predetermine   vt. 預先確定；預先決定；預先查明

Notes

1. A typical bond is an interest-only loan, meaning that the borrower will pay the interest every period, but none of the principal will be repaid until the end of the loan.

一個典型的債券是一種只付息貸款，這意味著每一個時期借款人將支付利息，但沒有本金被償還，直至貸款到期日。

2. A Yankee bond is a bond issued in the US by foreign banks and corporations and is denominated in US dollars. According to the Securities Act of 1933, these bonds must first be registered with the Securities and Exchange Commission (SEC) before they can be sold.

揚基債券是由外國銀行和公司發行的美國債券，並以美元標價。根據1933年的證券法，這些債券必須先在證券交易委員會（SEC）註冊，它們才可出售。

3. Eurobonds are attractive financing tools as they give issuers the flexibility to choose the country in which to offer their bond according to the country's regulatory constraints.

歐洲債券是有吸引力的融資工具，因為它們給予發行人靈活選擇，尤其是對於國家的法規約束提供其債券的國家而言。

4. A callable bond could be repurchased or called by the issuer before the bond matures. A put bond allows the holder to force the issuer to repurchase the security at specified dates before maturity. A convertible bond can be converted into a predetermined amount of the company's stock at certain times during its life, usually at the discretion of the bondholder.

可贖回債券在債券到期之前可由發行人回購或收回。看跌期權的債券持有者可以在到期前指定日期迫使發行人回購該證券。可轉換股票債券可在其生命週期中被轉換成在特定的時間預訂量的公司的股票，通常由債券持有人決定。

Section 5　Management of International Investments

Lesson 1　Sources of Risk in investing

This section identifies various risks of international investments and ways to managing them. Then we study the dynamic process of managing an international portfolio. The main risks of investing are as following: business risk, financial risk, market risk, inflation risk and liquidity risk.

Business risk is the risk that a firm will experience a period of poor earnings and resultant failure. The firm will not have cash to meet its operating expenses, to pay interest to bondholders or dividends to stockholders. It is the greatest for firms in cyclical or relatively new industries. Angel investors and venture capitalists are faced with huge business risk.

Financial risk is associated with the use of debt financing by companies. The companies must pay interest and principal to creditors otherwise they default on their debts. Companies that are purely equity financed have nearly no financial risk. Investors check the debt equity ratio, current ratio, and other factors to evaluate the financial strength of a firm that interests them.

Market risk refers to the fluctuation in returns caused by moves in the overall market. All securities are exposed to market risk, although it affects primarily common stocks. It includes a wide range of factors exogenous to securities themselves, such as recessions, wars, structural changes in the economy, and changes in consumer preferences.

Market risk caused by changes in the interest rate is often referred to as interest rate risk separately. Generally speaking, interest rates and security prices have an inverse relationship. For example, when market interest rate goes up, bond value goes down. This is a big concern for bondholders.

Inflation risk is the probability of loss resulting from erosion of all income or in the value of assets due to inflation shrinks the purchasing power of money, say US dollars. With rising costs of goods and services, the return from investments will be reduced when we consider it in the real term.

The real return can be approximately calculated by subtracting inflation rate from the nominal return. For example, if the rate of inflation is 5% over a year and the rate of return is 3% on a portfolio, then the investor has effectively taken a loss of 2% even though he/she has made a profit in absolute terms.

Liquidity risk stems from the lack of marketability of an investment that cannot be sold quickly enough and without significant price concession. A T-bill has little or no liquidity risk while a small OTC stock may have substantial liquidity risk. Venture capital investments are illiquid therefore investors must check and see the exit mechanisms in the business plan.

The risks discussed above faced by any investor even if he/she invests only in the domestic market. International investors are exposed to additional risks. The most important ones are currency risk and political risk. We introduce them in the following sections.

New Words, *Phrases and Expressions*
resultant　n. 合力；結果；[化學] 生成物；　adj. 結果的；合成的
bondholder　n. 公債證書所有者，公司債所有者
cyclical　adj. 週期的，循環的
exogenous　adj. 外生的；外因的；外成的
separately　adv. 分別地；分離地；個別地
subtract　vt. 減去；扣掉
stem from　起源於
concession　n. 讓步；特許（權）；承認；退位
mechanism　n. 機制；原理，途徑；進程；機械裝置；技巧

Notes
1. Business risk is the risk that a firm will experience a period of poor earnings and re-

sultant failure. The firm will not have cash to meet its operating expenses, to pay interest to bondholders or dividends to stockholders.

經營風險是指公司經歷一段時期的財務不佳以及由此產生麻煩的風險。該公司將不會有充足的現金，以應付其營運開支、支付利息給債券持有人或支付股息給股東。

2. Financial risk is associated with the use of debt financing by companies. The companies must pay interest and principal to creditors otherwise they default on their debts.

財務風險與公司使用債務融資有關。該公司必須支付利息和本金給債權人，否則它們會拖欠債務。

3. Inflation risk is the probability of loss resulting from erosion of all income or in the value of assets due to inflation shrinks the purchasing power of money, say US dollars.

通貨膨脹風險是指所有收入縮水，或資產價值產生損失，貨幣購買力下降的可能性，比如美元。

4. Liquidity risk stems from the lack of marketability of an investment that cannot be sold quickly enough and without significant price concession.

流動性風險源於投資的市場力缺乏，無法足夠快地出售，並且沒有顯著的價格優惠。

Lesson 2　Currency Risk and Its Management

What is currency risk? Currency risk is also called exchange rate risk. Investors who invest internationally must convert the foreign gains into their domestic currency. The return measured by domestic currency therefore is affected by the change in exchange rate. For example, a Chinese investor who buys a US stock denominated in US dollars must ultimately convert the returns from this stock back to Chinese Yuan. Due to the weak dollar in the past few years, Chinese investors returns from the US stock market were reduced a lot. At the same time, the weak dollar adds to US investors' dollar—denominated returns in foreign stocks.

How to manage currency risk? Currency risk is managed using either futures or forward currency contracts. Futures are standardized contract traded on futures exchanges while forwards are customized contracts traded over-the-counter. Currency forwards are sometimes referred to as currency swaps. No matter which one we use, the purpose is to lock in an exchange rate for an expected transaction in the future.

Investors that use currency futures to manage the exchange rate risk are called hedgers. They do need to sell the underlying foreign currency in the future and want to shift the risk of depreciation of foreign currency versus domestic currency. In contrast, there are speculators in the market. They are willing to assume the risk of change in exchange rate and they want to profit from it. They do not need or own the underlying currency.

The process of hedging currency risk using futures is simple. The investor sells appropriate contracts today (takes a short position). When the foreign currency depreciates, he/she get less domestic currency in the spot market but the gain from the futures position will help increase that amount. When the foreign currency appreciates, the futures position will incur a loss, which will reduce the amount of domestic currency he/she finally gets. However, it is considered the cost of hedging. As long as the exchange rate specified in the futures contract is acceptable to the investors, he/she should use futures to protect him/her against the bad results.

Here is an example provided by Chicago Mercantile Exchange Inc. to illustrate the use of currency futures to hedge Mexican peso foreign currency risk. A US hedge fund continues to find Mexican Treasury bill (CETES) yields attractive and decides to rollover an investment in CETES, whose principal plus interest at maturity in five months will be 850 million Mexican pesos (MP). Knowing that its all—in return is subject to US dollar versus Mexican peso exchange rate risk. The US hedge fund wanted to hedge its exposure of converting the CETES investment back into US dollars. After assessing the available alternatives, the US hedge fund chose to hedge with exchange-traded Mexican peso futures contracts.

The trading unit of the Mexican peso futures is 500,000 MP. Therefore, on May 1st, the US hedge fund sells 1,700 September Mexican peso futures at $0.089,50 per MP (equivalent to 850 million MP). Over the course of the next five months the US dollar exchange rate for the Mexican peso falls to $0.086,00 per MP. As the Mexican peso futures price falls, the hedge fund's trading account at its clearing member firm is credited the gains on the position by the exchange clearing house. The price move from US $0.089,50 to US $0.086,00 per Mexican peso on the short Mexican peso futures position represents a gain of US $0.003,50 per Mexican peso. This is equal to a profit of US $1,750 per contract times 1,700 contracts for a net position gain of US $2,975,000. (For the purpose of these examples, calculations do not include brokerage or clearing fees that may be associated with exchange transactions.) This US dollar profit when added to the US dollars resulting from conversion of the Mexican peso-denominated CETES principal and interest into US dollars at the lower dollar/peso exchange rate (of US $0.086,00 per Mexican peso), results in an effective rate equivalent to the Mexican peso futures price at the start of the hedge.

New Words, Phrases and Expressions

convert into　使轉變；把……轉化成；折合
denominate　vt. 為……命名；把……稱作……；　adj. 有特定名稱的
ultimately　adv. 最後；根本；基本上
standardized　adj. 標準的；標準化的；定型的
　　　　　　　v. 使合乎標準；按標準校準；制定標準
over-the-counter　adj. 不通過交易所而直接售給顧客的；不需處方可以出售的

swap　n. 交換；交換之物；　　vt. 與……交換；以……作交換；　　vi. 交換；交易

hedger　n. 種植樹籬者；兩方下註者；騎牆派；做套期保值的人

underlying　adj. 潛在的；根本的；　　v. 放在……的下面；為……的基礎；優先於

versus　prep. 對；與……相對；對抗

speculator　n. 投機者；思索者；倒賣者

depreciate　vt. 使貶值；貶低；輕視；　　vi. 貶值；輕視；貶低

incur　vt. 招致；引發；蒙受

rollover　n. 翻轉；（車）翻覆；延期付款

represent　vt. 代表；表現；描繪；回憶；再贈送；　　vi. 代表；提出異議

Notes

1. Currency risk is managed using either futures or forward currency contracts. Futures are standardized contract traded on futures exchanges while forwards are customized contracts traded over-the-counter.

貨幣風險通常使用期貨或遠期貨幣合約來管理。期貨合約在期貨交易所交易，而遠期貨幣合約一般在櫃臺交易。

2. As long as the exchange rate specified in the futures contract is acceptable to the investors, he/she should use futures to protect him/her against the bad results.

只要在期貨合約中指定的匯率是投資人可以接受的，他/她就可以利用期貨合約來保護自己，免受壞的結果傷害。

Lesson 3　Political Risk and Its Management

What is political risk? Political risk is also called geopolitical risk. It refers to the risk that a foreign government will alter its policies or other regulations so that it significantly affects one's investment. More broadly, it can apply to the risk that a nation will refuse to comply with an agreement to which it is a party, or that political violence will hurt an investment or business.

We can divide political risks to the following types which are showed in table 7.1. The first type is firm-specific risk and country-level risk. Firm-specific political risks are risks directed at a particular company. For example. HSBC's headquarters in Istanbul was wrecked by a terrorist attack in 2003. Country-level political risks are countrywide. For example, there was a civil War in the host country. The second type is government risk and instability risk. Government risks arise from a government authority. Instability risks, arise from political power struggles.

Table 7.1 Types of Political Risk

	Government risks	Instability risks
Firm specific risks	Discriminatory regulations	Sabotage
	Creeping expropriation	Kidnappings
	Breach of contract	Firm-specific boycotts
Country level risks	Mass nationalization	Mass labor strikes
	Regulatory changes	Urban rioting
	Currency inconvertibity	Civil wars

Source: www.transnational-dispute-management.com.

As we know the different types of political risks, the next important question is: how to manage political risk? The commonest way is called political risk insurance. Political risk insurance is defined to be the coverage for business firms operating abroad to insure them against loss due to political upheavals including war, revolution, confiscation, incontrovertibility of currency, and other such losses. The insurance is aimed not at preventing a loss but rather at assuring the investor that compensation will be received for all or part of the investment if a loss does occur.

Many countries offer some form of political risk insurance to their nationals. Examples include Australia, Belgium, Canada, Denmark, France, Germany, Japan, the Netherlands, Norway, Sweden, United Kingdom and the United States. The political risk insurance coverage varies across countries.

Political risk insurance can be obtained through private companies or through national or multilateral government insurance programs. The cost of government insurance is usually cheaper than that offered by private companies. Government policies tend to follow a standard form while with a private company, there is more flexibility and opportunity to negotiate the provisions of the policy. The rationale of the government companies is the promotion of foreign investments. Therefore, government policies are for a long term, 15 years or 20 years and they usually only insure new projects or the expansion of existing project. The private insurers write shorter term policies, normally on a 3-year basis, and they insure existing as well as new investments. Under government policies, the host government is informed of the coverage. It is believed that a foreign country will be less likely to take expropriatory action if they know it will lead to a direct claim and may cause potential conflict with the other government. However, disclosing the coverage of a private policy can nullify the policy.

An example of private political risk insurer is American International Group, Inc. (AIG), one of the world's largest insurance organizations. It was created 90 years ago when an American entrepreneur named C. V. Starr founded AIG'S earliest predecessor company in Shanghai. AIG now has operations in more than 130 countries and jurisdictions.

The political risk insurance is managed by AIG Global Trade & Political Risk Insurance Company (AIG Global). AIG Global works in conjunction with AIG World Source in the US and American International Underwriters (AIU) outside the US to meet the global insurance needs of its customers. For more than 20 years, the AIG companies have provided Political Risk Insurance to numerous US and foreign-based clients. There are political risk insurance services for equity investors, financial institutions, contractors, projects, importers and exports.

An example of government political risk insurer is the United States Overseas Private Investment Corporation (OPIC) established in 1971, OPIC is a federal program to insure private US investments in foreign countries. Currently OPIC does business in over 150 countries. It provides protection against three political risks: (1) inability to convert foreign currency; (2) expropriation of facilities by a foreign country; and (3) war or revolution. The program is guaranteed by the full faith and credit of the US government.

There are other political risk mitigation methods as following: Another way to tackle political risk is to form joint venture with a local company. A well-regarded local company acts as a buffer against the risk of state intervention. A third option is to hire lobbying firms. The purpose is to sway critical decisions in a company's favor.

All the three methods discussed above have their own drawbacks. The effectiveness is really case by case. A piece of suggestion to investors is to take preemptive actions. Investors should do their own homework about the political, economic, social and legal environment of the country where they want to invest in. Creating a political risk friendly environment is also valuable, which includes developing good relationships with the local workforce, constantly tracking the changes in political situation of the host country, establishing the sound company's reputation and corporate image.

Finally, notice that although emerging markets are labeled with high political risks, developed countries are not insulated. At various times, UK, France and Italy have raised concerns about nationalization. Even the benchmark for low political risk, the US, was tarnished by the recent financial crisis. As an example, Bear Sterns was propped up and pushed into the arms of a competitor, while Lehman Brothers was allowed to collapse. All was decided by Washington, some investors complained.

New Words, Phrases and Expressions
geopolitical　adj. 地理政治學的
violence　n. 暴力；侵犯；激烈；歪曲
headquarter　vi. 設立總部；　vt. 在⋯設總部
wreck　n. 破壞；失事；殘骸；失去健康的人　v. 破壞；使失事；拆毀
rioting　n. 暴亂、騷動、騷亂；（感情等的）爆發，放縱，發洩
upheaval　n. 劇變；動亂；動盪

confiscation　n. 沒收；徵用；充公
incontrovertibility　n. 無可辯駁性
expropriatory　adj. 徵用的，沒收的；剝奪的
nullify；　vt. 使無效，作廢；取消
predecessor　n. 前任，前輩
jurisdiction　n. 司法權，審判權，管轄權，權限，權力
mitigation　n. 減輕；緩和；平靜
preemptive　adj. 優先購買的；先發制人的；有先買權的
benchmark　n. 基準；標準檢查程序；　vt. 用基準問題測試（計算機系統等）

Notes

1. Political risk is also called geopolitical risk. It refers to the risk that a foreign government will alter its policies or other regulations so that it significantly affects one's investment.

政治風險也被稱為地緣政治風險，它是指外國政府將改變其政策或其他規定，使其顯著影響一個人的投資的風險。

2. Political risk insurance is defined to be the coverage for business firms operating abroad to insure them against loss due to political upheavals including war, revolution, confiscation, incontrovertibility of currency, and other such losses.

政治風險保險被定義為確保國外經營公司不因政治動盪造成損失的公司保險項目，這些政治動盪包括戰爭、革命、沒收、貨幣不可兌換以及其他類似的損失。

Lesson 4　International Investment Philosophies

This lesson introduces the philosophies followed by international fund managers. Individual investors are faced with exactly the same choices when they invest internationally and when they select portfolio managers. Therefore, the portfolio manager discussed below could be an individual investor.

A passively managed portfolio simply attempts to track the return on a selected market index, to capitalize on a long-term performance while keeping all costs at a minimum. The assumption is that market is efficient: all securities prices quickly and fully reflect all available information. There is no technique that can help investors consistently beat the market. The domestic index fund approach is observed index US, the UK, other European countries and Japan, where the security market is considered efficient. Recently, the trend toward global indexing is strongly felt among institutional investors.

To pursue active strategies, investors assume that they have some advantages over other investors. Such as superior analytical skills, information, judgment. The objective is to beat a benchmark imposed by the client. Active equity portfolio management includes the following respects：（1）Market timing. Managers attempt to earn excess return by temporarily increasing or decreasing the exposure in one or more markets or currencies.（2）Security selection. To achieve a given asset allocation strategy, mangers using fundamental analysis and technical analysis to actively select undervalued securities. (3) Sector/industry selection. Managers shift weights in the portfolio in order to take advantage of some sectors that are expected to do relatively better, and avoid or deemphasize those sectors that are expected to do relatively worse. In an international portfolio, the manager allocates funds across worldwide industries. (4) Regional/country allocation. Managers make choice of national markets and currencies. For example, emerging markets VS. developed markets. (5) Style selection. Managers choose to invest in companies with different attributes. For example, value stocks VS. growth stocks, large-cap stocks VS. small-cap stocks. (6) Currency hedging. Managers can actively manage their currency risk. For a global bond portfolio, mangers Can conduct active strategies through sector/credit selection, duration/yield curve management and yield enhancement techniques.

An individual investor can use top-down or bottom-up ways. In the top down approach, the manager first decides on the weight of each asset class (cash, bonds, stocks and others) in the portfolio and then selects individual securities to satisfy that allocation. For example, in a global equity portfolio, the manager decides on the percentage allocated to various countries then selects the best stocks in each market. Asset allocation is potentially the most profitable and important investment management decision.

In the bottom-up approach, the manager studies many individual securities then selects the best regardless of their national origin or currency denomination to build a portfolio. This implies that the portfolio is exposed to various currency and political risk. Some managers may combine the two approaches to have better control of risk.

There is another choice between quantitative and subjectivemodels. Investment is filled with numbers. Quantitative models are commonly used in the investment decision making process. However, the models require estimates about future situations and nobody could be consistently correct on his estimations. In other words, due to the uncertainties, quantitative models are subject to errors. Portfolio managers need to resort to subjective judgments especially when they are investing in the global market which is more complex than the domestic market.

New Words, Phrases and Expressions
consistently   adv. 一貫地；一致地；堅實地
analytical   adj. 分析的；解析的；善於分析的
deemphasize   vt. 使不重要；不再給予強調
potentially   adv. 可能地，潛在地
quantitative   adj. 定量的；量的，數量的
uncertainties   n. [數] 不確定性；不確定因素（uncertainty 的復數）
Notes
1. A passively managed portfolio simply attempts to track the return on a selected mar-

ket index, to capitalize on a long-term performance while keeping all costs at a minimum.

積極管理的投資組合只是試圖跟蹤選定的市場指數的回報率，它在利用其長期性能的同時把所有的成本降到最低。

2. Managers shift weights in the portfolio in order to take advantage of some sectors that are expected to do relatively better, and avoid or deemphasize those sectors that are expected to do relatively worse.

管理者對投資組合賦予權重，以充分利用預期做得比較好的一些部門，避免或淡化預期做得相對比較差的部門。

3. The models require estimates about future situations and nobody could be consistently correct on his estimations. In other words, due to the uncertainties, quantitative models are subject to errors.

該模式需要估計未來的情況，沒有人的估計能夠是一貫正確的。換句話說，由於存在不確定性，量化模型可能存在誤差。

Exercises

Ⅰ. Questions for discussion

1. Explain the motives for international investments.
2. Comment on the expected return and risk of the three types of markets.
3. Discuss the characteristics of international investments as compared with domestic investments.
4. Say something about Chinese investment environment.
5. What do you think about international arrangements on investment environment?
6. Explain FDI in your own words.
7. Explain and compare the two FDI modes.
8. Discuss the three basic forms of FDI in the host country and give some examples in China.
9. List the reasons for international securities investments.
10. Explain the risks involved in international investments.
11. Compare the methods of political risk management.

Ⅱ. Multiple choices

1. International traders _____ goods and—services may enter the foreign exchange market in order to finance their trade transactions.
  A. of   B. on
  C. in   D. with

2. We believe that the company's _____ to STDM, a private entity, which dominates the country's entertainment and hotel industries, is slightly overrated.
  A. dependence   B. influence
  C. risk   D. exposure

3. The Eurocurrency market has grown _____ the most important short—term

credit market.

  A. up   B. into
  C. to   D. out of

 4. The United States has been the largest home country in the world _____ international investment for quite a number of years.

  A. of   B. to
  C. for   D. on

 5. A very large loan for a single borrower to finance a major project, arranged by a consortium is often referred to as a _____ loan.

  A. consortiumed   B. syndicate
  C. syndicated   D. groupage

 6. Exchange restrictions tend to _____ outward foreign investment.

  A. promote   B. encourage
  C. facilitate   D. prevent

 7. International money marketsare international financial centers for raising _____ capital needed by international commercial and industrial institutions.

  A. long-term   B. medium-term
  C. short-term   D. share

 8. The modern Eurodollar market arose shortly after _____ .

  A. World War II   B. World War I
  C. the 1950s   D. the 1960s

 9. Many exchange dealers believe that interest rates are responsible for whether a currency is bought or sold _____ at a higher exchange rate or at a lower exchange rate.

  A. future   B. forward
  C. futures   D. forwardly

 10. Capital tends to flow out of a country where interest rates are _____ resulting in a fall in the exchange rate of the currency of that country.

  A. low   B. high
  C. stable   D. changeable

Ⅲ. Translate the following sentences into English

 1. 國際投資者有各種各樣的投資可供選擇，這些可供選擇的投資因其提供的利潤及攜帶的風險的不同而不同。

 2. 外匯匯率是指某一時間貨幣進行兌換的價格。匯率的變化受許多政治經濟因素的影響。

 3. 國際直接投資往往涉及生產要素的移動，而在國際證券投資裡，投資者只是購買某一外國公司發行的股票或債券，而這些股票和債券往往隨時可以按市場價格變現。

 4. 新加坡是亞洲最重要的離岸金融市場之一，只有外資銀行和其他金融機構這

類非居民才允許在新加坡從事離岸金融業務。

5. 隨著國際金融市場的證券化，許多國家的企業通過在國際金融市場上發行股票和債券而籌集了大量的資金以滿足其資本支出的需要。

Further Reading Comprehension

The Trap of Payday Loans Can Lead to Triple-digit Interest Rates

Michelle Singletary

If I said,「Don't rob Peter to pay Paul,」you would probably understand that I was warning against making a desperate move to fix a financial problem that often makes a bad situation worse.

Yet millions of people do just that when they get a payday loan.

These are small loans that a borrower promises to repay with the next paycheck or benefit check.

Stop and think about this.

If you can't pay your expenses with your current paycheck, how is borrowing from the next one going to help? Yes, it may solve a problem today. But if you can't repay the loan, you're likely to create a long tether to a financial product with expensive fees.

Yet, I understand why people get them. It's quick cash. It's easy to get if you have a bank account and income. And if you're in a financial jam, the fees can seem reasonable. A charge of $15 to borrow $100 doesn't seem extreme or exploitative to borrowers trying to avoid having a service turned off or catch up on their rent or mortgage.

But when fees are annualized, they often amount to triple-digit interest rates or more. I've seen payday loan contracts with four-digit interest rates. Payday lenders are required to tell you the finance charge and the annual interest rate (the cost of the credit) on a yearly basis.

Defenders argue that these loans provide a service for people who need short-term cash. And they are right. Many people feel they are being rescued. Until things go wrong. And they do, for a lot of folks.

The Pew Charitable Trusts says the average loan size is $375, but most people can only afford to pay $50 in a two-week period after paying other regular expenses.「Repeat borrowing is the norm, because customers usually cannot afford to pay the loans off on payday and cover their other expenses, so they repeatedly pay fees to renew or reborrow,」a 2013 report from Pew said.「Lenders depend on this repeat borrowing, because they would not earn enough revenue to stay in business if the average customer paid off the loan within a few weeks.」

After examining data from more than 12 million loans in 30 states, the Consumer Financial Protection Bureau found that more than 80 percent of payday loans are rolled over or are followed by another loan within 14 days. Monthly borrowers are disproportionately likely to stay in debt for a whopping 11 months or longer.

In a new report, the CFPB, which began supervision of payday lenders in 2012, focused on repeat payday loan borrowers. The agency noted that with a typical payday fee of 15 percent, consumers who took out a loan and then had six renewals paid more in fees than the original loan amount.

Think you can handle this type of loan?

I've counseled people who were stuck in a tormenting cycle of payday loans. One woman I was trying to help had a payday loan with an annualized interest rate of more than 1,000 percent. After several back-to-back loans, her debt obligation ate up most ofher paycheck.

Although lots of payday business is done online, storefront lenders continue to operate in mostly low-income neighborhoods. Organizations and agencies that fight and advocate on behalf of consumers have long understood the implication of the payday loan trap, especially for the most financially vulnerable.

Because payday lenders collect their money using post-dated checks or by getting customers to give them electronic access to their bank account, they don't have to look at a borrower's ability to pay when compared to existing expenses or existing debt, says Tom Feltner, director of financial services for the Consumer Federation of America.

Last year, the Office of the Comptroller of the Currency and the Federal Deposit Insurance Corp. imposed tougher standards on banks that offer short-term, high-interest loans similar to storefront payday loans. The institutions have to determine a customer's ability to repay. And the same should be true for Internet and storefront payday operations.

「Wehave to make sure regardless of what channel a borrower uses to take out a payday loan, there needs to be strong ability-to-repay standards.」Feltner said.

Come on, CFPB. Make a regulatory move. No payday loan should be made without assessing a person's ability to repay — and repay without repeated borrowing.

From: http://m.washingtonpost.com/business/the-trap-of-payday-loans-can-lead-to-triple-digit-interest-rates/2014/03/25/ca1853dc-b471-11e3-8cb6-284052554d74_story.html.

Chapter 8
International Business Law

Chapter checklist

In this chapter we will mainly introduce the international business laws. It can be divided into three parts: introduction to international business law, main contents of international business law and resolution of international business disputes. The more important and difficult part is: the main contents of international business laws.

When you have completed your study of this chapter, you will be able to

1. Define International business law system and explain the kinds of law system.

2. Receive a comprehensive understanding of the maincontents of international business law.

3. Explain the Resolution of International Business Disputes.

Section 1　An Introduction to International Business Law

Lesson 1　A Brief Introduction to International Business Law

International business is the economy of exchanging goods, services and intellectual property, conducted between individuals and businesses in multiple countries. In other words, international business maybe any domestic business operation that includes an international element. International elements create increased transaction cost and risk. Factors such as different in language, culture, economics, politics and laws bring about barriers and costs. The principal integral parts of international business—Goods, services and intellectual property funds in one country may find a ready market or more profitable market elsewhere. Thus today international business compromises a large and growing portion of the world's total business.

As the business world became more complex, and with the dawn of air travel and worldwide communication, a clearer and uniform set of modern rules governing international business are needed. In fact, the international business law has grown and shaped up significantly in the twentieth century, especially after World War Ⅱ. It is generally agreed that:

International business law is the body of rules and norms that regulates the cross-border transactions in goods, services and intellectual property between parties. Here 「parties」 include natural persons, legal persons, and international organizations. Under a few of circumstances, states may also play a unique role in the capacity of commercial not sovereign entity. Beside, states also play a unique role in regulating and supervising the international business between private parties in its capacity of a sovereign. In comparison with the traditional international business law, contemporary international business law covers much more extensively, such as law, maritime law, insure law, law of international technology transfer, industrial property law, international investment law, international financial law, international tax law, law of international dispute settlement.

National law is a major factor in private international law, because in dealing with conflicts problems, the national legal system takes center stage. Although many achievements have been obtained on the unification or harmonization of national legal system in the area of international business, great differences still exist from one state to another. These national legal systems may be classified in many ways: for example, by origin, cultural similarity, or political ideology. Generally, the two most widely distributed families are the Civil Law system and the common law system.

Civil Law System

The oldest and most influential of the legal system is the Romano-Germanic legal system, commonly called the Civil Law. The Civil Law System is characterized by the use of detailed codes that established both basic principle and details rules for regulating the conduct of individuals. And the grounds for deciding cases are found in these codes. The civil law codes only covers: the law of persons, family law, property law, and inheritance law, law of obligations, commercial law, labor law, and criminal law. Prominent examples and models for much of the civil law world are the French Civil Code of 1804 and German Civil codes of 1896. They were models for most contemporary civil codes in countries and regions such as France, Germany, Italy and Japan, etc.

Common Law System

The Common Law System is much less widely distributed than the Civil Law. The Common Civil Law's basis is court decisions, or precedent. That means the principles or rules of law on which a court rested a previous decisions are authoritative in all future cases in which the facts are substantially the same. The Common Civil Law is based on the system of judge-made rules that originated in medieval England around twelfth century and then was carried to England's colonies such as the United States, Canada, Australia, India, Hong Kong, etc.

Common and Civil law

To start at a basic level, the civil law system established in continental Europe is based on a comprehensive code, where as the common law system, established in England and carried on in its former colonies, evolves through case precedent. Civil law came from the Roman tradition and was codified in the sixth century; France codified the law into a civil, commercial, penal, civil procedure, and criminal procedure code. Other European countries, such as Germany and Switzerland, followed with a codification of their law. The colonization of Africa, Asia, and Latin American spread the civil law system. However, some countries such as Japan simply adopted law based upon the civil law model.

This explanation could be amplified by contrasting the role of judge and lawyer in civil and common law system takes on many of the functions of the lawyer in deciding what evidence needs to be developed or produced. In a common law jurisdiction, the judge is more neutral and rules more on requests by the parties' lawyers. Thus, in a common law system, the person one hires as a lawyer arguably plays a more critical role in the outcome. However, this difference should not be interpreted in the extreme because in civil law states, lawyers are still important

Lesson 2 Sources of International Business Law

Sources in legal sense refer to something such as a constitution, treaty, statute, or custom that provide authority for legislation andfor judicial decisions. The modern international business law is derived from a number of sources.: international treaties and conventions, international business usage and customs, national law, and other sources such as resolutions of international organizations and soft law.

International Treaties and Conventions

Legally, treaties are binding agreements between two or more states, and conventions are legally binding agreements between states sponsored by international organizations, such as the United Nations. In the field of international business, the purpose of some treaties is to liberalize trade between the contracting states such as GATT. Another group of treaties or conventions aims at unification of law. Most of unifying law treaties or conventions only provide rules for situations with an international dimension, for example, international transportation or international sale of goods. The provisions of the treaties or conventions become part of the nations law of the contracting states. For a specific treaty or convention, only the contracting states are bound by it. A number of treaties are, for instance, not yet in force or are at present only applied by a limited number of countries. For that reason many unifying law treaties or conventions achieve only a limited or a regional unification.

Economically, the following are the most important international conventions: The United Nations Convention on Contract for the International sale of Goods in 1980 (hereafter called CISG) in the area of the international sale of goods, Convention for the Unification of Certain Rules of Law Relating to Bills of Lading in 1924 (hereafter called the Hague Rules) in the area of carriage of goods by sea, The Paris Convention on the Protection of Industrial Property in 1883 and revised in 1979 (hereafter called the Pairs Convention) in the field of technical trade, WTO's Understanding of Rules and Procedures Governing the Settlement of Disputes in 1995 (hereafter called WTO's DSU) and Convention on the Recognition and Enforcement of Foreign Arbitral Awards in 1958 (hereafter called the New York Convention), etc.

International Trade Usages and Customs

International trade usages and customs stem from the tradition of law merchant or the *lex mercatoria* in the history of international trade. International trade usages and customs mean the general rules and practices in international trade activities that have become generally adopted through unvarying habit and common use. In fact, each sector of industry has developed its own practices and usages, which are adopted to the needs of the sector. Thus there are specific usages in the bank sector, grain trade, the oil industry, etc. International trade usages and customs are usually widely recognized in the specific industry, they are often incorporated into international sales contracts or standard clauses to specify the rights and obligations of the parties.

International trade usages and customs used to be oral rather than in writing. In recent years some non-governmental organizations, with the view to facilitating the use of trade customs and usages, have collected and complied them into sets of rules in a written form. For example, the WarsawOxford Rules by the International Chamber of Commerce in 1932, the International Rules for the Interpretation of Trade Terms (Incoterms), the Uniform Customs and Practice for Commercial Documentary Credits (UCP). These international trade usages and customs now are widely recognized and accepted in international business area.

National Law

Although there are increasing number of international treaties or conventions, and trade usages, national law is usually one of the most important sources of international business law in the foreseeable future. It refers to all rules and norms made by a state. Every state has established its own legal system governing domestic transactions. For instance, national law in China includes The Contract Law of the People's Republic of China, Trademark Law of the People's Republic of China, The Law of the People's Republic of China on Chinese-Foreign Joint Ventures, The Civil Procedure Law of the People's Republic of China, Arbitration Rules of China International Economic and Trade Arbitration Commission, etc. The international cross-border transaction linking with different states, lead to the possibility that different legal systems will govern the transaction.

National law will, in the long run, still be important source of international business law, chiefly because, though international business is universally recognized, it is not accepted by all the countries in the world; on the other hand, many rule sin international business law are customarily developed, and state legislature is needed necessarily to assure efficient enforcement and authority to the rules.

Chapter 8　International Business Law

Other resources

Resources of international organizations are sometimes binding. For example, all members of the United Nations are bond by the resolutions of the Security Council by virtue of the UN Charter. Resolutions and recommendations of the International Labor Organization are binding on the members if the members are obliged by a treaty to give effect to these resolutions and recommendations in their national legislations.

Soft law consists of provisions in model laws, principles to be found in legal guides, and in scholarly restatements of international business law. All these rules and principles are not legally binding and enforceable unless the parties to an international transaction decided otherwise. That is, parties are free to choose to follow the soft law or just ignore it. But soft law is capable of having or developing its legal effects. It often gives guidance for the conduct of the parties, and the interpretation of the agreement of the parties or an international convention.

International model law means the rules and norms worked out and passed by some international organizations for the free choice by individual nations. The most important international model laws are the UNDITRAL Model on InternationalCommercial Arbitration by United Nations Commission on International Trade Law (UNCITRAL Model Law). Principles of International Commercial Contract by the International Institute for the Unification of Private Law

New Words, Phrases and Expressions
uniform　adj. 統一的，一致的；　n. 制服，校服
supervise　v. 監督，管理，領導
sovereign　n. 主權，君主；　adj. 有主權的，至高無上的
unify　vt. 統一；使相同，使一致
harmonization　n. 調和化；融洽；一致
specify　vt. 指定；詳述；提出……的條件；使具有特性
inheritance　n. 遺傳；遺產；繼承
scholarly　adj. 學術性的；有學者風度的；學者的
prominent　adj. 顯著的；傑出的；突出的
institute　n. 協會；學會；學院；（教育、專業等）機構
treaties　n. 條約
precedent　n. 先例；前例
customs　n. 慣例
codification　n. 編纂，整理；法典編纂
foreseeable　adj. 可預知的；能預測的
recommendation　n. 推薦；推薦信；建議
legislation　n. 立法；法律

Notes

1. This explanation could be amplified by contrasting the role of judge and lawyer in civil and common law system takes on many of the functions of the lawyer in deciding what evidence needs to be developed or produced.

而在決定需要開發或生產什麼樣的證據時，大陸法和英美法體系中律師呈現出許多職能。通過對比法官和律師的職能，這種解釋得以廣泛化。

2. International trade usages and customs are usually widely recognized in the specific industry, they are often incorporated into international sales contracts or standard clauses to specify the rights and obligations of the parties.

國際貿易慣例和習俗，通常在特定的行業被廣泛認可，它們往往被納入國際銷售合同或標準條款，來規定雙方的權利和義務。

3. National law will, in the long run, still be important source of international business law, chiefly because, though international business is universally recognized, it is not accepted by all the countries in the world.

從長遠來看，國家法律仍然是國際商法的重要來源，主要是因為雖然國際業務是舉世公認的，但它並沒有被世界上所有的國家接受。

Section 2　Main Contents of International Business Law

Lesson 1　International Contract Law

Contract is playing a more and more important role in today's international sale of goods, and considered as basic tool for international business transaction. For instance, a successful business transaction has to be based on at least sales contract that provides the transfer of title and risk of the goods and the obligations of parties, transportation contract the states the transfer of possession of the goods, contract of payment that stipulates the effective payment, and insurance company, etc. Therefore, a successful business transaction is chiefly decided by a good contract.

In common law countries, contract under English law means a promise or set of promises, for breach of which the law gives a remedy, or the performance of which the law in some way recognizes as a duty. The nature of contract, under civil law, is a「meeting of minds」or「mutual assent」. Article 1101 of「French Civil Code」states:「Contract is a mutual assent with one person or more is obligated to give a thing, to do or not to do a thing to one person or more persons.」China Contract Law has the similar provision, and Article a states「A contract in this Law refers to an agreement establishing, modifying and terminating the civil rights and obligations between subjects of equal footing, that is, between natural persons, legal persons or other organizations.」

Form in contract law means whether a contract is made in writing or orally. Traditionally, many countries have required that a contract be in writing. The English Statute of Frauds of 1677 required a signed writing to enforce a wide variety of contracts, including contracts for the sale of goods. The same requirement reappears in the UK Sale of Goods Act of 1893, the US Uniform Sales Act of 1906. Today, however, most oral sales contracts quite enforceable. The civil countries of Europe for the most part never had any writing requirement at all. Under the CISG, contracts for the sale of goods need not be in writing. Article 11 states: 「A contract of sale need not be concluded in or evidenced by writing and is not subject to any other requirements as to form.」

Legal System of International Business Contract includes United Nations Convention on contracts for the International Sale of Goods (CISG), UNIDROIT Principles of International Commercial Contracts (PICC) and the convention on contracts for the international sale of goods.

United Nations Convention on contracts for the International Sale of Goods (CISG)

The domestic laws of various countries often continued to develop along different lines. Thus, legal rules pertaining to the creation and completion of contracts differ from country to country. In the more recent past decades, as international merchants have sought to take advantage of global sales opportunities, the differences have sometimes created problems. Consequently, countries, with the help of international organization, came together to create the CISG to standardize certain international contract principles. Countries may choose to be parties to this Convention, though private parties continue to have the choice of opting out or opting into the agreement. CISG came into force on 1 January 1988 with the original eleven Stats. By the end of 2009, the number of Contracting States has reached 71 including Australia, Canada, China, France, Germany, Italy, Russia, Singapore, Swiss, Switzerland, United States, Japan, and other major countries in world trade covering countries from every geographical region, every stage of economic development and every major legal system. Thus CISG is regarded to be one of the most successful instruments of international sales law harmonization.

UNIDROIT Principles of International Commercial Contracts (PICC)

In 1968 to celebrate the 40th anniversary of the foundation of UNIDROIT, President of UNIDROIT proposed to undertake a similar initiative as 「Restatements of the Law」 in the United States of America, at the international level. In 1980 a special working Group set up with the task of preparing the various drafts chapters of the PICC. The members of the Group were leading experts in the field of contract law and international trade law. The working Group concluded the last reading of the different draft chapters in February 1994, after which the text was submitted to a special Editorial Committee, chaired by E. Allan Farnsworth and composed of Rapporteurs on the various chapters. In May 1994, the Governing Council gave its formal imprimatur to the UNIDROIT PICC and recommended their widest possible distribution in practice. At present, the UNDROIT PICC exists in many languages. The most important ways in which the UNIDROIT PICC can be used as a model for national and international legislators; as a means of interpreting and supplementing existing uniform law; as rules governing the contract; as a substitute for the domestic law otherwise applicable-are set out in the Preamble.

The convention on contracts for the international sale of goods

Early attempts at constructing an international law of sales actually began in the late 1920s with the work of the International Institute for the Unification of Private Law, or UNIDROIT, an organization of European lawyers that was working closely with the League of Nations. It successfully developed two conventions in 1964. However, the efforts was primarily a European one (the United States and many other countries did not participate in drafting these documents), and the conventions never received wide acceptance. In 1966, the United Nations created the U. N. Commission on International Trade Law, or UNCITRAL. UNCITRAL consists of thirty-six representatives from nations in every region of the world. Supported by a highly respected staff of lawyers, it is headquartered at the U. N. Vienna International Centre in Austria. UNCITRAL has drafted several widely accepted legal codes for international business, including the CISG.

Validity and Formation of Contract

Under the common law, a valid contract is an agreement that contains all of the essential elements of a contract. The civil law countries have similar stipulations to that except for the legally sufficient consideration. A valid contract contains a number of elements as following:

Invitation

An invitation offer is not capable of being turned into a contract by acceptance. It is used to invite others to make offer.

The offer

An offer is a proposal by one person toanother indicating an intention to enter into a contract under special terms. Requirement of Offer: first is the requirement of offer including: The contract law of most nations hold that an offer must be addressed to one or more specific persons; second is under article 11 of the CISG an offer must be sufficiently definite and indicates the intention of the offeror to be bound; Third is an offer becomes valid when it arrives the offeree.

As a general rule, an offer isn't binding on the offeree. If the offeree doesn't accept it, he has no obligation to give a notice to the offeror. In common law countries an offer generally isn't binding on the offeror, due to the lack of consideration not being signed and sealed, and the offeror can withdraw the offer at any time before the offeree's acceptance. Under Germany law an offer is binding on the offeror, while under French law the offeror can revoke the offer before the acceptance of the offeree. Article 14 of the CISG states: Until a contract is concluded an offer may be revoked if the revocation reaches offeroee before he has dispatched an acceptance; However, an offer can't be revoked, if it indicates, whether by stating a fixed time for acceptance to rely on the offer as being irrevocable and the offeree has acted in reliance on the offer.

Under the United Kingdom law an offer lapses: On the death either of the offeror or the offeree before acceptance; By non-acceptance within the time prescribed for acceptance by the offeror; When no time for acceptance is prescribed, by non-acceptance within reasonable time. What is a reasonable time depends on the nature of the contract and the circumstances of the case.

Revocation of offer means that the offeror notifies the offeree before acceptance of invalidity of the offer so as to be free from it. Under English common law: an offer may be revoked at any time before acceptance. Revocation does not take effect until it is actually communicated to the offeree. Article 16 of CISG states: Until a contract is concluded an offer may be revoked if the revocation reaches the offeree before has dispatched an acceptance; However, an offer cannot be revoked: if it indicates, whether by stating a fixed time for the offeree to rely on the offer as being irrevocable; or if it was reasonable for the offeree to rely on the offer as being irrevocable and the offeree has acted in reliance on the offer.

The acceptance

A contract isn't formed until the offer is accepted by the offeree. The acceptance is the offeree's manifestation of intention to be bound to the terms of the offer. Requirements of acceptance: An acceptance must be made by the offeree; An acceptance must be made within the period of validity; An acceptance must match the terms of the offer exactly unequivocally. Otherwise it is considered a counteroffer an thus a rejection of the original offer.

Under the common law, a contract is formed when the acceptance is dispatch is the time the letter is put into the hands of the postal authorities. This is known as 「Mail box rule」. Under the most civil law countries, the receipt theory is used. The difference between the two relates to the allocation of risk when an acceptance is lost or delayed. For example, a buyer sends a seller an acceptance through the mails and the acceptance is lost. If the dispatch theory were applied, a contract would have come into existence at the same time the acceptance was mailed, and the seller would be required to perform. Under the receipt rule, however, no contract would exist, and a perception that more fairly allocates responsibility for loss or delay.

Battle of the forms

In the commercial reality, a contract is often formed after several rounds of negotiation. The chains of correspondence or exchange of documents between sellers and buyers oftenmake it difficult to tell which one is offer and which one is offer and which one is acceptance. Most of the countries follow The Mirror Image Rule. The rule required that the terms in an acceptance match the terms of the offer exactly and un-equivocally. Under this rule, a purported acceptance that contains different or additional rule present special problems when both parties negotiate back and forth via standard business forms. According to this rule, if buyer's purchase order form and seller's order confirmation differ in any term, there will be no acceptance.

Consideration in Common Law

Consideration is one of the three elements of contract formation in common law, the other two being offer and acceptance.

Definition

Consideration is some benefit received by a party who gives a promise or performs an act, or some detriment suffered by a party who receives a promise. It may also be defined as 「that which is actually given or accepted in return for a promise.」 Put in plain English, consideration is the price you pay to purchase another person's promise.

General rules on consideration

Consideration required for all simple contracts. Consideration is necessary for the validity of every contract not under seal; Consideration must be have a value that is recognized by the law but need not be equal to the promise; Consideration must be present or future and cannot be past; Consideration must be possible to perform. A promise to do the impossible would not be accepted as consideration; Consideration must be legal, an illegal consideration makes the whole contract invalid; Consideration must move from the promisee. This means that the person to whom the promise is made must furnish the consideration; Consideration must not be too vague. Consideration must not be so vague that value cannot be placed on it; Performance of an existing contractual duty is not valuable consideration.

Exceptions to the consideration requirement

The consideration requirement is a classic example of a traditional contract law rule. Modern courts and legislatures have responded to this potential for injustice by carving out numerous exceptions to the requirement of consideration.

Contract under seal does not need consideration; Promissory Estoppel. Traditional contract law is basically designed to protect bargains that people make and that satisfy all legal requirements for a binding contract. Around the turn of this century, however, the courts were confronted with cases in which persons relied on promises made by others that did not amount to contracts because they lacked consideration required for a contract.

Lesson 2　Laws of International Sales of Goods

United Nations Convention on contracts for the International Sale of Goods (CISG)

In the late 1920s, the international Institute for the Unification of Private Law (UNIDROIT) began early attempts at constructing an international law of sales and developed two conventions in 1964. However, the conventions never received wide acceptance because, the United States and many other common law countries did not participate in drafting these documents, and the conventions were primarily a European one. In 1966 the

United States created the U. N. Commission on International Trade Law (UNCITRAL). UNCITRAL consists of 36 representatives from nations around the world. UNCITRAL has drafted several widely accepted legal codes for international business, including the Convention on Contracts for the International Sale of Goods (CISG). And it is a convention or agreement among nations that is binding once the legislature of a country adopts it. The CISG has already been adopted by many countries and is being considered by still others. Translations are available in Arabic, Chinese, English, French, Russian, Spanish, and other languages.

The CISG applies if the following three conditions are met: The contract if for the commercial sale of goods. It is between parties whose places of business are in different countries (nationality or citizenship of individuals is not a determining factor. The places of business are located in countries that have ratified the convention.

The following types of sales have been specifically excluded from the convention: Consumer goods sold for personal, family or household use; Goods bought at auction. Stocks, securities, negotiable instruments, or money; Ships, vessels, or aircraft; Electricity; Assembly contracts for the supply of goods to be manufactured or produced wherein the buyer provides a 「substantial part of the materials necessary for such manufacture or production」; Contracts that are in 「preponderant part」 for the supply of labor or other services; Liability of the seller for death or personal injury caused by the goods; Contracts where the parties specifically agree to 「opt out」 of the convention or where they choose to be bound by some other law.

Interpreting of the CISG

The underlying goal of CISG is the creation of a uniform body of international commercial sales law. In deciding questions governed by the Convention, Article 7 (2) directs a court to look to the following sources, in this order: the Convention, the general principles on which the Convention is based, and the rules of private international law.

The convention

CISG implies that a court may use only the plain meaning of the language in the Convention. At present, the number of cases that have interpreted CISG is small, but undoubtedly this will change quickly. As it does, the courts will have to keep in mind the Convention's admonitions: that CISG is an international treaty, that its purpose is to establish uniform international rules, and that the courts are bound by the principle of good faith in interpreting the Convention. The directive that it can be interpreted to promote uniformity in its application compels courts to examine and follow the decisions of the courts in other contracting states.

General Principles

CISG calls for courts to look to the general principles on which the Convention is based when interpreting its provisions, but it gives no list of general principles. It does sell out the mechanism for determining them. They must be derived from derived from particular sections within the Convention, and then extended, by analogy, to case at hand. In choosing this particular mechanism the drafters rejected the adoption of general principles derived from public or private international law. As well as from domestic law codes. This limitation on the sources that the courts may turn to in creating general principles was consciously made, and reflects the drafters' concern for uniformity and consistency, both in the drafting and in the evolution of the Convention.

Rules of Private International Law

The rules of private international law are the third and final source for interpreting the Convention. They may be used, however, only when CISG itself does not directly settle a matter, or when the matter cannot be resolved by the application of a general principle derived from the Convention itself.

Statements and Conduct of the Parties

A contract is sometimes said to be formed only when the parties have a 「meeting of minds」 or a 「common intent」. This comes from the idea, commonly accepted in many civil law countries, that parties are bound by a contract only when they subject their 「will」 to its terms. This theory is called subjective intent approach. This shortcoming has led some courts, notably those in the common law countries, to reject completely the use of subjective intent in the interpretation of a contract.

Negotiations

When a court is to determine intent—be it the party's subjective intent or a reasonable person's objective understanding—Article 8 (3) of CISG directs that 「due consideration」 be given 「to all relevant circumstances」, including (1) the negotiations leading up to the contract, (2) the practices which the parties have established between themselves, and (3) the parties' conduct after they agree to the conduct. In addition to considering prior and contemporaneous conducts, CISG lets a court consider the parties' subsequent conduct. Again, this is contrary to the practice in some domestic tribunals. CISG, however, does reflect the most widely followed practice, and as is the case for evidence, the parties are free to insert a provision in their contract excluding its consideration.

Practices and Usages

Both Article 8 (3) and Article 9 (1) of CISG state that parties are bound by 「any practices which they have established between themselves」. Article 9 (1) also allows a court to consider any usages which the parties agree to, and Article 9 (2) lets it consider 「a usage of which the parties knew or ought to have known and which in international trade is widely known to, and regularly observed by parties to contracts of the type involved in the particular trade concerned」. Trade usages are not considered unless a contract specifically adopted them and they did not violate statutory rules. In the Western world, on the other hand, flexibility and freedom of contract are more important than certainty. In some circumstances, trade usages apply in the West even when they contradict a statutory provision or the contract is silent. The delegates ultimately agreed to let a court consider international trade usages, but only if they are 「widely known」and「regularly observed」.

UNIDROIT Principles of International Commercial Contracts (PICC)

In 1968 to celebrate the 40th anniversary of the foundation of UNIDROIT, President of UNIDROIT proposed to undertake a similar initiative as「Restatements of the Law」in the United States of America, at the international level. The UNIDROIT Governing Council-the Institute's highest scientific organ-decide in 1971 to include in the Work Programme of the Institute the initiative. After the New York Programme was approved by the Members of UNIDROIT, the President set up a small Streering Committee, composed of Professors Rene David (University of Aix-en-Provence), Clive M. Schnitthoff (City University of London) and Tudor Popescu (University of Bucharest), in representation of the civil law, the common law and the former socialist systems, was set up with the task of making preliminary inquiries. In 1980 a special working Group set up with the task of preparing the various draft chapters of the PICC. The members of the Group were leading experts in the field of contract law and international trade law.

The working Group concluded the last reading of the different draft chapters in February 1994, after which the text was submitted to a special Editorial Committee, chaired by E. Allan Farnsworth and composed of Rapporteurs on the various chapters. In May1994, the Governing Council gave its formal imprimatur to the UNIDROIT PICC and recommended their widest possible distribution in practice. At present, the UNDROIT PICC exist in many languages.

Scope of the UNIDOIT PICC

In the Preamble to the UNIDROID PICC it is stated that they「set forth general rules for international commercial contracts」. The scope of the UNIDROIT PICC is supposed to be limited to「international」contracts. The UNIDROIT PICC do not provide any express definition, but, as pointed out in the Comment, the assumption is that the concept of 「commercial」contracts should be understood in the broadest possible way, so as to include not only trade transactions for the supply or exchange of goods or services, but also other types of economic transactions, such as investment and/or concession agreements, contracts for professional services, etc.

General Principles of the UNIDROIT PICC

There are a number of principles underlying the UNIDROIT PICC, and it is highlighting at least the most significant of these. Freedom of Contract: One of the most fundamental ideas underlying the UNIDROIT PICC is that of freedom of contract. This principle is expressly stated in Art 1.1 Which reads:「The parties are free to enter a contract and to determine its contract.」It is clear that the parties enjoy the widest possible autonomy, in the sense that they may in each individual case either exclude the application of the UNIDROIT

PICC in whole or in part or modify their content so as to adapt them to the specific needs of the king of transaction involved.

Openness to Usages

Another essential element of the UNIDROIT PICC is its marked openness to usages. The reason for this is to be found in the general objective pursued by the UNIDROIT PICC to provide a regulation sufficiently flexible to permit of its constantly adapting to the ever-changing technical and economic conditions of cross-border trade.

Favor Contracts

Another principle underlying the UNIDROIT PICC is the so called favor contracts, i. e. The aim of preserving the contract whenever possible, thus limiting the number of cases in which its existence or validity may be questioned or in which it may be terminated before time. The reason behind this is the acknowledgement that despite shortcomings which might arise in the course of the formation or performance of the contract it is normally in the interest of both parties to do all possible to keep their original bargain alive than to renounce it and to look for alternative goods or services elsewhere on the market.

Observance of Good Faith and Fair Dealing in International Trade

Paragraph 1 of Art 1.7 Clearly states:「each party must act in accordance with good faith and fair dealing in international trade,」and this clearly indicates that under the system of the UNIDROIT PICC it is throughout the life of the contract, including the negotiation process, that the parties' behavior must confirm to good faith and fair dealing.

The purpose of the UNIDROIT PICC

The most important ways in which the UNIDROIT PICC can be used as a model for national and international legislators; as a means of interpreting and supplementing existing uniform law; as rules governing the contract; as a substitute for the domestic law otherwise applicable-are set out in the Preamble. As stated in the Introduction, there may well be other uses for the UNIDROIT PICC, first and foremost that of constituting a guide for drafting international contracts: as a model for National and International Legislators; as a means of Interpreting and Supplementing existing International instruments; as a guide for Drafting Contracts; as rules Governing the Contract.

Comparison of the UNIDROIT PICC and CISG

In comparison with UNIDROIT PICC, CISG has at least the following shortcomings: (a) CISG is less comprehensive in contents than UNIDROIT PICC. (b) Some provisions in UNIDROIT PICC are more reasonable and suitable to international trade practices. (c) The trade-encouraging principle is better embodied in UNIDROIT PICC

UNIDROIT PICC as a means of interpreting and supplementing CISG: According to Article 25 of CISG, the non-performance that substantially deprives the aggrieved party of what it was entitled to expect under the contract, provided the other party could not reasonably have foreseen such result, amounts to fundamental breach of contract, and the defaulting party have the right to terminate the contract. However, how to terminate the fundamental breach of contract remains to be a problem. Article 7.3.1 of UNIDROIT PICC indicates as further factors to be taken into account in each single case whether strict compliance with the non-fulfilled obligation is of essence under the contract, whether the aggrieved party has reason to believe that it cannot rely on the other party's future performance, and finally whether the defaulting party would suffer disproportionate loss as a result of the preparation or performance if the contract is terminated.

International Rules for the Interpretation of Trade Terms (Incoterms2000)

In today's international trade practices, most of the transactions are carried out by means of trade term. International trade is, in contrast to national trade, characterized by long distance between parties, more comprehensive scope, more complex links, more risks, etc,. In order to save time and reduce trade cost, merchants in European countries began gradually to use trade terms as a shorthand methods of expressing shipping terms as well as allocating the risk and loss. Trade terms always express in the form of abbreviated symbols, such as FOB or CIF. The most common method to defining trade terms is to incorporate them into the contract by reference to some independent source or publication. The most important source of the Interpretation of Trade Terms revised in 2000 (hereafter called Incoterms2000).

Contents of Incoterms 2000

Incoterms2000 include 13 trade terms that are classified into 4 groups—E, F, C, and D—according to the relative responsibilities of each party and to the point at which the risk of loss passes from sellerto buyer. Incoterms apply to a contract of international sale of goods only if the parties have incorporated them into their contract.

The present edition of Incoterms lists the following trade terms:

E term refers EXW term (works, mill, factory, mine warehouse, etc.) Under this term, the seller need only make the goods available at this factory (or mill, farm, warehouse, or other place of business) and present the buyer with an invoice for payment. The buyer must arrange all transportation and bear all risks and expenses of the journey from that point.

F terms include FCA (free carrier), FAS (free alongside ship), and FOB (free on board). Generally, under F terms the buyer pays the cost of the main international carriage. The seller places the goods in thehands of a carrier named by the buyer and at a time and place named in the contract. The risk of loss passes from seller to buyer at that time. The buyer arrange the transportation and pay all freight costs, however, if it is convenient and the parties agree, the seller may pay the freight and add that amount to the invoice price already quoted.

C terms include CFR (cost and freight), CIF (cost, insurance and freight), CPT (carriage paid to), and CIP (carriage and insurance paid to). C terms are also shipment contracts. Under CFR terms the seller should contract for transport and pay freight charges to the named port of destination; arrange for loading goods on board ship, usually of seller's choice, and pay costs of loading; obtain export license' notify buyer of shipment, etc. The buyer should purchase document of title and take delivery from ocean carrier; pay import duties, etc,. I f the seller intends to arrange ocean transportation, but will be delivering the goods to a road or rail carrier, inland waterway, or to a multimodal terminal operator for transit to the seaport, the seller may wish to quote CPT. CIP terms are the same as under CPT, with the added requirement that the seller procure insurance to cover the buyer's risk of loss.

D terms include DAF (delivered at frontier), DES (delivery ex quay), DEQ (delivered ex quay), DDU (delivery duty unpaid), and DDP (delivered duty unpaid). Contract with D terms of sale are destination contracts. If the seller in China is willing to enter into a destination contract,   then he must be willing to accept far greater responsibility than under any other terms.

Passing of the Risk and Property

Passing of Risk

The loss of goods through fire, theft, or other means can occur at any time: prior to delivery, during transit or inspection, or after delivery. In most cases, the loss will be covered by insurance.   Even so, it is important to determine whether the buyer or seller is responsible for obtaining the insurance.

Common law

Sales of Goods Act Article 20 of Sales of Goods Act states: Unless otherwise agreed, the goods remain at the seller's risk until the property in them is transferred to the buyer, but when the property in them is transferred to the buyer the goods are at the buyer's risk whether delivery has been made or not. But where delivery has been delayed through the fault of either buyer or seller the goods are at the risk of the party at fault as regards any loss which might not have occurred for such fault.

Civil law

As a represent of civil law, CISG contains provisions that allocate the risk of loss in Article 66-70: CISG allows the parties to decide the time of passing of risk; The effect of passing of risk. Article 66 states that loss of or damage to the goods after the risk has passed to the buyer does not discharge him from his obligation to pay the price, unless the loss or damage is due to an act or omission of the seller; Where the goods need to be moved, Under Article 67, if the contract of sale involves carriage of the goods and the seller is not bound to hand them over at a particular place, the risk pass to the buyer when the goods are handed over to the first carrier for transmission to the buyer in accordance with the contract of sale. Also the risk does not pass to the buyer until the goods are clearly identified to the contract, whether by makings on the goods, by shipping documents, by notice given to the buyer or otherwise.

The passing of risk of the goods in transit

Usually, the determination of time of passing of risk of the goods in transit is extremely difficult, because neither the seller nor the buyer knows anything about the goods at disposal of the carrier. The article 67 of CISG provides three solutions to solve this problem: (a) The risk in respect of goods sold in transit passes to the buyer from the time of the conclusion of the contract. (b) If the circumstances so indicate, the risk is assumed by the buyer from the time the goods were handed over to the carrier who issued the documents embodying the contract of carriage. (c) If at the time of the consideration of the contract of sale the seller know or ought to have known that the goods had been lost or damaged and did not disclose this to the buyer, the loss or damage is at the risk of the seller.

Fundamental Breach of Contract's Effect on the passing of risk

According to the article 70 of the CISG, if the seller commits a fundamental breach of contract in accordance with the CISG, it in due course of the trade does not effect the passing of risk to the buyer, but the buyer has the remedies available to the damages caused by the seller's fundamental breach.

Passing of Property

The property in the goods means the ownership of the goods, as distinguished from their possession. In international trade it is extremely important to know the precise moment of time at which the property in the goods passes from the seller to the buyer. In case of the destruction of the goods by fire or other accidental cause it is necessary to know which party has to bear the loss.

Internationally, however, different countries have different provisions to this problem.

Common law

Specific Goods

Under Sales of Goods Act, specific goods are goods identified and agreed upon at the time the contract is made. The property of specific goods passes to the buyer at the time when the parties intend it to pass.

Unascertained Goods

Sales of Goods Act provide that the property in unascertained goods does not pass until the goods are ascertained. Unascertained goods are goods define by description only, e. g. 100 tons of coal, and not goods identified and agree upon when the contract is made. The property in unascertained or future goods sold by description passes to the buyer when goods of that description and in a deliverable state are unconditionally appropriated to the contract, either by the seller with the assent of the buyer or by the buyer with the assent of the seller.

Civil law

Article 4 of the CISG provides that Convention governs only the formation of the contract of sale and the rights and obligations of the seller and the buyer arising from such a contract. In particular, except as otherwise expressly provided in this Convention, it is concerned with: (a) the validity of the contract or of any of its provisions or of any wage; (b) the effect which the contract may have on the property in the goods sold.

Remedies for Breach of Contract

Remedies for breach of contract are provided by the laws of almost every country, which are intended to give the parties the benefit they expected to obtain from thecontract and to put the parties into the economic position they would have been in had the breach not occurred. The remedies available to a buyer or seller under CISG are drawn from both common law and civil law systems. These remedies include: suspension of performance, avoidance of the contract, damages, specific performance, seller's right to cure, seller's additional time to perform and price reduction.

General Principles of Remedies

The Doctrines of Compensation

The doctrine of expectation interest: This doctrine means that both parties at the time of the conclusion of a contract have expectation interest lost by the party in breach. The doctrine of reliance interest: This means that when the defendant's promise to perform his contractual obligations, the claimant has acted to his compensate the defendant. The doctrine of restitution interest: A claimant who claims the protection of his restitution interest does not wish to be compensated for the loss which he has suffered; rather, he wishes to deprive the defendant of a gain which he has made at the claimant's expense.

The Doctrine of Limitations on Damages

A person may not recover for all losses flowing from a breach. There must be limitations on damages. Non-breaching party has the duty to mitigate damages. Damages must be established with a reasonable degree of certainty. Damages must be either foreseeable to the breaching party or would naturally flow from the breach.

Remedies of the buyer and seller

Buyer's Remedies

A seller may breach a contract in a number of different ways, the most common are: failing to make an agreed delivery, delivering goods that do not conform to the contract, and indicating an intention not so fulfill the obligations under the contract.

The remedies that are unique to the buyer are: To compel specific performance. Specific performance is used when a court requires a party to the contract to performs, or carry out its part of the bargain. The usual remedy in civil law countries is that of specific performance, while the usual legal remedy in contract cases in common law countries is an award for money damages. To avoid the contract for fundamental breach or non-delivery:

Under common law and civil law, if the breach is material or fundamental, the buyer need not take delivery nor pay for the goods, nor find a buyer to take them. To reduce the price: Price Reduction for the buyer is used when the seller makes only a partial shipment, or if the goods are nonconforming. To refuse early delivery and excess quantities: Article 52 of CISG provides: 「 (1) If the seller delivers the goods before the date fixed, the buyer may take delivery or refuse to take delivery. (2) If the sellers deliver a quantity of goods greater than that provided for in the contract, the buyer may takes delivery of all part of the excess quantity, he must pay for it at the contract rate.」

Seller's Remedies

The remedies that are unique to the seller are: to compel specific performance, to avoid the contract for a fundamental breach or failure to cure a defect, and to obtain missing specifications. Clearly, the seller's remedies are rather similar to those the buyer.

Remedies available to both buyers and sellers

The remedies available to both buyers and sellers are: suspension of performance, avoidance in anticipation of a fundamental breach, avoidance of an installment contract, avoidance, and damages.

Obligation of the seller and buyer

The discussion on the obligation of the seller and the buyer is to be based on principally the CISG because it usually governs most of the international business transaction in China.

Seller's obligations

Under the CISG a seller is required to deliver the goods, hand over any documents relating to them, and ensure that the goods conform with the contract.

Buyer's obligations

The buyer has two primary obligations under CISG in a sale contract: to pay the price, and to take delivery of the goods.

Lesson 3 The International Technology Trade Law

Intellectual Property Rights

Fundamental of Intellectual Property Rights

Intellectual property refers to creation of the human mind, and in essence, useful information or knowledge. For the purpose academic study, intellectual property is generally classified into two main types: copyright and industrial property. Industrial property is itself divided into two categories: patents and trademarks. Besides know-how or trade secrets is also included in the realm if intellectual property rights.

Copyrights

Copyrights belong to the authors of any work that can be fixed in a tangible medium for the purpose of communication, such as literary, dramatic, musical or artistic works; sound recordings; films; radio and television broadcasts; and computer programs. Theoretically a copyright is classified into pecuniary rights and moral rights. The 「pecuniary rights」 inherent in a copyright allow an author to exploit a work for economic gain. Whereas moral right is an author's personal rights to keep others from tampering with his works. Moral rights are not recognized in the copyright laws of common law legal system such as the U K and the United States, they claim that the author can bring an action against the distortion, mutilation or modification of his or her work according to contract law and tort law. A copyright applies only within the territory of the state gaining it. About the duration for which copyrights may be granted, the TRIPS Agreement sets the minimum time period of the life of author plus 50 years.

Patents

Patents are 「a statutory privilege granted by the government to inventors, andto others deriving their rights from the inventors, for a fixed period of years, to exclude other persons from manufacturing, using or selling a patented product or from utilizing a patented method or process」. There are two basic reasons why governments grant patents. One is that patents are a confirmation of the private property rights of the inventor. The second is that a patent is a grant of a special monopoly, which prevents others from utilizing the patentee's invention without his consent, to encourage invention and industrial development across the state.

Trademarks

Trademarks are used by merchants and others to identify themselves and their products from similar goods or services supplied by others. Now trademarks have become one of the core factors in the competitiveness of enterprises. In the broad sense, there are different kinds of trademarks.

Know-how or Trade Secrets

Know-how is practical expertise acquired from study, training, and experience. Unlike other forms of intellectual property, know-how is protected by contract, tort and trade secrecy laws, rather than by particular statutory enactments.

International Intellectual Property Organizations and Treaties

World Intellectual Property Organizations

The World Intellectual Property Organization (WIPO), created in 1967, is responsible for administering the International Convention for the Protection of Industrial Property (the Paris Convention) and the Berne Convention for the Protection of Literary and Artistic Works (the Berne Convention). There are now more than 170 parties to this organization, including China. One of WIPO's more important tasks is to facilitate the transfer of technology, especially to and among developing countries. The council for Trade-Related Aspects of Intellectual Property Rights (TRIPS Council) was created in 1995 with the adoption of the Agreement Establishing the World Trade Organization (WTO Agreement).

The Council is responsible for overseeing the TRIPS, which is an annex to the WTO Agreement.

Intellectual Property Treaties

Most countries relating to intellectual property rights are regulated internationally by multilateral treaties. Most of these treaties are administered by the World Intellectual Property Organization and the Council for Trade-Related Aspects of Intellectual Property Rights. The principal comprehensive agreement establishing general intellectual property obligations for most countries of the world is the Agreement on Trade-Related Aspects of Intellectual Property Rights. (TRIPS Agreement), which came onto effect in1995 with the adoption of the Agreement Establishing the World Trade Organization (WTO Agreement). The goal of the TRIPS Agreement is to create a multilateral and comprehensive set of rights and obligations governing the international trade in intellectual property rights. To accomplish this goal, the Agreement establishes a common minimum of protection for intellectual property rights applicable within all the WTO members.

International licensing agreement

According to the Foreign Trade Law of People's Republic of China 2004, in the international licensing agreement, if the intellectual property right owner is involved in any one of such practices as preventing the licensee form challenging the validity of the intellectual property right in the licensing contract, conducting coercive package licensing or incorporating exclusive grant-back conditions in the licensing contract, which impairs the fair competition order of foreign trade, the Ministry of Commerce of People's Republic of China may take measures as necessary to eliminate such impairment.

Regulations on International licensing

Intellectual property rights may be transferred from one country to another in various ways. In addition to an owner's personal use abroad, theowner may assign part or all of the rights, license another to work them, or establish a franchise, or a government may grant a compulsory license so that a third party may exploit the rights.

Compulsory Licenses

A compulsory license is to work a patent or use a copyright without the consent of the owner but based on a license that is issued by a government. Compulsory licenses are common in most countries of the world, especially in developing countries. They arise when the owner of the intellectual property, in particular patents or copyrights, refuses or is unable to work the property in a particular country within a certain period of time. In such a case, a third party may apply for a compulsory license, which will be issued by the government without the consent of the owner.

Regulations of International Franchising

Franchising constitutes a rapidly expanding form of doing business abroad. Now it even becomes a business model in which many different owners share a single brand name. A parent companyallows the franchisee to use the company's strategies and trademarks; in exchange, the franchisee pays an initial fee and royalties based on revenues. The parent company also provides the franchisee with support, including advertising and training, as part of the franchising agreement. For the franchisor, the franchise is a faster, cheaper form of expansion than building 「chain stores」 to distribute goods and avoid investment and liability over a chain. For the franchisee, the franchisor's success is the success of the franchisees. The franchisee is said to have a greater incentive than a direct employee because he or she has a direct stake in the business. Under the intellectual property laws, franchise is a specialized license in which the licensor franchiser permits the licensee to work the licensor's intellectual property under the supervision and control of a franchiser.

A license allows a licensee to use an intellectual property, usually a trademark, for the licensee's own purposes. Depending on the licensing agreement between them, the licensee may use the intellectual property as a component of its own products; it may sell the products derived from it under the licensor's name or under its own name. Sometimes the licensee may even sell the intellectual property or the products derived from it in direct competition to the licensor. However, a franchisee has more limited rights. The key difference is that a franchisee is regarded as a unit or element of the franchiser's business.

Three types of franchise have evolved since the beginning of the twentieth century: distributorship, chain-style business, and manufacturing or processing plan. A distributorship franchise exists when a manufacturer licenses a dealer to sell its product; the automobile dealership is an example. A chain-style franchise is an arrangement in which a franchise operates under a franchiser's trade name and is identified as a part of the franchiser's business chain. Most of the foreign fast-food restaurants are operated in this way. A manufacturing or processing plant franchise exists when a franchiser provides the franchisee with the formula or the essential ingredients to make a popular product. The franchisee then wholesales or retails the products according to the standards established by the franchiser. Examples of this kind are Coca-Cola and other soft-drink firms.

Lesson 4　The Regulations of International Trade

Regulations of international trade are to limit the access that a particular good enter to a market. Regulations may be imposed on the inbound end, by imposing a regulation on imports, or on the outbound end, by imposing a restriction on exports. The international trade regulation policy include: Tariffs Regulation, Non-Tariffs Regulation, Antidumping, Subsidies and Countervailing Duties, Safeguards etc. Such regulation policy can divided into two different forms: From government conduct; From WTO rules.

Tariffs Regulation

Tariffs is a duty (or duty) levied upon goods transported from one customs area to another either for protective or revenue purposes. The customs duties can be divided into different categories according to different standards or from different angles: import duties, export duties and transit duties, specific duties, ad valorem duties etc., ordinary duties, preferential duties etc., revenue duties and protective duties, agreement tariffs and bound tariffs etc., principle duties and additional duties.

Non-Tariffs Regulation

Non-Tariffs Regulation refer to policy measures used by the government, otherthan tariffs, that can potentially affect trade in goods. Non-Tariffs measures can serve legitimate public policy goals, and it can take many forms, including import quotas, discriminatory product standards, buy-at-home rules for government purchases, and administrative red tape to harass importers of foreign products.

Quantitative Restrictions

Countries can used border measures to limit or restrict the importation of goods in two ways. First, countries can increase the price of imports by levying duties or charges on imports. Second, countries can limit the quantity of imports by imposing import quotas or other quantitative restrictions on importation. Quantitative Restrictions can take on several different forms, including: an outright prohibition or ban, a quota, indicating the precise amount of a good that maybe imported or exported, a requirement for a license to import/export good, voluntary export restraints.

Technical Barrier to Trade

The TBT Agreement governs three types of measures: technical regulations, standards and conformity assessment procedures. Technical Regulation, as the TBT Agreement defines, is a document which lays down product characteristics or their related processes and production methods, including the applicable administrative provisions, with which compliance is mandatory. It may also include or deal exclusively with terminology, symbols, packaging, marking or labeling requirements as they apply to a product, process or production method.

Antidumping

Dumping is what happens when a producer exports a product at a price less than the normal value of the product. According to Art. Ⅵ. 1 GATT and Art. 2.1 Antidumping, dumping occurs when the export price of a good is less than its normal value (NV). The difference between the two prices is the dumping margin. Calculating Normal Value by: use of the home market price as NV, circumstances for disregarding home market price as the basic for NV, use of export price in a third country as the basis for NV, constructed NV, non-market economies.

Based on Art. Ⅵ. 1 GATT, in order to impose an antidumping duty, a WTO member must prove: dumping, existence, or threat, of injury to the domestic industry (or the material retardation of the establishment of such industry), a causal link between the dumping and injury (or material retardation). Of these three requirements, the most complicated is the determination of dumping, which is extremely technical, in nature.

Subsidies and Countervailing Duties

Subsidy

According to WTO SCM Agreement, a subsidy as follows: a financial contribution by a government or any public body; which confers a benefit; to a specific recipient. It can be divided into three difference categories: prohibited subsidies, actionable subsidies and non-actionable subsidies.

Prohibited Subsidies

Prohibited Subsidies are 「red light」 subsidies that are illegal under all circumstances. Two forms of subsidies are expressly prohibited under Art 3. SCM: local content subsidies and export subsidies. Local content measures have been outlawedsince the inception of the GATT (Art III. 5); hence, outlawing subsidies conditional upon the use of domestic value added should not come as a surprise. Art III SCM makes clear that it does not matter whether this sole condition or one of several condition. The attitude of the world trading system towards export subsidies, on the other hand, has remarkably evolved since the inception of the GATT. The GATT initially followed a much more lenient approach than the SCM-Agreement on this score. Art VI GATT implicitly recognized that subsidies may be a legitimate instrument of domestic public policy by specifying that countervailing duties to offset the subsidy can only be imposed if the effect of subsidization has resulted in material injury to a domestic industry.

Actionable Subsidies

Actionable subsidies are 「yellow light」 subsidies, whose legality depends on the circumstances that arise from their effect. The vast majority of government measures that amount to a 「subsidy」 under the SCM Agreement fall under the category of actionable subsidies. Such subsidies are not outright illegal. Instead, actionable subsidy is impermissible under WTO law only if it causes one of the three types of adverse effect to the interests of other WTO members included in Art 5 SCM: (1) injury to the domestic industry; (2) nullification or impairment of benefits; and (3) serious prejudice. Therefore, to prevail in WTO litigation that an actionable subsidy results in one of these three adverse effect.

Non-Actionable Subsidies

The now-expirednon-actionable subsidies were once referred to as 「green light」 subsidies that are legal under all circumstances. The SCM Agreement, as originally drafted, provided for a third category of subsidies non-actionable subsidies. Three types of subsidies were deemed non-actionable subsidies: regional aid, environmental subsidies, and subsidies for research and development (R & D) purposes. They were initially contracted for a 5 year provisional period. In the absence of agreement to keep this category in place, non-actionable subsidies ceased to exist as of January 1th, 2000. Consequently, a scheme which qualifies as subsidy under the SCM Agreement, nowadays, either a prohibited or an actionable subsidy.

Countervailing Duties

Recourse to multilateral dispute settlement is not the only option available to a WTO member when its domestic producers are confronted with competition from subsidized imports within its own market. In such a circumstance, a WTO member may also make resource to a specific contingent protection instrument designed to tackle unfair trade caused by subsidization: a countervailing duty (CVD).

The process of imposing a CVD is undertaken unilaterally through a domestic administrative procedure (similar to that of anti-dumping); it is generallymore expeditious than WTO dispute settlement. However, success through the CVD channel only results in the imposition of higher duties for the subsidized good within one's own market. A CVD is not effective at countering the negative impact that one's domestic producers may face in third-country markets from subsidized foreign competitors.

Safeguards

WTO members can lawfully impose safeguards in the form of tariffs, quotas, or tariff quotas if they can show that, as a result of unforeseen developments, imports have risen, and caused injury to the domestic industry producing the like product. Actually, the unforeseen developments requirement does not figure anywhere in the WTO agreement on Safeguards (SG); it is reflected in Art GATT only. To impose safeguards, four requirements that must be cumulatively met: increased quantities of imports, as a result of unforeseen developments, that have caused or threaten to cause serious injury to the domestic industry producing the like product, in the sense that a causal link between the increased quantity of imports and the serious injury (threat of injury) is (can be) established.

Unlike the other two instruments of contingent protection (AD and CVD), safeguard measures are not imposed against imports from a particular trading partner. Instead, they must be applied in a non-discriminatory manner to all imports. A number of other important difference, as compared to AD and CVD measures, exist: safeguard measure must be progressively liberalized after a year; WTO ambers imposing a safeguard must agree on adequate means of compensation with affected trading partners once the safeguard exceeds three years; safeguards also cannot be renewed over and over again, but instead must be terminated after a fixed period of time and not be re-introduced for another fixed period of time.

New Words, Phrases and Expressions
enforceable   vt. 實施，執行，強迫
valid   adj. 有效的，合法的，正當的
offeror   n. 要約人
offeree   n. 受要約人
lapse   n. [法]（權利等的）失效，消失

revocation   n. 廢止，撤回
remedy   n. 救濟方法
compensate   vi. 補償，賠償；抵消
intellectual property   n. 知識產權
addition   n. 增加；加法；附加物
instrument   n. 儀器
distributorship   n. 分銷權
domestic laws   國內法
sought to   尋求

Notes

1. As it does, the courts will have to keep in mind the Convention's admonitions: that CISG is an international treaty, that its purpose is to establish uniform international rules, and that the courts are bound by the principle of good faith in interpreting the Convention.

同樣的，法庭將必須牢記公約的告誡：CISG 是一項國際條約，其目的是建立統一的國際規則，而且在解釋本公約時，法庭要遵循誠實信用原則。

2. This limitation on the sources that the courts may turn to in creating general principles was consciously made, and reflects the drafters' concern for uniformity and consistence, both in the drafting and in the evolution of the Convention.

這種對於法庭可能轉向有意識地創造一般性原則的限制，體現了起草者對公約的起草階段和演變過程中統一性和一致性的關注。

3. The most important ways in which the UNIDROIT PICC can be used as a model for national and international legislators; as a means of interpreting and supplementing existing uniform law; as rules governing the contract; as a substitute for the domestic law otherwise applicable-are set out in the Preamble.

UNIDROIT PICC 是可以作為國家和國際立法模式的最重要的方式；作為一種解釋和補充現有統一法律的方式；規定合同；代替國內法，否則要在序言中載明。

4. The usual remedy in civil law countries is that of specific performance, while the usual legal remedy in contract cases in common law countries is an award for money damages.

在大陸法系國家，通常的補救方法是依照具體表現，而在英美法系國家，通常對於合同的法律補救辦法是金錢賠償。

5. Copyrights belong to the authors of any work that can be fixed in a tangible medium for the purpose of communication, such as literary, dramatic, musical or artistic works; sound recordings; films; radio and television broadcasts; and computer programs.

版權是屬於作者的可以固定在一個有形的媒介中用來溝通的作品，如文學、戲劇、音樂或藝術作品；錄音、電影、廣播和電視廣播、計算機程序。

Section 3　Resolution of International Business Disputes

　　The growth of international commercial arbitration (ICA) is in part a retreat from the vicissitudes and uncertainties of international business litigation. More positively, ICA offers predictability and neutrality as a forum (who knows which court you may end up in) and the potential for specialized expertise (most judges know little of international law). ICA also allows the parties to select and shape the procedures and costs of dispute resolution. That said, ICA procedures are relatively informal and not laden with legal rights.

　　One of the most attractive attributes of ICA is the enforceability in national courts of arbitral awards under the New York Convention. Roughly 120 nations participate in the New York Convention. Another major advantage of ICA is the support of legal regimes that give arbitration agreements dispositive effects. One of the least attractive attributes of ICA is the minimal availability of pre-trial provisional remedies. In addition, many arbitrators focus on splitting the differences between the parties, not the vindication of legal rights which in courts might result in 「winner takes all」.

Arbitration agreements

　　The agreement to arbitrate is the foundation stone of international commercial arbitration. It serves to evidence the consent of the parties to submit to arbitration; and this consent is indispensable to a process of settling disputes which depends upon the agreement of the parties for its very existence. Arbitration agreements, traditionally called compromise, come in two basic types. The first, and more common, is an agreement to submit future disputes to arbitration; The second is an agreement to submit existing disputes to arbitration.

　　Whether a valid agreement to arbitrate exists depends on the specifies of the arbitration clause, not the entire business agreement. The arbitration clause is severable, and issues of validity (such as fraud in the inducement of the arbitration clause and unconscionability directed to it. Many courts will stretch the limits of the New York Convention in order to uphold an arbitration clause. When there is a battle of forms, the same judicial bias towards arbitration often be found. But in most cases, the disputes must 「arise under」 the business transaction to be arbitrable and legal claims falling outside the transaction remain in court.

　　The closure or misdescription of an arbitration center designated in the agreement (e. g., the New York chamber of commerce) is no barrier to arbitration. A substitute arbitrator will be appointed by the court if the parties can not agree. The U. S. Supreme Court held in Commonwealth Coatings (1968) that arbitrators are subject to 「requirements of impartiality」 and must 「disclose to the parties any dealings that might create an impression of possible bias」. That said, most U. S. Courts are loathe to intrude in proceedings or vacate an arbitration award on disclosure grounds.

There is a split of opinion as to whether the implied ground of 「 manifest disregard of the law 」 bars enforcement of an arbitral award in U. S. courts under the New York Convention. Article V of the New York Convention does not recognize manifest disregard of the law as a basis for denial of enforcement. Another issue concerning he New York Convention is whether to adjourn U. S. enforcement proceedings if parallel proceedings to vacate the award have been commenced in the country of arbitration. Despite the risks of forum shopping and delay, adjournment can be appropriate, depending upon the circumstances. Obtaining compulsory non-party discovery in private commercial arbitrations has been denied.

Cases in the United States have pointed out that parties cannot refer a dispute to a court while an arbitration is in progress or block enforcement of an award in the United States in reliance upon the fact that the award, although binding in the country where rendered, is under appeal there. Interim orders of arbitrators, such as records disclosures, may be enforceable 「 awards 」 under the New York Convention. After the arbitration is concluded, a party may not able to block enforcement of the award in reliance upon in the United States Foreign Sovereign Immunities Act, but a court may decline to enforce in reliance upon the Act of State Doctrine. One court granted enforcement, under the Convention, of a New York award rendered in favor of a non-citizen claimant against a non-citizen defendant.

When arbitral awards are annulled at their situations, courts in enforcing jurisdictions have taken different positions on the enforceability of the award. French courts enforced an improperly vacated award to the detriment of claimant who had prevailed in a second arbitration. One U. S. Court refused to honor the clearly legitimate annulment of an arbitral award by an Egyptian court because the parties had agreed not to appeal the award. The Second Circuit, on the other hand, recognized the annulment of two arbitral awards vacated by a Nigerian court and refused enforcement. The New York Convention does not address the treatment of annulled arbitral awards.

The United States Supreme Court indicated in First Options (1995) that questions of the arbitrability of disputes may be arbitrated, but only if the parties have manifested a clear willingness to be bound by arbitration on such issues. Silence or ambiguity should favor judicial review of arbitrability issues. Arbitration clauses that adopt the UNITRAL Rules meet the requirement of clarity to arbitrate arbitrability because Article 21 conveys jurisdictional issues to the tribunal.

International Business Arbitration Procedure

The arbitration procedure is relatively straightforward. In its simplest form, the parties begin the process by the selection of an arbitrator or arbitrators. If they have agreed upon a place for arbitration to take place, the location decided; otherwise the arbitrator or tribunal may fix the location. The parties then exchange statements of claim and defense and any further statements, including any pleas as to the jurisdiction of the arbitration tribunal. The next step in the process is hearing itself, where the arbitrator or arbitration will hear the evidence of the parties, any expert testimony, and final arguments. The arbitrator or arbitration tribunal will then deliberate and issues an award. Generally, if the parties have selected an arbitration institution, rules of the arbitration institution shall apply to the process of arbitration, unless the parties agree to be bound by other arbitration rules. For example, UNCITRAL rules covers the appointing authority, notice requirements, representation of the parties, challenges of arbitrators, evidence, hearings, statements of claims and defenses, pleas to the arbitrator's jurisdiction, provisional remedies, experts, default, rule waivers, the form and effect of the award, applicable law, settlement, interpretation of the award and costs. Arbitration Rules of China International Economic and Trade Arbitration Commission (2005) cover similar items of matters.

Appointment of Arbitrators

An arbitration tribunal consists of one or three arbitrators. When only one arbitrator has to adjudicate on a dispute the parties usually have no difficulty over this appointment. In case of three arbitrators, each of the party, within 15 days upon receiving the notice of arbitration, appoints an arbitrator and a third arbitrator is then appointed jointly by the parties or by the arbitration institution based on the authorization of the parties. The third arbitrator is the presiding arbitrator. The nominated arbitration institution is also competent to appoint an arbitrator if one of the parties refuses to co-operate. The arbitrator appointed by one party cannot consider himself as an agent of that party; he must have an independent and impartial view. If this independence and impartiality are compromised the arbitrator must refuse or hand back the appointment. Arbitrators must, in any event, disclose to the parties all instances, such as conflict of interests, financial, etc. That could jeopardize their independence and impartiality. And the parties are entitled to challenge the independence and impartiality of arbitrators with cause.

Arbitration Proceedings

Parties have a wide range of options for the setting up of arbitration proceedings. If they don't make use of these options, it is up to the arbitration institution todirect the arbitration proceedings. Arbitration Rules of China International Economic and Trade Arbitration Commission (2005) require a fundamental condition for the arbitration procedure that every party is entitled to a equal and fair trial. A breach of the fundamental rule of the right to a equal and fair trial results in the award set aside by the court of China.

The parties state the claims and defenses, submit evidence and debate. The arbitration can order an expert assessment, a local visit, hear parties or witnesses and decide which evidence will be considered sufficient proof. If one of the parties does not co-operate without legal reasons the arbitrators has jurisdiction to make an award by default. The arbitrators are required to make their award within six months, and at the due request of arbitrators the Director of China International Economic and Trade Arbitration Commission may grant an extension.

Enforcement of Arbitral awards: New York Convention

The enforcement of an international commercial arbitration award is obviously of critical importance to the winner in the award. Parties are deemed to be bound by the arbitral award unless the award has been annulled. As for the international commercial arbitration, often the losing party gives effect to the arbitral award voluntarily. Because there is frequently social and commercial pressure from the commercial sector to abide by the arbitral award: an enterprise known for defaulting on arbitral awards looses credibility. However, the arbitral award can be enforced where necessary through a court order. Necessary, this requires a mechanism by which the award of a private adjudicator becomes enforceable by public authorities——the award must be recognized and enforced by states. The award is enforceable where the losing party has assets located or where that party resides. To overcome the problems of enforcing an international arbitral award in another country than that of the seat, an important international treaty has been adopted the United Nations Convention on the Recognition and Enforcement of foreign Arbitral Awards (New York Convention 1958).

The New York Convention commits the courts in each Contracting State to recognize and enforce arbitration clauses and separate arbitration agreement for the resolution of international commercial disputes. Where the court finds an arbitral clause or agreement, it 「shall... refer the parties to arbitration, unless it finds that the said agreement is null and void, inoperative, or incapable of being performed」 (emphasis added). The convention also commits the courts in each Contracting State to recognize and enforce the awards of arbitral tribunals under such clauses or agreements, and also sets forth the limited grounds under which recognition and enforcement may be refused. Grounds for refusal to enforce include: incapacity or invalidity of the agreement containing the arbitration clause 「under the law applicable to」 a party to the agreement, lack of proper notice of the arbitration proceedings or the appointment of the arbitrator, failure of the arbitral award to restrict itself to the terms of the submission to arbitration, or decision of the matters not within the scope of that subdivision, composition of the arbitral tribunal not according to the arbitration agreement or applicable law, and non-finality of the arbitral award under applicable law.

In addition to these grounds for refusal, recognition or enforcement may also be refused if it would be contrary to the public policy of the country in which enforcement is sought; or if the subject matter of the dispute cannot be settled by arbitration under the law of that country. Courts in the United States have taken the position that the public policy limitation on the Convention is to be construed narrowly and to be applied only where enforcement would violate the forum state's most basic notions of morality and justice. Recourse to the limitations of the Convention, in order to defeat its applicability, has been greeted with judicial caution.

Whether the New York Convention applies generally turns upon where the award was or will be made, not the citizenship of the parties. A growing number of the courts in developing nations are issuing injunctions against arbitral proceedings before they commence. Many of these injunctions seem deliberately intended to protect local companies. Parties who proceed to arbitrate after such an injunction has been issued do so at their peril. Subsequent enforcement of the award under the New York Convention in the enjoining nation will almost certainly be voided on grounds of public policy. Hence enforcement can only proceed in non-enjoining jurisdictions, assuming that their public policy permits this agreement.

New Words, Phrases and Expressions
unconscionability   n. 顯失公平
jurisdiction   n. 管轄範圍內
arbitrable   adj. 可裁決的，可仲裁的
default   n. 違約；缺席；缺乏
misdescription   n. 錯誤（或虛假）的描述
credibility   n. 可靠性，可信性；確實性
adjudicator   n. 評判員
restrict   vt. 限制；約束
limitations   n. 局限性；（限制）因素；邊界
injunction   n. 禁制令
nominated   v. 提名……為候選人

Notes

1. In case of three arbitrators, each of the party, within 15 days upon receiving the notice of arbitration, appoints an arbitrator and a third arbitrator is then appointed jointly by the parties or by the arbitration institution based on the authorization of the parties.

在有三名仲裁員的案例中，每一方，自收到仲裁通知之日起 15 日內，指定一個仲裁員，然後再由雙方共同指定或仲裁機構基於當事人的授權任命第三名仲裁員。

2. In addition to these grounds for refusal, recognition or enforcement may also be refused if it would be contrary to the public policy of the country in which enforcement is sought; or if the subject matter of the dispute cannot be settled by arbitration under the law of that country.

除了這些拒絕理由，如果違背其執行請求國的公共政策，或者爭議的標的物不能被仲裁根據該國法律解決，承認或執行也可能被拒絕。

Exercises

Ⅰ. Noun explanation

(1) International business and International business law

(2) International trade usages and customs

(3) Definition of contract

(4) Revocation of offer

(5) Intellectual property

(6) Arbitration agreements

Ⅱ. Think about the following questions

(1) Elements to a valid contract.

(2) Sources of international business law.

(3) Calculated standard of Normal Value.

(4) How to arrange the International Business Arbitration to avoid risks.

Ⅲ. Translate the following sentences:

A. The remedies available to both buyers and sellers are: suspension of performance, avoidance in anticipation of a fundamental breach, avoidance of an installment contract

B. To impose safeguards, four requirements that must be cumulatively met: increased quantities of imports, as a result of unforeseen developments, that have caused or threaten to cause serious injury to the domestic industry producing the like product, in the sense that a causal link between the increased quantity of imports and the serious injury (threat of injury) is (can be) established.

Further Reading Comprehension

New publication on the application of substantive law now available from the ICC Store Paris, 21 October 2014

The International Chamber of Commerce (ICC) Institute of World Business Law has released The Application of Substantive Law by International Arbitrators, the eleventh volume of the Institute's Dossier Series on international commercial law and arbitration.

Lawyers, arbitrators, magistrates, academics and corporate counsel can now get a copy of the new publication which will help them solve some of the complex questions they face: How should the contents of the applicable law be proved by the parties? Are the arbitrators free to ascertain the applicable rules of law independently from the parties?

Dossier XI assembles in eleven chapters the salient points that were brought up during panel discussions of the Institute's 33rd Annual Meeting. The conference was co-organized and co-chaired by Fabio Bortolotti, Partner, Buffa, Bortolotti Mathis Studi Legali Associati, Turin and Pierre Mayer, Partner, Dechert LLP, Paris, both Council Members of the ICC Institute.

「When international arbitrators apply the rules of a national legal system, they need to deal with a number of questions on how the applicable law should be proven and on the arbitrators' right to inquire autonomously about its contents.」said Mr. Bortolotti.

「In case the applicable domestic law does not correspond to expectations of parties engaged in international commerce, arbitrators are put before a difficult choice between a flexible approach based on commercial practice, usage and reasonable expectations of parties involved in the type of commerce at stake, and a rigorous application of the rules of law,」 he added.

Mr. Mayer said:「This new publication provides a perspective on the right balance between these conflicting needs, which can only be found case by case, taking into account all the relevant circumstances. It offers a comprehensive analysis of the role of the arbitrators and the expectations of the parties in case of contradiction between the applicable law and the needs of international business.

「It offers a thorough picture of practical issues related to the application of substantive law when international arbitrators apply the rules of a national legal system.」

The Dossiers of the renowned ICC Institute of World Business Law are the outcome of the Institute's annual meetings, where experts from around the globe come together to discuss topical issues of international commercial law including arbitration.

The next Annual Meeting will gather international trade and dispute resolution professionals in Paris on 24 November with a conference on「Addressing Issues of Corruption in Commercial and Investment Arbitration」. The proceedings from this event will be compiled in another edition of the Institute Dossier and participants attending the conference will be entitled to a free copy.

The Dossiers Series analyses important issues which are frequently faced by arbitrators and counsel in international disputes. The ICC Institute of World Business Law is a think-tank that provides research, training and information to the legal profession concerned with the development of international business law. Its main objectives are to foster wider knowledge and the development of the law and practices of international business.

附錄 1
國際經貿專業英語的特點

自 2001 年加入世界貿易組織以來，中國對外貿易事業發展迅猛，近年來一直保持世界貿易大國和出口大國的領先地位。隨著全球經濟一體化趨勢的發展，中國「一帶一路」倡議的落實，對外經貿事業必將在未來持續全方位、深層次發展，同時，隨著「大貿易」的拓展、貿易運行平臺的豐富，國際合作與交流必將在更大範圍內、更寬領域裡日趨頻繁。在此背景下，既能掌握紮實的國際經濟貿易理論及專業知識技能，又能熟練使用英語進行國際經貿業務溝通的高層次複合涉外經貿人才將會備受企業青睞。

用於國際經濟貿易合作與交流的英語通常被稱為國際經貿英語（International Business English）[①]。作為專門用途英語（English for Specific Purposes，ESP）的一個分支，國際經貿英語既涉及基礎英語的語言共核，又涉及國際貿易、國際金融、國際商法、國際工商管理等領域的專業知識。但它並非基礎英語與專業知識的簡單結合。特定的語言使用場合、語言使用者和使用目的賦予其鮮明的語用特點：準確、程式化、簡潔和禮貌。這些特點具體體現在國際經貿英語的詞彙、語義、句式、語法、篇章和修辭等多個層面。瞭解這些特點，對於我們進一步學習國際經貿英語有著非常重要的意義。

1 國際經貿英語的語用特點

與通用英語（General English）不同，專門用途英語通常具有兩個特點：①學習者具有明確的學習目標：為滿足特定行業的需求，學習者需達到在某些學科內使用英語的能力。②特定的內容或專門化的內容。特定行業與專門化內容決定了專門用途英語所特有的語言使用者、語言使用場合與使用目的。國際經貿英語作為一種專門用途英語，也具有這兩個特點。

語言的主要功能是有效地溝通與交流。國際經貿英語作為英語的一種社會功能變體，也不例外。具體講，它是指在跨文化國際經貿領域中，由生產商、代理商、消費者、投資者等為進行和完成相關經貿交易活動而使用的一種英語變體。經貿英

[①] 也稱商務英語、商貿英語、外貿英語等。

語在詞彙、句法、篇章上具有特定表現形式、表達程式、言語規範以及言語規律。依照語用學的「合作原則」與「禮貌原則」，經貿英語在語用方面表現出準確、程式化、簡潔與禮貌的特點。

1.1 準確

準確是運用國際經貿英語的根本。無論是貿易談判還是書信往來，國際經貿英語都力求用詞準確、內容清楚、說明具體、敘述完整。因為經貿活動涉及貿易雙方的責任、權利、利益、風險等利害關係，同時，紀要、備忘錄、函電等文字材料往往又是各種貿易單證、合同合法有效的依據，因此，這些材料的格式、內容甚至標點符號的任何錯誤、紕漏都可能導致意想不到的後果。

例如：Although we are keen to meet your requirements, we regret that we are unable to comply with your request to reduce the prices as our prices are closely calculated. Even if there is difference between our prices and those of other suppliers, you will find it profitable to buy from US because the quality of our products is far better than any other foreign makers in your district. （儘管我們熱切地想滿足你方的要求，然而，由於我們的價格已經精打細算，的確不能再降。即使別人的產品價格優惠一些，購買我們的產品依然是有利可圖的，因為在貴地區銷售的外國產品中，我們產品的質量最上乘。）該段文字有兩個層次。第一層次的主旨是拒絕減價，通過 Although 和 as 引導的條件狀語從句和原因狀語從句，闡述了說話人主觀上良好的出發點與實際困難之間的矛盾，委婉而禮貌地表達主旨。第二層次則是勸說對方購買自己的產品，even if 和 because 兩個從句同樣闡明了價高但質優這一事實，有效地打消了對方的顧慮。全段文章思維縝密，邏輯性強，言之有理。

1.2 程式化（規範）

使用程式化語言是國際經貿英語的第二個顯著特點，也是其實現準確表達的途徑之一。由於涉及貿易雙方的責任與風險、權利與義務，在具有法律效力的文件（包括各種合同、協議以及信用證等）以及貿易信函中的語言有較強的規律性，套語較多。這使其區別於其他語言變體，如廣告語言。前者較多使用專業、半專業詞彙，句型一定；而後者則非常靈活，重在創新，講究文採。

例如：This Sole Agency Agreement is entered into through negotiation between ××× Corporation (here after called Party A) and ××× Company (here after called Party B) on the basis of equality an d mutual benefits to develop business on the terms and conditions set forth below ［經×××總公司（以下簡稱甲方）與×××公司（以下簡稱乙方）通過友好協商，本著平等互利發展貿易的原則，達成獨家代理協議，茲訂立條款如下］。

不過，同法律文件和單證相比，商業信函的程式要簡略一些，主要表現為一些套語、固定搭配及句式結構的頻繁使用。例如：We refer to your letter of (dated)…，Referring your letter of (dated)…，With reference to your letter of (dated)…，（我們收到貴方×月×日的函），又例如：We have for acknowledgement your letter dated…，We acknowledge receipt of your letter dated…，We acknowledge your letter dated…（上述三種結構均可譯為：茲談及貴方×月×日的函）

1.3 簡潔明瞭

近年來，隨著國際互聯網上電子郵件的廣泛使用，人們進一步要求作為信息載體的語言要簡明、實用，一方面能快捷地傳遞信息，另一方面能相應降低通信費用。所以如今的國際經貿英語越來越呈現口語化和非正式化的傾向。其特徵是：

（1）開門見山，直奔主題。

例如：In reply to your letter of May9, we are glad to book your trial order for 500 pcs of car air conditioning corn pressers.（應貴方5月9日的來函，我方欣然接受你方訂單，試訂500臺汽車空調壓縮機。）

（2）刪除不必要的、華而不實的詞句和無關痛癢的修飾成分。

例如：We can not at present entertain any fresh orders for magnet clutches, <u>as we have</u>（owing to）heavy commitments.（由於我們承攬過多，目前不能接受任何新的電磁離合器訂單。）上述句子的畫線部分如果用括號中的內容來替代，就會更加簡單明瞭。

1.4 注重禮貌

商務活動通常是在雙方反覆協商、妥協中達成的。貿易雙方為解決分歧，取得意見一致，往往會遵循禮貌原則（Politeness Principle）和合作原則（Cooperative Principle）。禮貌原則、合作原則較好地解釋了言外之意和間接言語行為。在經貿往來函電中，雙方務必在字裡行間流露出熱情、周到和善解人意的語氣，盡量用客觀、肯定和積極的態度去談論問題，既要維護自身的利益也要顧及對方的面子，切不可一言不慎，傷害雙方的合作關係。表達禮貌的方式如下：

（1）使用請求式代替命令式。在外貿往來函電中，在商務英語中，如果採用命令式，容易冒犯對方，引起對方的不滿和反感，從而影響貿易的順利進行。而請求式則能避免居高臨下，體現雙方的平等地位。因此，常常用「I（We）wish…」「I（We）hope…」「I（We）think…」「I'd like to…」「Please…」「We are glad/pleased…」「We shall be obliged/grateful/appreciated if」等句型來提出要求或表達意願，語氣委婉，觀點鮮明。例如：We hope you will reconsider our offer and fax US your order our confirmation at your earliest convenience.（希望貴方重新考慮我方報價，早日來電訂貨，以便我們確認。）

（2）靈活使用主動、被動語態。通常，漢語具有人稱傾向，主動句式較為常見；而英語多使用被動語態以體現客觀、理性。漢語注重主題，除非強調某人受損或受益，很少出現被動態；而英語注重接受者的感受，被動語態更為普遍。語態不同，語義、語氣也隨之發生變化，並最終產生不一樣的交際效果。在國際商務信函中，常常用主動句明確地表達交易雙方的立場和觀點，句式簡潔明瞭。例如：Part A hereby appoints Part B as its exclusive sales agent in Singapore. 又如：The Seller hereby warrants that the goods meet the quality standard and are free from all defects. 但在某些情況下，如在表示否定含義時，常採用被動語態。因為這樣可避免語氣咄咄逼人，引起反感，從而有利於雙方在融洽的氣氛中建立貿易關係。

（3）採用變通的方式。在商務信函和談判中，為了達成協議，雙方都會為找出解決問題的辦法而提出一些建議和看法。在意見不一致的情況下，任何斷然拒絕的方式，都可能會使貿易雙方失去談判的興趣，嚴重的話，還會導致雙方貿易關係破裂。因此，有必要在語言表達方式上加以靈活變通，以促成交易。例如可通過轉換角度或用副詞 unfortunately、hardly、really、scarcely 等來弱化否定語氣。例如：I'm really sorry that our clutches can hardly satisfy your requirements.（對我方離合器不能滿足你方要求，甚感抱歉。）在此句型中，「really」一詞使語氣更加委婉，讓對方感覺到態度的真誠，而「hardly」則極好地避免了絕對化，雙方都易於接受。

2　國際經貿英語的語言特點

特定的語言使用者、語言使用場合以及使用目的決定了國際經貿英語的語用特點：準確、程式化（規範）、簡潔與禮貌。這些特點具體體現在國際經貿英語的詞彙、句式與篇章、修辭等多個層面。

2.1　國際經貿英語的詞彙特點

準確性是國際經貿英語的語用特點之一。在詞彙層面，國際經貿英語通過使用大量專業術語和專業詞彙，力求準確傳遞雙方信息。這類詞彙語義明確、固定、規範。不過，由於國際經貿英語涉及面廣泛，如工業生產、經濟、貿易、法律等諸多領域，這類詞彙數量多且龐雜。如經濟方面的 economic lot size（經濟批量）、progressive tax（累進稅）；國際貿易方面的 Letter of Guarantee（銀行保函）、standby credit（備用信用證）；金融方面的 foreign exchange market（外匯交易市場）、financial and operating ratios（財務與經營比）等。這類詞彙通常要結合專業課程，通過學習專業知識來掌握其語義的內涵與外延。此外，各專業領域都有相關的專業術語辭典，提供常用的釋義，以確保各方能準確理解，避免不必要的誤解和爭議。

除專業術語和專業詞彙外，經貿英語中還有部分半專業詞彙。它們是由普通詞彙轉化而來的，其特點是既能作為專業術語使用，同時又是普通語言詞彙的一部分。這類詞彙不但使用頻率高，而且絕大多數一詞多義，其語用的準確性常以下面的方式實現：

（1）以語境（上下文或搭配）判斷。半專業詞彙看似和普通英語相似，但由於所用語境不同，意義早已改變。例如，「gas」在普通英語中指氣體，如 such gases as hydrogen and oxygen（如氫氣和氧氣之類的氣體），但在經貿英語中，其語義便不再單一，如 gas station（加油站）、gas cooker（煤氣爐）、gas meter（煤氣表）、gas in a coal mine（瓦斯）、liquefied natural gas（液化天然氣）等。又如 Top quality cigars are being **sold at a premium.**「premium」的普通意義是「額外費用、津貼」,但是在國際經貿英語文本中,該詞就具有語用意義。如1:Just as in that year, when those catastrophes were followed by substantial increase in insurance premiums, insurance are already lobbying for rate relief. 在這一例句中,「premium」一詞是指保險業務中的「保險金」;如2:Some observers contend that, because of the mobility of capital, globalization limits the union workers to achieve a「union wage premium」, thus decreasing the bargaining power of workers vis-à- vis capital. 在此句中,「premium」則應指金融英語中的「溢價」。

　　(2) 以詞項重複避免歧義。半專業詞彙常以詞項重複來避免歧義,使表達準確無誤。詞項重複是指同義詞或近義詞的詞項重複,每組的詞語皆為同義詞或近義詞上下義詞和概括詞的復現等。重複出現的詞項起著補充和完善的作用,能夠避免歧義,起到強調以達到準確無誤的目的。詞項重複在經貿類文書如協議、單證、合同等中較為常見。例如:terms and conditions(條款)、losses and damages(損失)、null and void(無效)、all and any(全部)、modification or alteration(修改或變更)等。

　　此外,半專業術語還包括許多典型套話,又稱行話。它們是在經貿往來中逐漸形成的一種被大家熟悉並普遍接受的表達方式。這些商務套語的格式、用詞用句相對固定,商務人士通過這些程式化的行文用語來傳遞需要表達的內容。

　　國際經貿英語還吸收了不少外來詞、古體詞。外來詞大多來自拉丁語和法語。例如,拉丁語:as per(按照)、re(事由)、ex(在前)、ad valorem(duty)從價(稅)、ad hoc(arbitration)專門(仲裁);法語:force majeure(保險用語,「不可抗力」)、en route(在途中)、vis-à- vis(和……相對)。常見的古體詞主要是以here、there、where 為詞根的詞,例如 hereby、hereinafter、thereby、thereof、whereas、whereby 等。使用古體詞增加了經貿文本正式、嚴肅的意味。不論外來詞,還是古體詞,它們的使用也體現了商務英語語體行文準確、簡潔的要求。

　　國際經貿英語詞彙中還大量使用縮略詞。縮略詞實質是詞的結構的簡化,從語用角度講,縮略詞的使用可以實現交流的簡便、高效和快捷。如 IMF(International Monetary Fund,國際貨幣基金組織)、GSP(Generalized System of Preferences,普惠制)等。

　　專業術語、半專業詞彙、行話、古體詞、外來詞以及縮略詞等的使用,體現了國際經貿英語準確、程式化(規範)、簡潔與禮貌的語用特點。

2.2 國際經貿英語的句法特點

　　「準確、程式化、簡潔與禮貌」的語用特點同樣體現在經貿英語的句法層面。

　　一般講,英語的句型——簡單句、並列句和複合句——在表達語義時有不同的側重。簡單句句子結構簡單明瞭,表達語義突出。並列句以並列連詞連接兩個句子,其語義具有相同的重要性,且彼此發生關聯。主從復句以從句語義為輔,重點突出主句的語義。

　　作為專門用途英語之一的國際經貿英語,涉及的大多是國際商務和經貿領域,

專業性強，且關乎雙方的責任與義務、利益和風險，這要求其文本在行文上需嚴謹，文意嚴密、細緻。所以，在句法層面上，國際經貿英語文本在靈活使用簡單句和並列句的同時，較多地採用了英語長句，以表達該專業領域嚴謹、細緻且複雜的概念和關係。

英語長句是在基本句型的基礎上使用非謂語動詞短語、邏輯性定語、並列句和主從復句等，擴展句子的修飾、限定和附加成分，從而可以表達非常嚴謹、細緻、複雜的概念和關係。英語長句雖冗長，但是信息量大；雖讓人眼花繚亂，但邏輯層次清晰。

例1：Microeconomics examines the behavior of basic elements in the economy, including individual agents and markets, their interactions, and the outcomes of interactions.（個體經濟學考察經濟主體的各種行為，包括個體行為人和市場，他們之間的相互影響以及相互影響的後果。）這個長句有兩個層次：①主幹部分是 Microeconomics examines the behavior of basic elements in the economy，其餘的為次要成分。②第二個層次是伴隨狀語：including individual agents and markets, their interactions, and the outcomes of interactions.

例2：As an important sub-field of economics, the basic focus of finance is the working of the capital markets and the pricing of capital assets, and its methodology is the use of close substitutes to price financial contracts and instruments, especially applied to value instruments whose characteristics extend across time and whose payoffs depend upon the resolution of uncertainty.（作為經濟學的重要分支，金融依靠替代品對金融合約和工具，特別是那些具備跨期和基於解決不確定性問題來進行結算的工具進行定價，以實現其資本市場營運與資本定價的功能。）這個句子有三個層次：①主幹層次是由 and 連接的並列句；②在 and 連詞之前是一個使用並列結構的複雜單句，其主幹部分是 the basic focus of finance is the working of the capital markets and the pricing of capital assets；and 之後是帶非謂語結構的複雜單句，其主幹部分是 its methodology is the use of close substitutes to price financial contracts and instruments。③第一個單句帶有一個介詞結構 As an important sub-field of economics；後一個單句帶有一個非謂語結構 especially applied to value instruments whose characteristics extend across time and whose payoffs depend upon the resolution of uncertainty.

2.3 國際經貿英語的篇章特點

「準確、程式化、簡潔與禮貌」的語用特點還體現在國際經貿英語的篇章層面。

國際經貿英語涉及的領域廣泛，其文本也種類繁多，如在國際商務中的合同文本、對外經貿函電、信用證等。在篇章層面上，都有一定的行文結構要求，以符合該專業領域交往的常見邏輯與規範。

以貿易雙方的通信為例。首先在格式方面，除遵循一般的通信格式外，貿易雙方通常會對來往信函加編號或分類，以使來往信函能得到及時關注、高效處理，也方便日後查閱。一般以信函討論的主題為依據分類，或在信函上標明主題詞，或是

進行編號。編號常和日期一起提及。Your ref：（你方編號）Our ref：（我方編號）常置於信函內公司抬頭下方。主題詞是在稱呼語下方，以名詞短語的形式進行標註。

例：

Dear Ms. Philips,

Translation Services & Fees

Thank you for your letter of 22 January enquiring about our translation service.

同時，貿易雙方之間信函往往會根據主題而有約定俗成的內容格式。如建立貿易關係的信函內容包括：介紹從何處獲得對方公司信息，介紹本公司業務，表達合作願望；報價函的內容包括：介紹公司經營範圍/提及對方詢價函，報價，對達成交易的期望；還盤函包括：感謝對方的報價函，對某商品感興趣以及對其報價的看法，對達成交易的期待等。信函內容僅為經貿業務，任何無關內容均不涉及。

2.4 國際經貿英語的修辭特點

一般認為以「準確、程式化、簡潔和禮貌」為語用目標的國際經貿英語不會涉及修辭。但是，國際經貿英語中也存在著一些經貿市場特色詞彙和表達，它們被用於描述經濟貿易發展趨勢及市場方面的變化，形象而生動，具有擬人的修辭特點。如，描述價格下跌的詞彙有 decline、decrease、drop、fall、dip、dive、plunge、slip、go down；又如 jump、leap、rise、recovery、surge、increase、climb、soar、grow（描述價格上漲）、end high、low、down、steady、firm、soft、quiet、mixed（描述市場收盤情況）。此外，還為敦促對方接受報價或盡快下單而使用一些誇張的用法，如「Our quotations have been accepted by other clients. A greet of orders have been bargained and large inquiries are **pouring in** our company so we cannot cut down our price.」「For your suggestion, we have seriously considered, although we acknowledge that the price can be reduced, but your price is too low. We are ready to compromise. Our new offer is ×××, which is **our bottom line**, I hope you can consider accepting.」

又如：「Our user has regular demand for this commodity. Please quote us **your most favorable price.**」

「Thank you for your letter. For your proposal, we agree to accept your new price. We believe that **this cooperation will give us a special harvest.**」

附錄 2
國際經貿英語的學習方法

1 如何聽懂財經新聞廣播

對於金融界、財經界人士而言，獲取信息是極為重要的。能否在風雲變幻的商場上得到第一手信息是決定一個人能否迅速、正確地做出商業決策的重要因素之一。新聞作為一個十分重要也十分常見的信息通道，其重要性不言而喻。而英文財經廣播因其時間限制，更需要做到語言嚴謹、簡潔、專業，同時還要保證內容廣泛，涵蓋經濟、金融、商業、管理等與日常生活息息相關的各方面的信息。如何打下堅實的聽力基礎，熟練地從英語財經新聞中獲取必要信息，對有志於從事國際經濟貿易的人而言相當重要。

1.1 財經新聞報導的特徵

財經報導的主要功能是傳遞信息（transaction of information）。及時瞭解、準確掌握英文財經報導的信息內容，除需具備一定的專業知識外，還應該對這類報導的語言運用規律和特徵有所認識。總體而言，英文財經報導從個別字詞的選用到全篇內容的陳述都體現了新聞報導慣有的及時、客觀、新穎等特點，另外，由於專業上的要求，財經報導在形式和內容等方面又體現出自己獨有的特徵。

語言學上一般把語言的功能分為五類：信息功能、表情功能、指示功能、寒暄功能（phatic function）以及美學功能。對於財經新聞而言，毫無疑問，其主要功能是信息功能。為達到在最短時間內向目標受眾傳遞最大量的信息，新聞必須以最短的篇幅包含盡可能多的信息。同時，由於時間的限制，要字斟句酌也不現實。因此，財經新聞與一般的新聞相比較，前者更具即時性，否則就無法反應瞬息萬變的財經世界，更無從談及傳遞信息了。這些在句法上主要表現為慣用擴展的簡單句、常省略名詞性從句中的關係代詞、較多使用數字和比較結構、大量使用引語四大特徵。

（1）慣用擴展的簡單句

例如：「NEW YORK - Stocks roared, with the technology-laden Nasdaq market posting its biggest gain ever, after the Federal Reserve threw a lifeline to Wall Street with an unexpected interest-rate cut to keep the world's largest economy from slowing too much.」

（紐約——股市上揚，高科技股集中的納斯達克股市在美聯儲向華爾街出乎意料地拋出突然宣布降息這根救生索後迎來了其最大一次漲幅。美聯儲此次降息旨在防止美國經濟發展進一步滯緩。）從上例可以看出，其基本結構是一個簡單句（stocks roared），但通過使用從句和介詞短語作為修飾成分，將這一簡單句予以擴展。例句40個單詞，構成一個段落，句中有插入語、時間狀語從句及目的狀語，這些成分都是為了補充說明相關的情況和消息。這是因為英文財經新聞稿沒有太多的時間進行語言上的推敲，插入成分通常比較多。上文中 with the technology-laden Nasdaq market posting its biggest gain ever 即插入語，從句中的 with 引導的介賓結構也多達15個單詞，其目的是要說明美聯儲的做法及最終目的，是個相對重要的信息，但並未單獨列句，而是以介賓短語作為方式狀語及不定式作為目的狀語這樣的附屬結構體現出來。大量使用含有冗長修飾成分的簡單句便成了財經新聞英語的一個重要特徵。

（2）名詞性從句中常常省略關係代詞

財經新聞英語具備口語化傾向這一特徵，該特徵除了表現在句式上多採用擴展的簡單句外（如前例所示），還表現在臨時插入的從句前省略了連接詞，顯得與主句的關係不夠緊湊，有口語化的特點。例如：「The time for release of the statistics, normally in the afternoon, was moved up to 8:50 a.m., ministry officials said, to allow Japanese markets to react first to the numbers, as is the practice in most industrialized countries.」（Bloomberg Sep. 18, 1995）（政府部門官員說，原本在下午公布的經濟數據將提前到上午8:50公布，是為了讓日本國內市場首先對這些數據做出反應，這也是許多工業化國家採取的慣例。）這句話中的定語從句「normally in the afternoon」=「which is normally in the afternoon」，同位語「as is the practice in most industrialized countries」=as that it is the practice in most industrialized countries，連接詞「which」和「that」都被省略了，因此句子更傾向於口語表達特點。

（3）比較結構在財經新聞英語中被普遍使用

與一般新聞不同，財經新聞經常涉及價格、百分比、總值、平均值等各類經濟統計數據，而且這些數據之間的比較或參照因為可反應經濟情況在一段時間內的變化趨勢，往往成為財經報導的重要組成部分。因此，比較結構在英文財經報導的句型中佔有很大的比例，這也構成了財經新聞英語句法上的第三大特點。比較機構就其比較雙方的關係來劃分，可分為三種主要類型：一是不對等比較，即比較的一方大於另一方，根據不同的語境，可用的形容詞副詞比較級包括 less、higher、lower、narrower、smaller 等，這些形容詞副詞比較級的運用可形象而貼切地反應經濟在不同階段不同地區的區別或經濟發展變化的趨勢。例如：The corn price in the corn-growing provinces in northeastern China is comparatively high, at 1,000 RMB (US $120.96) per ton in Heilongjiang, higher than those of harvests in 1999 and 2000.［中國東北各玉米主產省份玉米價格相對走高，黑龍江省玉米價格達1,000元人民幣每噸（120.96美元每噸），比1999年和2000年都高。］二是用形容詞或副詞最高級作為限制語，達到強調的效果。例如：The blue-chip Dow Jones industrial average climbed 299.60 points, or 2.81 percent, to 10,945.75, its eighth-largest point gain ever and its strongest close in two months. （藍籌股道瓊斯平均指數攀升了 299.60 點，或 2.81%，達 10,945.75 點，為歷史第八大漲幅，也是兩個月來道瓊斯指數最高收盤價。） 三是通過介詞連接比較的兩個方面，強調發展變化的趨勢。常用的介詞包括 against、opposite，詞組有 compared with、in contrast to 等。例如：Secretary Evans acknowledged that oil surplus capacity, which has dropped considerably compared with the 1980s and 1990s. （秘書長伊萬承認石油順差額與 20 世紀八九十年代相比有很大幅度的下降。）

（4）大量使用引語，增強報導客觀真實性

新聞報導特別是財經新聞，客觀真實是最為重要的。現代人已與經濟生活密不可分，人們關注財經新聞，因為它能給予及時有效的信息。財經新聞因此格外追求客觀性效果，給讀者以權威感，因而它的正文中往往採用大量的引述。這些引述可能來自政府官員、市場人士、分析評論專家等，可能是直接引述，也有可能是間接引述。其所有的目的都是最大限度地體現其客觀性和生動性。例如：Winfried Vahland, CEO of VGC Company expressed optimism about the development trend in the second half of 2007,「Skoda Octavia, the new fashion of Volkswagen had a good start and continuous orders. Another style of Magotan which led market trend would be launched after a few weeks and laid a good foundation for the sales in the second half of the year.」「大眾汽車集團（中國）總裁範安德博士表達了他對 2007 年下半年市場的樂觀態度：「斯柯達歐雅——大眾的新車款，已經有了一個好開端和源源不斷的訂單。另一款引導市場導向的新車型邁騰也會在幾星期後進入市場，這必然會給今年下半年的銷售業績打下堅實的基礎。」」在這個句子中，只有剛開始部分是報導者的闡述，接下來部分全是直接引用當事人的話來佐證報導者的觀點，凸顯報導的真實性。

1.2 財經新聞聽力學習策略

在對財經新聞英語特點進行分析的基礎上，大家也要瞭解自己在財經新聞英語聽力學習中會遇到的困難，以便根據自己的特點進行專門練習。英語新聞聽力理解是個信息處理的過程。從信息的輸入角度看，可以分為接收信息和理解信息。信息的交流有六個因素，即信息源、轉換器、信道、收信者、信宿和噪聲。學習者收聽英語新聞的過程實際上就是對信息的「接收」和「建立意義的過程」，即接收語音信號，與原已儲存在大腦中的語言知識、背景知識建立起聯繫，並進行解碼、辨認、分析和歸納的過程。這一環節被稱為「建立意義的過程」（construction process）。一般來說，在聽英語新聞時，聲音信號是一連串接著一連串傳入大腦的，所以大腦的解碼、辨認、分析、歸納和意義建立的過程是動態的過程。在這個動態過程中，語言知識、背景文化知識以及解碼、辨認、分析、歸納和理解速度制約著學習者對新聞的理解，成為制約聽力的主要因素。語言知識包括詞彙量、語音知識和語法知識，它是決定聽力理解成敗的關鍵。在這一因素中，詞彙量起了決定作用。尤其對財經新聞英語來說，學生不僅需要充足的普通詞彙，還應具備相當數量的專業詞彙，如經濟、金融、保險等專業術語。有時即使學習者對某個句子在文法上不存在問題，但由於缺少背景文化知識，同樣也理解不了其含義。財經新聞英語涉及範圍廣，包

括天文、地理、科技、文化、政治、軍事、經濟、生活等方方面面，它們對聽力理解的影響極大。解碼、辨認、分析、歸納和理解速度慢也是影響聽力的重要因素。英語新聞播放的句子是連續的，在詞與詞、句與句之間好像是有停頓，音段與音段之間好像有界限，但實際上並不十分明顯。從理論上講，一條完整新聞內容的播出是一種不斷變化的信號，由詞素、詞、短語、從句、句子和語段等部分組成，它們從耳朵傳輸到大腦的過程是連續性的，同時也是一次性的。更有甚者，音素之間、詞與詞之間是前後依賴的，這就產生了「平行傳遞」（paralled transmission），即幾個音素幾個單詞信息幾乎平行地傳遞入耳朵。這種傳遞方式增加了學習者對語言解碼、辨認、分析、歸納和理解的難度。

以下聽力學習策略能對同學們的學習起到相當大的幫助：

（1）培養良好的聽力技巧

首先，嚴把語音關。聽力的基礎是語音，一個語音不標準的人是無法完全聽懂語音純正的英語的。學習者應以此為基點，訓練自己的語音，如連讀、弱讀、同化、爆破等語音現象，並熟悉英式發音和美式發音的區別。在聽的同時可要求自己模仿發音、跟讀新聞，加深一些特定的發音現象在大腦中的印象以幫助培養對語音的辨識能力和敏感度。

其次，記錄要點。這要求學習者有篩選關鍵詞的能力，以關鍵詞的形式記錄下主旨。可以鍛煉自己用一兩個詞記下一句話的意思，用一兩個字母代表一個詞，利用聽力材料中間隔停頓的時間迅速鞏固之前的短期記憶內容，幫助自己在聽力中加快筆記速度，指導自己學習、總結和探索一些速記方法。英文速記的方法可以是首字母或縮略詞，如：a. m. morning；etc. and so on，也可以是常用符號，它們均可以簡化筆記方式、節約記錄時間。

最後，做好聽前準備。瞭解背景知識或熟悉聽力語篇話題有助於提高外語聽力理解能力。這一點在新聞聽力學習中尤為重要。如果有了足夠的相關信息激活，則容易產生聯想，就更能把握新聞內容。新聞報導力求準確，往往羅列大量事實、數據，詳細地報導事件的來龍去脈，因此新聞裡的人名、地名、組織名、時間、內容、目的、結果、原因等均應引起同學們的高度注意。做好聽新聞之前的準備工作，不僅能幫助聽者預測新聞內容，減少聽力學習的盲目性，更可以讓同學們在一開始就聽得懂或基本聽得懂新聞大意，有利於保持大家的學習興趣，克服畏難心理。

（2）課堂訓練和課下累積相結合

課堂的核心任務是將學生聽新聞主題與細節訓練相結合。要培養學生聽新聞主題，抓住主旨的能力，首先需要教師設計關於主題的問題供學生回答練習。老師可以訓練學生的猜測能力，除通過導語信息猜測下文內容外，對於稍長的新聞材料也可採取先播放新聞的前半部分，鼓勵學生猜測接下來可能發生的事情，並在聽完材料後驗證是否準確。教師也可以對剛聽過的新聞材料做一定的篇章結構分析，使學生清楚新聞報導的題材和句型特點。在主題討論之後，開展對細節信息獲取能力的訓練。根據題材的不同，要分清細節主次，判斷出哪些是所需的關鍵信息，哪些是次要信息，並完成相關練習題目。另外，對涉及聽力材料的背景及專業知識還要多做

介紹,有意識地擴大學生的知識面,加深學生對英語國家現狀及文化的瞭解,培養他們的學習興趣。最後可以引導學生獨立歸納要點並對新聞進行口頭復述,進一步強化學生的綜合概括能力及語言表達能力。課後學生應主動拓寬瞭解英語新聞的途徑。如多讀英文期刊、報紙或瀏覽英語學習網站,堅持收聽收看英語廣播、電視,包括 VOA、BBC 及中央電視臺國際頻道等。同時同學們應該多瞭解當今世界熱點問題,把握新聞報導的句型結構和篇章結構,大量接觸世界各國的政治、經濟、軍事、外交、宗教、歷史、地理、國際關係、社會習俗和風土人情等知識,提高個人閱讀新聞報導的水準和興趣。應同時要求自己不斷累積財經專業詞彙,如財經新聞報導常出現的專業術語、動詞以及常見的地名、各國首腦及主要頭銜、各國主要行政機構、國際組織等一系列專用名稱,形成讀—聽—讀的反覆循環。

(3) 利用網路資源,建立財經新聞英語語料庫

語料庫 (Corpus) 是按照一定的語言原則,運用隨機抽樣的方法,收集自然出現的連續的語言運用文本或話語片段而建成的電子文庫。語料庫提供給學生的語言數據具有兩大特徵,一是高質量輸入,二是大數量輸入。語料庫的應用能夠提高學習者的語言直覺,鍛煉其處理語言變體 (variation) 的能力,能夠幫助學習者掌握地道的語言。語料庫是語言教學潛在的豐富資源。目前世界各國主要媒體都在網上開設了網站,80%以上的新聞是以英語為載體的。同學們應當熟練掌握網路工具及操作技巧,密切跟蹤網上浩如菸海的財經新聞,敏銳捕捉財經新聞英語變化的新動向,為自己提供一個豐富、真實的語言學習環境,指引自己基於大量的語料庫數據觀察、概括和歸納語言使用現象,自我發現語言規則及語用特徵。該語料庫的構建應緊密聯繫當前現實各國經濟發展前沿和熱點問題,在一段時間內有意識收集、整理和記憶財經專業術語,甚至包括國家名、經濟事項,在此基礎之上,把題材類似的新聞集結起來,在一段時間內集中收聽、反覆收聽,熟練地掌握一個題材以後,再轉向另外一個題材。

(4) 精聽泛聽相結合

精聽時,同學們在聽的過程中需解決播音員的語速快和大腦對語音、詞彙和句子的解碼、辨認、分析、歸納和理解的速度慢的矛盾。精聽的目的是進行連音及省音適應性訓練,突破語音關,熟悉語音特點,習慣快速自然話語的聽覺形式,並提高大腦言語聽辨能力和聽辨速度,同時掌握大量的相關詞彙。英文廣播財經新聞的語速一般在每分鐘 150 個單詞左右,播音時間每次 10 分鐘。在基礎階段,有條件的學習者可把當天的新聞用電子設備錄制下來,再反覆聽,借助辭典,直到把每一個詞、每一個短語都聽懂。沒有條件的學習者可以反覆聽當天的新聞。以美國的 VOA 和英國的 BBC 等電臺為例,它們的新聞一般都滾動式播出,即每個整點時間都重複當天第一次播出的內容。大家可以重複收聽,從孤立音素、詞彙、語句的聽辨,到連續語音、詞彙、語句的辨別,把連續、快速的語言形式儲存在大腦中,形成語言習慣,以提高解碼、辨認、分析、歸納和理解的速度。泛聽,就是大量輸入。大量輸入總的來說是以廣泛接觸為宗旨,自然輸入、自然吸收,不強記硬背,更不要求接觸的全輸入、輸入的全吸收。每天保證用足夠時間大量地聽,保證輸入的密集性,

不斷適應語速、語流、提高反應速度，並運用大腦已儲存的語音知識、詞彙知識、語法知識和背景文化知識對所輸入的內容進行解碼、辨認、分析、歸納和理解。精聽在初級階段尤為重要，泛聽也是必不可少的學習手段。精聽和泛聽是相輔相成的。

（5）建立視、聽、說、寫四位一體學習體系

英語學習本來就是一個系統工程，聽、說、讀、寫、譯相輔相成，缺一不可。因此，全方位、多角度建立一個系統的學習體系至關重要，也適用於整個英語學習過程。學習者可以根據聽力材料的不同，運用圖片、PPT、新聞視頻等直觀手段創設情境，幫助個人理解語篇大意、句子要點和段落重點。例如：VOA 中相關國際股市波動的報導，可以通過多媒體手段，幫助理解股市相關領域的生詞，掃除聽力障礙。此外在練習英語新聞聽力的同時，可以先設計好問題。在聽力訓練之後，檢查自己是否可以復述新聞內容，並對新聞展開評論，甚至可以對一段難度較大的材料進行背誦和默寫，在此基礎上反覆精聽，做到字字聽懂、句句明白。

另外，在平時的聽力練習中，學習者可以用便於自己識別的符號或關鍵詞語邊聽邊寫，對重要信息進行記錄，鍛煉自己的速記能力。也可以利用語料庫中語速較慢的材料，逐句播放、逐句聽寫，增進自己對材料的理解和精聽的能力。

總而言之，財經英語聽力的學習、提高是一個漫長的過程，需要同學們的耐心和堅持不懈的努力。同學們可以在瞭解財經新聞用詞、語句特點後，根據自身需求和各自面臨的重點難點，在老師的幫助下制定適合自己的學習方案。通過有組織有系統有計劃的嚴格訓練，相信同學們在財經英語聽力方面能有明顯的進步和長足的發展，為同學們今後的學習與發展打下良好的基礎。

2 如何閱讀財經類英文雜誌和報紙

西方財經貿易類報刊，較有代表性的有 Financial Journal（《金融時報》）、Wall Street Journal（《華爾街報》）、Business Week（《商業周刊》）、The Economist（《經濟學人》）和 Fortune（《財富》）等。其他的綜合性報刊有 Washington Post（《華盛頓郵報》）、New York Times（《紐約時報》）、The Times（《泰晤士報》）等，此類經貿報刊中的文章不僅是國際經貿專業學生學習專業英語的途徑之一，而且是各企業界經濟界人士瞭解西方乃至全球財經貿易發展動態的窗口。它們除了具有一般新聞報導文體的特徵外，還具有內容高度專業化、背景複雜化及其反應在語言上的複雜和高深等特點。要讀懂這些報刊上的文章並非易事，除了需要讀者具有一定的英文基礎外，還需要具有較豐富的時事、經濟、政治、歷史和文化背景知識，同時需要瞭解這類報刊中的文章的語言特徵，並掌握一定的閱讀技巧。

2.1　財經英語報刊文章特點淺析

（1）財經英語報刊文章結構特點

通常一篇完整的英語財經報刊文章由標題、導語和正文組成。標題具有引導讀

者選擇信息的功能，在報刊文章寫作中的重要地位不言而喻。所謂導語就是導讀之語。一般報紙上的重大新聞報導和評述都有導語。導語主要是對主題進行提煉、概括，起到引導讀者的作用。一篇好報導應該是在利用導語將讀者吸引住以後，通過同樣引人入勝的文字和具體內容牽動讀者的心，從而使讀者把文章一氣讀完。讀者看完導語段後，就知道了文章的關鍵信息，接下來就是報刊文章的正文，即向讀者詳細交代信息和評述觀點的文章主體部分。報刊文章正文一般採取「倒金字塔」結構來寫作，就是先開門見山，把主要的內容說出來，以便抓住讀者，然後才逐步做進一步的解釋和說明，介紹有關背景知識等。在先後排列順序上，一般是把比較重要的和與主要內容關係密切的材料排在前面，次要的和關係不太密切的排在後面。報刊文章的這種篇章結構是和讀者的需要密切相關的。報刊種類繁多，而讀者時間有限。如果報導的開頭是與主題關係不大的開場白，讀者就不會再往下看。如果在報導一開始就點出主要內容，讀者很容易做出是否再往下看的決定。即使不再往下看，讀者也瞭解了主要內容，報導的基本任務也完成了。以下將通過對財經報刊文章各部分的介紹來讓同學們對財經新聞報刊文章有所瞭解。

①標題。標題即文章的題目，用來簡明扼要地向讀者揭示文章的主要內容，使讀者在最短的時間內獲得盡可能多的信息。由於報刊版面有限，標題應簡潔搶眼、表意確切和生動活潑。其標題一般具有以下特點：A. 為了達到簡潔的效果，標題中常出現名詞短語、形容詞短語、動詞不定式和分詞短語等特殊用法。例如 Government Web Sites Invaded（政府網站被黑客入侵）（用過去分詞表被動）。B. 有些標題由一個簡單句構成。就句型來看，有陳述標題、疑問標題、和引述標題等。標題中可以使用逗號、冒號、引號、破折號和問號，但一般不用句號。例如：*Economic Troubles Cloud New Market's Future*（經濟困境使新市場前景蒙上陰影）（陳述標題）、*Is the dollar too Weak or Too Strong?*（美元是太疲軟還是太堅挺?）（疑問標題）*surplus is a「Time Bomb」EC warns Japan*（歐共體警告日本，順差猶如定時炸彈）（引述標題）。C. 為了使標題顯得生動和有吸引力，還可以借用成語或習語。例如：*No Pain, No Gucci*（沒有付出就沒有Gucci）（引用了 No Pain, No gain 的習語）、*A Tale of Two Debtors*（兩個債務國的故事）（借用了狄更斯小說 *A Tale of Two Cities* 標題）。D. 有時為了突出重點，在版面允許的前提下，標題可以做成複合結構，就是說不但有主標題還有副標題，這在財經英文雜誌和報紙中經常出現。副標題通常位於主標題的下方，印刷字體小於主標題，以示區別。副標題的功能主要是對主標題進行補充和解釋。例如：*Why no longer no alluring? Investors are intriguingly wary of America's recovery*（為什麼不再如此有吸引力？投資者對美國的復甦抱有懷疑）、*Accounting for change. The need for radical audit reform in America grows ever more pressing*（會計要改革。完全徹底的審計改革在美國勢在必行）。

②導語。導語一般分為概括式和延緩式。概括式導語，又稱硬導語，它要求開門見山、一語中的，用簡潔的語言將最重要的新聞主題告訴讀者，多用於時效性強的硬新聞。延緩式導語，又稱軟導語，注重文字駕馭技術，多以通過具有文學化風格的懸念來吸引讀者，多用於軟新聞。一般來說，報紙類文章採用概括式導語，顯

然雜誌類文章相對於報紙類文章來說採用的是延緩式導語。按照英文報刊編排的習慣，一般消息導語要求在 35 個單詞以內。作為雜誌採用的軟新聞，導語可長一些。但萬變不離其宗，導語就是導語，只是起到引導的作用，不可本末倒置。

③正文。從結構安排來看，前面已經提到，英語經貿報刊文章正文一般採用「倒金字塔式」即按照重要性遞減的順序來組織材料，把主要的內容先說出來，然後才逐步做進一步的闡釋和論證。正文有以下幾大特點：A. 為了突出報導的及時性和全面性，記者要在盡可能短的時間內發出盡量多的消息。為了使句子包含最大限度的信息量，記者通常會盡量拉長一個句子。另外，財經類文章本身內容複雜，需要有與之相應的表達形式，而且其作者是既具有良好的語言駕馭能力同時又熟悉財經領域專業知識的專業寫作記者，讀者又多為受教育程度較高的經貿管理人員、政府經貿官員以及專業研究人員等，使得財經雜誌和報紙的文字具有高深的特點。所以英文財經報導中的句子偏長，邏輯關係複雜，結構比較鬆散，插入成分通常比較多。例如：「World Bank Managing Director Richard Frank said in a statement that the ＄52 billion international rescue package assembled by the Clinton administration——which includes ＄17 billion from the International Monetary Fund——would meet Mexico's short-term financial crisis, which blew up after the government devalued the peso in December.」（世界銀行行長理查德‧弗蘭克在一項聲明中說，由美國克林頓政府組織籌集的 520 億美元的國際援助貸款，其中包括國際貨幣基金組織提供的 170 億美元貸款，將用於解決墨西哥目前面臨的短期金融危機。這次危機是墨西哥政府在 12 月將墨西哥比索貶值後引發的。）這一句中「that」引導的賓語從句中插入了兩個定語從句，都是為了補充說明相關的消息，包含了大量的信息。B. 財經類報紙雜誌經常涉及各類經濟數據、市場價格及其他有關的統計數字，這些數據之間的比較是財經報導的一個重要組成部分。因此，比較結構在英文財經報導的句型中佔有很大的比例。

④詞彙特點。從詞彙來看，英文財經報紙雜誌正文也有一些特點。包括：A. 專業名詞和專業術語。英文財經報紙雜誌文章屬專業性文章，因此其語言的一個最大特徵即是對專業名詞和專業術語的大量使用。其不僅涵蓋了經濟、貿易、投資、貨幣金融、管理、行銷、商品市場等各方面的基本專業詞彙，而且許多常用詞被賦予特殊的含義。例如：long（多頭）、short（空頭）、call（看漲）、put（看跌）等。另外，隨著科學技術的發展和經濟貿易方式的不斷發展變化，新的詞彙也層出不窮。如果不瞭解這些詞彙在專業上的特殊含義，就可能無法真正透徹理解有關的報導和消息。B. 外來語、俚語、口語的使用。財經英文報導為了體現新穎、通俗等特點，也經常用一些外來語、俚語、口語表達形式以及比較生動、靈活的小詞。例如：The challenge for Chun is to rein in the chaebol without damaging their absolutely essential contributions to the Korean economy——and without stepping on the toes of the chaebol's enormously powerful chieftains.（全斗煥總統所面臨的挑戰就是既要控制住那些財團，同時又不要傷害到它們對韓國經濟絕對重要的貢獻，也不要得罪財團中勢力強大的頭頭腦腦們。）其中「step on the toes of」是俗語，意為「觸犯、得罪」。C. 比喻和象

徵。比喻和象徵等修辭手段也常見於英文財經報紙雜誌中，可以使得文章更加生動且具有趣味性。比如：Since he was made manager, the company has been running in the black.（自從他當上經理之後，公司一直盈利。）be in the black 是 be in the red 的反義詞。這兩個術語都來自記帳時所用的墨水的顏色，黑字象徵盈餘，紅字象徵虧損，俗稱「赤字」。又比如 Pretty soon, experts say, cell phones and the Internet will go together like peanut butter and jelly.（專家稱，很快手機和因特網將像花生醬和果醬一樣互相依存了。）這裡將手機和因特網的關係比喻為花生醬和果醬的關係，即為一種互為補充、相互依存的關係，文章既顯得通俗易懂，又顯得生動活潑。D. 財經報紙雜誌文章中用詞的另一個突出特點是形容經濟數據及市場行情變動的詞通俗、形象。比如表示「趨平」意思時，會用到 level off（持平）、reach a plateau（達到平臺階段）、remain constant（保持不變）、hold steady（持穩）、stand at（數量維持在）、hover（徘徊）。表示波動時，會用到 fluctuate wildly（劇烈波動）、seesaw（波動）、move up and down（上下波動）、rise and fall（興衰）、booms and busts（盛衰）、ups and downs（起伏）。

2.2 英文財經報紙雜誌閱讀策略

通過上文的學習，大家對今後學習中會遇到的英文財經文章的特點就有了一個大致的瞭解。為了幫助同學們更好地學習，早日能順暢流利地從英文財經報紙雜誌中獲取有用信息，這裡列出一些非常有用的學習方法。

（1）端正心態，培養長期閱讀英文財經報紙雜誌的興趣和習慣

俗話說，興趣是最好的老師。閱讀英文財經報紙雜誌一開始一定是一件枯燥而又困難的事情。但是，作為新時代的大學生，面對國家對於大批具有國際視野、通曉國際規則、能參與國際事務與國際競爭的國際化人才的需要，時刻關注世界經濟動態，既是大學生學習的需要，也是擴大財經視野、培養國際化思維的重要途徑。因此，我們要提高對閱讀英文財經類報紙雜誌的必要性和重要性的認知。在日常生活與學習中，同學們可以通過建立閱讀分享小組，定期共同閱讀、共同分享、參與小組討論和舉行辯論會等方式來培養閱讀興趣，增強自己閱讀英文財經類報紙雜誌的能力。

良好的心態、堅定的意志和良好的閱讀習慣是提高英文財經報紙雜誌閱讀效率的基本保障。同學們應該堅持沉著冷靜地閱讀和思考，自覺抵制將英語下意識地翻譯成母語的習慣，加快閱讀速度。除此之外，同學們也要找到最適合自己能力和習慣的每日閱讀時間表，用自己喜歡和習慣的方式去獲取信息，閱讀英文財經類報紙雜誌的能力就將在潛移默化中提高。同時，同學們畢業後將從事與國際經貿相關的工作，如果要在繁忙的工作間隙這種相對有限的時間裡盡可能多地搜集到信息，養成每天定時定量閱讀英文財經報紙雜誌的習慣也會對大家有積極的作用。

（2）瞭解專業背景，豐富專業詞彙

所有財經報紙雜誌的文章內容都離不開當時或者前後一段時間裡特定的政治、經濟和社會背景。讀者如果對這些背景知識缺乏瞭解，即使有很好的英文功底，也

不可能完全準確地領會、理解文章中所要傳達的相關經濟信息。如果能夠理解相關的政治和經濟方面的背景知識，就能在一定程度上幫助我們對文章進行理解和掌握，甚至在一定程度上能彌補語言能力上的不足。同學們上課時應該認真學習自己的專業知識，通過各種手段來豐富自己的專業知識。同時，除了平時關注經貿有關知識外，與專業相關的管理、金融、法律等專業知識也需要涉獵。因為財經英文報紙雜誌上的文章都有很強的綜合性，通常包括不少的專業背景知識，所以，同學們一定要不斷學習以掌握這些專業基礎學科知識，從而提高自己對專業背景的瞭解和財經相關內容的敏感性，活躍自己的思維。除此之外，同學們還應該緊跟當今世界發展形勢，廣泛涉獵世界經濟、政治等方面的各種知識，把握時代潮流。

許多同學剛開始閱讀英文財經報紙雜誌時覺得很困難，有相當大一部分原因是因為個人掌握的專業詞彙不足。由於專業詞彙量的限制，不能夠準確、快速、全面地理解文章內容，或者由於對一些詞彙沒有完全掌握，只有片面瞭解，導致對文章內容理解的錯誤。因此，同學們必須通過各種渠道來拓展自己的詞彙量，比如通過前文介紹過的收聽財經新聞，同學們可以找到新聞原文，在完成聽力任務之後對照字典將不會的專業名詞查找出來，詳細瀏覽它們的專業意義，從而達到拓展自己詞彙量的目的。由於英文財經類報紙雜誌涉及的業務範圍十分廣泛，所以業務詞彙也顯得很豐富。同學們也可以先選擇一個特定的主題，根據該主題搜集英文財經報紙雜誌文章來閱讀，並在閱讀過程中找出自己不認識、不熟悉的專業詞彙，通過查字典等方式熟悉它們。這樣一來，同學們就能掌握與特定主題相關的一系列專業詞彙了。這樣一來，同學們不僅拓展了詞彙量，掌握了相關主題的核心詞彙，而且能更加深刻地理解該特定主題。不過同學們一定要注意：必須要準確理解並掌握自己所知道的每一個專業詞彙和用法。只有做到了這點，才能準確、全面地獲取英文財經報紙雜誌中的信息。隨著閱讀量的增大，詞彙量也會不斷擴大，閱讀速度就能不斷加快，閱讀過程也就能更加簡單。

（3）把握文章總體結構，剖析閱讀長句

英文財經報紙雜誌中文章的主題非常清晰，結構很完整，邏輯也很嚴密。因此閱讀此類文章，不能僅僅局限於對單個段落、句子以及某個詞彙的理解，而忽略了整體結果和邏輯思路。那樣可能引起不完整、片面地理解作者想要表達的意思，甚至有可能會曲解作者的真實意圖。從整體對結構以及思路進行把握，還可以大大提高閱讀效率，迅速找到重要的信息。通常而言，想要把握整體結構，我們應該根據前文分析過的英文財經報紙雜誌各部分入手。

在分析財經新聞的時候提到過，財經類文章需要在較少的句子中包含盡可能多的信息，因此在英文財經報紙雜誌中也會遇到一些結構複雜的長句，給讀者的閱讀帶來一定障礙。但是，如果同學們能夠靜下心來，認真分析和把握句子結構，也能領會到作者的意思。一般而言，分析長句的基本方法有如下這些：

第一，找出句子主幹部分。同學們可以從句子的語法結構以及相應的語法功能來進行句子的語法成分分析。分析一個句子，必須要有整體想法，不能局限於對個別詞彙的理解。通過對句子主幹成分進行全面分析，就是句子主謂成分，然後進行

其他輔助成分的確定，比如定語、插入語等。

第二，明確句間關係。明確句子之間的關係是並列關係還是主從複合關係。很多長句一般都會包含並列句和複合句。並列句的分句也非常重要，沒有什麼主次之分，大部分是使用一個冒號和一個逗號，也有由 but 或者 or 這些並列詞進行連接的。複合句的從屬分句是主句的一個成分（定語、表語、同位語等），主句與從句用關聯詞結合在一起，如 that、who、which、when 等。

第三，化整為零，分析長難句。分析長難句的重點是進行分解，即分解成比較好理解的單元。首先找出主句和它的關鍵部分，然後分析與主要部分相關的修飾成分，再分析主句別的成分相關的一些修飾成分，對它們之間的邏輯關係進行判斷。例如：Factories will not buy machines unless they believe that the machine will produce goods that they are able to sell to consumers at a price that will cover all cost.（廠家一般情況下是不買機器的，除非它們相信那些機器創造的產品的售價能夠保本。）這是由一個主句「factories will not buy machines」以及四個從句組成的長句。在這裡，unless 引導條件狀語從句，第一個 that 引導賓語從句，而後面的 that 引導定語從句。總而言之，句子的主幹，就像主語、謂語以及賓語是主要信息的表達，而別的成分，不管多少，不管多麼複雜，都是輔助成分。只要能夠抓住句子的主幹成分，就能夠理解文章的意思。

（4）運用相關閱讀技巧

閱讀方法分為精讀（intensive reading）、泛讀（extensive reading）、略讀（skimming reading）和瀏覽（scanning reading）四種，每種閱讀方法都能在不同方面提高英語閱讀效率。例如閱讀英文財經報刊時，因為內容較多，篇幅較長，同學們可以適當採用泛讀、略讀、瀏覽等方法，而在閱讀一些專業雜誌的時候，為了能夠更多更好地理解作者想要表達的意思，同學們可以運用精讀的方法。不論如何，在瞭解各種閱讀方法的優劣勢之後，同學們可以根據所閱讀文章內容的特點和自己的閱讀習慣，選擇合適的閱讀方法組合以提高閱讀效率，增強自己的閱讀能力。

（5）利用中文信息幫助閱讀

無論是中文的財經報紙雜誌還是英文的財經報紙雜誌，都是用文字來傳遞信息的。對於一些有重大影響的財政、經濟、政治與金融大事，中外媒體均會予以報導。有些比較權威的英文財經報紙雜誌也會有中文版，同學們可以在閱讀原版文章時參照中文幫助理解。在剛開始閱讀英文報導時，如果感覺有難度，不妨先閱讀中文報導，瞭解相關信息和知識後再閱讀英文報導，這樣會比較輕鬆地讀懂。

英文財經報紙雜誌在傳遞信息的同時，也有其獨特的語言特色。提高英文財經報紙雜誌的閱讀能力不是紙上談兵，也不是一朝一夕就能煉成的。同學們可以根據前文所介紹的內容制訂自己的閱讀計劃，堅持閱讀，保持對世界經濟政治的敏感度，幫助自己提高英文財經報紙雜誌的閱讀效率，培養國際化意識。

3 如何用英文撰寫商務信函

　　商務信函就是用於商貿往來以處理商貿業務的信函。從事商貿往來，開展商貿業務，經常需要寫作商務信函。根據實際需要，寫作規範實用的商務信函，是商務人員必備的基本功。作為商務人員，如果不能熟練地寫作商務信函，就無法開展各種商貿業務，在商貿活動中將會寸步難行，甚至有被老板炒魷魚之虞。

3.1 商務信函的分類

　　寫作商務信函，首先需要瞭解其種類和功用。不同種類的商務信函具有不同的功用和特點。由於商貿業務範圍極廣，因此，商務信函的種類較多。根據其在商務活動中的功用，可分為兩大類：一類是交易磋商函，另一類是爭議索賠函。

　　（1）交易磋商函主要用於商貿活動的交易磋商。常見的交易磋商函包括：①推銷函：主要用於介紹、推銷產品和服務。雖說是商務信函，但兼有了廣告的一些功能。在寫法上，在語言的運用上，平實中也使用些廣告措辭，往往針對對方心理，宣傳自己產品的特點和優勢，目的是引起對方的購買慾望，具有一定的推銷能力。②諮詢函：主要用於向對方詢問有關商品的信息和情況，如是否有貨、有什麼樣的貨（規格型號、質量標準等）、什麼價格、怎樣提貨、售後服務等事宜。寫法上直截了當，想知道什麼就詢問什麼，突出最主要的未知信息。③報價函：主要針對需方價格等方面的詢問給予回復，為需方提供有關商品價格等方面的信息。寫作上有的放矢，針對性要強，切忌答非所問，讓人不得要領。④訂購函：主要用於訂購商品。訂購商品前，交易雙方多通過當面洽談或電話洽談，達成購買意向，訂購函則是白紙黑字，將訂購意向變成書面函件。因此，訂購函兼有合同的功用，往往成為合同糾紛案件仲裁審理的重要憑證。訂購函的寫作與合同寫作有相同之處，都要求考慮周詳，表述精確嚴密，沒有歧義和漏洞。

　　（2）爭議索賠函主要用於貿易糾紛、合同違約等情形，常見的爭議索賠函包括：①催款函：主要用於催收欠款。催款函寫作有一定難度，難在既要觸動對方，達到收回欠款的目的，又要顧及對方的臉面，給對方臺階，維持原來的合作關係。這就要求在措辭上反覆推敲，注意掌握好分寸。根據對方拖欠款項的嚴重程度，決定措辭的語氣輕重。初次催款，不妨語氣輕些，隨意一些，提醒一下對方即可；如果多次催收，對方仍無動於衷，一味賴帳，措辭則可嚴厲一些，加大催款的力度。②索賠函：主要用於協商解決合同違約、商務糾紛的問題。在合同履行的過程中，不履行合同約定應盡的義務是常有的事，如買賣合同，供方不能按時交貨，質量不合要求等；需方中途退貨、拖欠款項等。一旦出現了合同違約、合同糾紛，就需要寫作索賠函，向對方提出賠償要求。索賠函的寫作也有較大難度，因為這涉及雙方的切身利益，關係到雙方矛盾、分歧的解決。這就要求作者有較強的業務能力、處事能力和寫作功力，寫法措辭上要求做到有理有節。有理就是要提出充分的對方違

約的事實依據、法律依據；有節就是提出的處罰的力度要合情合理、輕重得當。

3.2 商務信函的寫作技巧

儘管商務信函種類較多，功用不同的商務信函具有不同的特點，但寫作上仍有共同之處，需要注意掌握以下幾個寫作技巧：

（1）格式方面

第一，格式規範，依規寫作。商函是商貿活動的憑據，具有一定的嚴肅性甚至法律效力，特別是訂貨函、索賠函常可作為處理商務糾紛、合同違約的文字憑據，因此，儘管商務信函寫作時經常不加標題，也不編發文字號，但並不是說對其格式規範可以掉以輕心，也要熟悉其格式規範，做到依規寫作。一份規範的商務信函，格式上包括以下幾個部分：①信頭：大多數商務信函都使用專用的信箋，信箋眉首處已印好信頭，包括發函公司的名稱、地址、郵政編碼、電話、傳真等項目，亦可將這些項目印在信箋下方，以橫線與信函內容相隔。如使用普通信箋，也可以沒有信頭。②稱謂：這是對商務信函行文對象的稱呼。若是單位，應用全稱，不宜使用簡稱，以示對收函方的尊重。若是個人，則依對方具體身分、職稱或職銜而定。如不知對方具體身分，則可以用「Miss」「Mr.」「Mrs」稱之。③問候語：若收函者是個人，應先問候一下以示禮貌。也可寫幾句客套話，但要注意自然得體。④正文：包括開頭、主體、結尾三個層次。開頭交代發函緣由，即為什麼發函、發函的起因或背景，寫作上要注意言簡意明，直入主題。主體寫清發函的事項，需要詢問、商洽的具體事宜，寫作上要注意直陳其事，要言不煩，切忌行文拖沓，空發議論。結尾提出希望和要求，要注意使用商詢、期盼的語氣，切忌語氣生硬，令人閱之不快。⑤祝頌語：看對象而定，要注意禮貌得體。⑥附件：隨信發送的文件，如報價單、樣品圖表、收據、欠款清單等。位於祝頌語之下另起空兩格。⑦落款：這是商務信函生效的標誌，右下方寫清發函方、發函日期並加蓋印章。

第二，格式範文，見下面的模板。

From：×××

To：×××

Date：×××

Dear Ms. Cunningham：

　　You and your company have been recommended to us by Charles Lewis of East Asia Building Material Supplies. Charles mentioned that your company provides high quality goods and services at a reasonable cost, and I am writing to inquire about establishing business cooperation between you and my company, China Merchandise Company. China Merchandise Company is one of the largest international exporters of Chinese goods. We have clients throughout the world, especially in the United States. Chinese ornamental merchandise represents one of our most popular products. As such, we would appreciate it if you could send us your latest sales catalog for our review. We believe that establishing business cooperation with Ornamental Decorations and Supplies will be mutually beneficial for your company and ours.

　　I am looking forward to receiving your catalog and doing business with you in the future.

Sincerely yours,

Mai Yang

Manager

　　(2) 內容方面

　　第一，一事一函，目的明確，言簡意明。

　　商務信函用於處理商貿業務，解決商務往來中的各種問題，因此，商務信函的寫作首先要求做到有的放矢，目的明確。與此相應，一份商函，只解決一個單一的商務問題，或者詢價或者報價，或者推銷或者訂貨，或者催款或者索賠，不要把好幾個問題混在一份商函中，把商函變成大雜燴。否則將不便於集中解決問題，也不便於對方回復。

　　俗話說，時間就是金錢。多數商務人士每天有很多的商務信函要閱讀和處理，有很多事情需要解決。簡單明瞭的信函對於他們繁忙的頭腦猶如一陣清風吹過，而囉唆繁瑣的信件必定讓讀者費盡頭腦，並有可能喪失閱讀的興趣而延誤時間。因此英語商務信函必追求內容清晰明了，切中要點，最好只有簡單的寒暄然後直入主題，讓客戶在最短時間就可以明白商務信函所要闡述的意思並快速做出反應。

　　第二，內容清晰準確。

　　內容清晰應該包含兩個方面：第一是在擬文前知道自己要寫什麼，第二是在讀者看到之後知道商務信函寫了什麼。如果一封信件中想要表達多個內容，最好每一個單獨的意思用單獨的段落來表達，這樣更便於讀者快速瞭解信件內容。文字的表達中還要注意不同文化的差異，以免發生誤解。盡量表達準確，避免使用一些含混不清、模棱兩可的詞彙。英語被一百多個非英語國家的商務人員在商務交往中使用，而這些商務人員並不是每個人都精通英語，因此在選詞上應該盡量選擇常規的表達方式和簡單易懂的詞彙，避免使用複雜句式和晦澀難懂的詞語。盡量用最少的文字表達最為完整的意思。但也不可一味追求簡潔，書信內容也要具體詳實、完整。商務信函以協商處理商貿業務問題為目的，語言表達上要求言簡意明，直陳其事，這樣才能提高商務活動的效率。商務信函寫作切忌拐彎抹角，拖泥帶水，含糊其詞，讓對方讀後如墜雲裡霧裡，不得要領，影響到商務問題的解決。在最近幾年裡，商務信函的格式也發生了一些變化，不要求過於正式，但務必簡潔，也可以使用一些約定俗成的英文縮寫。縮略語的使用也可使文章看上去更加簡潔，但對於商務英語縮寫的掌握一定要準確，以免引起麻煩。一些涉及時間和數字的表達不妨直接用阿拉伯數字表達，這樣不僅一目了然，並且可以引起重視。要注意在英語句子中，以數字開頭的，應使用單詞而不是數字。涉及金額的表述一定要明確易懂，沒有歧義，並注意在表述金額時加上金額的單位如 RMB、USD 等。與商務信函莊重典雅的文風相應，商務信函有一套常用的習慣用語，對這套習慣用語，寫作時要盡量熟練使用。

(3) 其他方面

第一，看清對象，措辭得體。由於商貿活動非常複雜，業務範圍極廣，商務信函對象的身分也非常複雜，或是公司客戶，或是供貨商，或是欠款人，或是合同違約方，這就要求看清收函對象，根據收函對象決定措辭和語氣，做到禮貌得體。如推銷函應熱情真誠；訂貨函要禮貌尊重；催款函和索賠函要講究分寸，既要表明催款索賠的鮮明態度，又要考慮對方的處境，顧及對方的面子。只有看清收函對象，到什麼山上唱什麼歌，對什麼人說什麼話，措辭得體，才能妥善處理各種商務關係，從而達到發函的目的。

第二，注意禮貌用語的使用。為了建立友好的貿易夥伴關係，禮貌是非常重要的。商務信函中應多使用禮貌用語和請求句式，少用命令句式。如「Will you？」「Please」等。字裡行間更應體現中國的外交禮節，做到不卑不亢、得體大方。對待客戶要一視同仁，即使是面對非常小的公司，在交流時也要做到以禮相待。拒絕對方時要盡量使用委婉的方式。由於不同的文化差異，語言的表達方式有很多不同。盡量避免中國式表達，注意英、美的表達差異。例如英語的表達多用物做主語，而漢語的表達多用人做主語。除了文字表達的方式不同，還要注意到對方的風俗習慣、宗教信仰等。例如穆斯林不吃豬肉、狗肉、骨頭等，我們應避免在信函中提及相關詞彙。在發送商業信函時，最好使用正規的公司信紙，帶有公司抬頭地址等，便於讀者在最快時間瞭解信件的來源並且有助於在客戶中形成對公司的印象。除此之外，在信件的末尾要署名或者蓋公司公章。及時回信也是對對方的一種尊重。信件的整體結構要清晰、整潔。文章應使用統一字號，字號的大小也要有所選擇，以易於閱讀為準，如果信件內包含表格，應盡量安排在一頁之內，便於對方閱讀和比較。商務信函一般不過多使用斜體、彩色字體，文章的標題或重要內容可加粗或加下畫線來標註，標點要使用正確以免引起歧義。文章最好在標題上就讓讀者可以瞭解信件的主要內容，在結構上就可以看出信件包含幾層意思。如果發電子郵件附件時，文件題目應註明時間，並且應和文章內所標註時間一致。

第三，熟悉商業心理。完成一封成功的商業書信，不僅要求寫信人精通英語和外貿知識，還應該通曉收信人的心理。貿易往來中，追求利益是無可厚非的，在商務信函磋商過程中遵循互利互惠的原則的同時應多從客戶的角度行文，突出積極的因素。在句式的使用上多加揣摩，可以將對方所得利益放在句首，這樣讀者對自己所得利益一目了然並且印象深刻。把積極因素放在句首更能激發對方的興趣並得到認同，當然對於其他事實情況不可避而不談，但可以弱化一些。商務信函應起到信息溝通的作用，只有在發信後得到了對方應有的反應，信函才能體現其價值。在起草商務信函時，只有站在對方的立場上思考和處理問題，才能更有效地溝通。少使用否定的句子，多從正面強調積極的因素，弱化消極的信息，有利於更好地傳達信息。

第四，仔細檢查，發送前確認行文準確無誤。在將信函發給客戶之前，一定要反覆檢查是否有錯，因為商務信函往往涉及貿易雙方的義務的責任以及企業和個人的利益。一些不準確和錯誤的信息可能帶來無法預料的麻煩和損失，如果在寫信函

時遇到了不確定的信息，一定要積極核實。商務信函質量的好壞會直接影響企業的形象。商務信函寫得好，可能贏得新的業務，有助於促進和發展與外貿夥伴的友好關係，也可以及時消除誤會並解決問題；反之，則可能引發矛盾，引起經濟糾紛，失去客戶。

4 如何進行國際經貿活動專業翻譯

從語言的社會功能變體來講，經貿英語是在跨文化國際經貿領域中，主要由生產商、代理商、消費者、投資者等使用的，以進行和完成相關經貿交際活動為目的，在詞彙、句法、篇章上具有特定的表現形式、表達程式、言語規範及言語規律的相對穩定的一種功能變體。在世界經濟全球化和一體化程度日益加深的今天，經貿英語翻譯作為貿易往來的橋樑，越來越重要，能否順利地引進國外資金和先進技術以及出口本國產品和進行國外投資，關鍵在於經貿英語翻譯的正確與否。如何做好國際經貿活動英語翻譯，已經成為國家經濟發展的當務之急。

4.1 培養專業翻譯所需素質

國際經貿活動英語譯者具備較高的語言素質、文化素質、國際貿易知識與翻譯理論及技巧的熟練掌握程度等專業水準、職業素質和複合型翻譯人才標準，將是保證順利進行經貿英語翻譯，加強國際經貿領域溝通、交流與合作的前提與基礎。經貿英語翻譯中需要具備的素質主要包括語言素質、文化素質和創新素質。

首先，翻譯者需要具備一定的語言素質。語言是說話者傳達信息的載體，譯者想要使用一種語言把另一種語言所要表達的內容忠實地表達出來，就必須具備熟練運用兩種語言的能力。本國語言和外語素質的高低直接影響著譯文質量的高低，所以譯者必須具有較高的語言素質。趙彥春先生認為：「翻譯不是靜態的代碼轉換，而是以關聯為準則，以順應為手段，以意圖為歸宿，盡量使譯文向原文趨同的動態行為。」對於語言素質，譯者需要有良好的外語語言聽、說、讀、寫、譯能力，以及豐富的專業外語詞彙、句型。對原文的閱讀理解能力決定了譯者翻譯水準的高低。所以，為了提高英語閱讀理解能力，同學們必須做好以下四個方面的工作：一要掌握足夠的外語詞彙，特別是國際貿易相關的專業詞彙量，並確保能在口筆譯中靈活運用；二要掌握系統的英語語法知識，減少低級錯誤或者不犯錯，以避免在國際貿易中引來不必要的爭議，造成不可挽回的損失；三要大量閱讀國際貿易的相關英語原文文本，提高自己的語言感悟能力，例如一見到 consignor 就要快速反應它所對應的中文是發貨人，另外還要懂得區分一些易混淆的詞彙，以避免不必要的損失；四要通曉本國語言。正如茅盾先生說的：「精通本國語文和被翻譯的語文是從事翻譯的起碼條件。」總之，英語語言基本功和漢語語言的通曉程度是經貿英語翻譯人才所必備的素質。語言素質越高，相應的翻譯質量也就越高。

其次，譯者還需要有一定的文化素質。現代翻譯尤其是經貿方面的翻譯大都是

文化翻譯。在經貿英語翻譯中，譯者關注更多的是國際慣例條款、談判內容，而忽略文化方面的因素。但翻譯的跨文化交際的屬性，要求譯者必須具備較強的文化素質。翻譯是一種努力使譯文全方位靠近原文的跨文化、跨語言的交際行為。文化的內涵包含多個方面，涉及面較廣，有物態文化層、制度文化層和行為文化層等。經濟領域遠比其他領域容易受到文化素質的影響。較早提出跨文化交際能力概念的西方學者提出了有效的跨文化交際的七要素：①向對方表示尊重並對其持積極態度的能力；②描述性、非評價性和非判斷性的態度；③最大限度地瞭解對方個性的能力；④移情能力；⑤應付不同情景的靈活機動能力；⑥相互交往能力；⑦能忍受新的和含糊不清的情景，並能從容不迫地對其做出反應的能力。根據跨文化國際貿易活動的特殊規律，想要在經貿活動中有效交際，至少應具備三方面的能力。其一是文化學習能力，即對國際貿易夥伴的文化背景有足夠的學習以及認知能力。比如，中國現在與東盟各國的貿易聯繫緊密，當我們與越南合作的時候，如果我們因為對其文化不夠瞭解而不小心冒犯了胡志明，或許會導致整個合作失敗。要知道胡志明在越南人民心中的地位不亞於毛澤東在我們心中的地位。如果我們沒有文化學習的能力，對文化缺乏足夠的敏感以及不具備善於學習對方特殊文化知識的能力，就會認識不到兩國之間的差異，進而以自己的文化標準期待對方，從而造成雙方談判失敗。譯者對潛在的商業夥伴的文化敏感度越高，就越能預料雙方存在的差異，這樣才能把差異帶來的消極因素變為有利因素。其二是文化理解能力。如果說文化學習能力僅僅是正確認識文化的過程，那麼文化理解能力就是對文化的內容進行必要的解釋。它主要指對商業夥伴採取不評價不判斷的態度。譯者需要從雙方的談話中領會和推測商業夥伴話語中的真正意思，這樣就可以準確順利地完成國際談判。其三是行為變通能力。俗話說，計劃趕不上變化。縱使是一位久經沙場的譯者，也無法完全掌握談判的動向。這就要求譯者要以變應變，對不同的交際場合採用不同方式。雖然任何交際都是人際交往，但是不同交際之中又存在著一定的差異性。就好像並非所有中國人都喜歡吃麵一樣，並非所有的西方人都有相同的思維模式。這就是產生了差異性。但是作為譯者，面對不同的文化和差異，最好的辦法就是適應。適應差異的同時，自己也可以從中自我反思，更好地解決問題。

最後，創新素質也是翻譯者需要具備的一項重要素質。翻譯是一門藝術，是一種基於源語的再創作。現代社會的發展越來越傾向多樣化、技術化和信息化。經貿英語翻譯人才不僅需要具有紮實的翻譯基礎知識，還必須具備與國際市場發展相適應的創新能力。翻譯人員在國際貿易活動中的身分是商務談判雙方的媒介與橋樑，任務是促成各類國際經貿活動的交融或交鋒，靈巧地解決語言或文化障礙而讓國際商務得以順利實現。但是現今國際市場競爭激烈，翻譯人員不能只是一味地守舊，在適當的時候還應站在「目的論」與「功能論」的翻譯視角，發揮自主創新精神。這樣會給譯員的翻譯增添不少色彩，同時也能更好地表達談判方的意願，適應時代的潮流。

4.2　掌握翻譯理論和技巧

翻譯既是一門科學，又是一種技能。翻譯這種人為的過程，是在一定的技巧下

進行有規律可循的跨文化、跨語言的交際與合作。對翻譯理論的掌握和理論水準的提高是譯者準確、快速完成外貿活動的關鍵。從語言方面而言，同學們不僅應該熟悉中國翻譯理論，還需要更多地瞭解西方的翻譯理論。如嚴復的「信達雅」和西方翻譯理論家奈達的「功能對等論」、紐馬克的「文化翻譯理論」等。而且翻譯從來就不是僵硬、機械的語言過程，而是靈活的、有規範性的人類參與的交際行為。不同的文體，有不同的翻譯要求及表達的標準。涉及國際商務活動的翻譯大多會涉及經貿領域所屬的專業詞彙，而且具有一定程度上的嚴謹性、科學性與通用性。這些都與翻譯理論與技巧息息相關。眾所周知，要做好一名經貿英語譯者，光靠理論是不行的，還需要從事豐富的翻譯實踐活動，獲得更多的翻譯技巧。而對翻譯技巧的嫻熟運用，又通過大量的經貿翻譯實踐反過來予以強化與活化。把理論與實踐在經貿英語翻譯活動中巧妙地結合起來，才能快速有效地提高譯者的翻譯水準，增強經貿英語翻譯的敏感與精確。翻譯技巧一般包括以下幾方面：

（1）準確理解原文，熟悉術語，瞭解專業知識

經貿英語的用法和日常用法不大相同，我們平常所熟悉的普通名詞在經貿英語中的意思經常會有一些變化，如何翻譯這些術語和習慣表達法，以及這些詞應該和哪些詞搭配都是非常重要的。近年來隨著經濟發展的全球化趨勢，新的詞語層出不窮，在翻譯時我們必須根據上下文去把握和理解其真正的含義。例如：We live in an era in which information, goods and capital speed around the globe, every hour of everyday.（我們生活在這樣一個時代，信息、貨物和資金每時每刻在世界上流動。）在這句話中，「speed」的本意是「速度」的意思，但在這裡根據上下文意思應該翻譯為動詞「流動」。又如：Since World War II, the US has been the world's largest economy and in most years, the world's largest exporter.（自從第二次世界大戰以來，美國一直是世界上最大的經濟大國。在大多數年代中，它又是世界上最大的出口國。）這個句子裡「economy」的本意是「經濟」，根據上下文，這裡翻譯成「經濟大國」。Any claim by the Buyers regarding the goods shipped shall be filed within 15 days after arrival of the goods at the port of destination specified in the relative Bill of Lading and supported by a survey report issued by survey or approved by the seller.（買方對於裝運貨物的任何索賠，必須於貨到提單規定的目的地15天內提出，並須提供經賣方同意的公證機構出具的檢驗報告。）這裡「claim」的意思變為「索賠」，短語「Bill of Lading」的意思是「提單」。

（2）掌握文體特點，不同文體的翻譯原則不同

文體不同其語言風格就不同，比如經貿合同的語言非常精練、正式、有很多附加修飾成分。例如：It shall be subject to ratification or acceptance by the sign a tory States.（本公約須經簽字國批准或接受。）英語用了shall，漢語翻譯時為了表達這種非常正式的文體用「應……」「須……」等詞。Kindly tell US what steps you are going to take in the way of compensation for the damage.（敬請告訴我方將以什麼方式來補償這次損失。）「Kindly tell US」使得說話的語氣顯得十分客氣。We would appreciate receiving details regarding the commodities.（如能告知該商品的詳細情況，則不勝感

激)。We must make claim on you for compensation with regard to this shipment, for it is not equal to the sample submitted, and contains large lumps, which will all have to be extracted before we are able to sell it. (此次所寄來的貨品與樣品大不相同,且有混入不少大塊,在銷售以前不得不全部清除。為此只好向貴公司提出損失賠償。) We hereby add confirmation to this cred it and we undertake that documents presented for payment in conformity with terms of this cred it will be duly on presentation. (茲對此證加保兌並保證於提示符合此證條款的單據時履行付款。) 在這些句子的翻譯中,都要盡可能地保持原文體現的正式、客氣的禮貌用語和風格。

(3) 研究、把握英、漢句子結構的差異

儘管英、漢兩種語言的主要句子結構大致是相同的,但其中仍存在較大的差異,翻譯時需要慎重考慮。例如:If any of the joint ventures wish to assign its registered capital, it must obtain the consent of the other parties to the venture. (合營者的註冊資本如果要轉讓,必須經合營各方同意。) 英語的主語是 any of the joint ventures「任何合營企業」,翻譯成漢語時卻將原文的一部分賓語變成了主語,因為英語中的主語要考慮能否作動詞所表達的動作的執行者,漢語相對而言比較隨便。We hand you here with an order for 100 tons bar iron as per the particulars on the annexed specifications, and shall be glad to hear that you are able to complete the same. (茲隨函寄上 100 噸鐵條訂單一份。條件按附帶規格中的具體規定。如貴公司能供貨,當不勝感激。) 此例,英文是一個句子,裡面有一個並列謂語 and shall be glad to hear 和一個賓語從句 that you are able to complete the same。如果將其翻譯成一個長句,漢語將顯得不倫不類,而將其分成三個句子處理就順當得多。

(4) 領會詞類轉換,把握句子、段落的聯繫

就句子內部而言,英語重形合,句子各成分聯繫緊密,漢語重意合,結構鬆散,更多地依賴句內各成分的關係。漢語重視句子間的聯繫,而英語不太重視句子間的聯繫。例如:The economic growth rate has been noticeably affected by the chaotic state of the market. (經濟增長的速度受到市場混亂的影響,這是顯而易見的。) 英語中 noticeably 是副詞,修飾謂語動詞 affected,翻譯時將其當成一個句子來處理反而符合漢語習慣。During the half-hour talk, the two sides exchanged views on the choice of terms of payment, but they made no mention of the mode of transportation. (在半小時的商談中,雙方就付款方式交換了意見,但沒有提到運輸方式。) mention 是名詞,譯成漢語時要變為動詞。

(5) 適當增詞、減詞,平衡斷句與並句的關係

英、漢語的句子結構、表達方式、修辭手段不盡相同,所以為使譯文更加精練,更符合漢語表達習慣,有時需要省略部分詞語;而有時為了使譯文更準確,可按意義、修辭和句法的需要在譯文中加入原文無其詞而有其意的詞,使譯文更加通達。同時也可將原文的一句話分成幾句,或根據具體情況把兩句話合成一句,即將原文的順序做適當的調整,這樣做的目的是使譯文的意思完整準確。We are pleased to have received your invitation to the symposium on Internet. (非常高興收到了參加因特網

會議的邀請。）中文譯文省去了代詞「我們」。But considered realistically, we had to face the fact that our products were less than good.（但是現實地考慮一下，我們不得不正視這樣的事實：我們的產品並不理想。）譯文增補冒號，將同位語從句分開處理。The government tried unsuccessfully to curb inflation through out the country.（政府試圖控制全國範圍內的通貨膨脹，但沒有成功。）譯文將原文中的一個副詞翻譯成一個句子。

　　在現今國際競爭日益激烈、全球化進程日趨加快的情況下，複合型的經貿翻譯人才是保證國際貿易活動成功的關鍵。只有以高素質的經貿英語翻譯人才作為後盾，企業才能在國際貿易活動中占得先機。譯者只有擁有堅實而合格的語言素質、文化素質、創新素質、國際貿易知識以及英語翻譯理論和翻譯技巧等，才能真正而全面地適應貿易國際化、全球化對經貿英語翻譯人才的需求，才能使我們的國際經貿活動有理、有利、有節地發展。

附錄 3
國際經貿專業英語詞彙英漢對照表

A

absolute advantage 絕對優勢
absorption 吸收
absorption approach 吸收論
absorption instrument 吸收工具
accounting cost 會計成本
accounting profit 會計利潤
adverse selection 逆向選擇
advertising 廣告
affreightment 租船運輸
agent 代理商
aggregate demand 總需求
aggregate supply 總供給
agreement of the Parties 協議選擇原則
allocation of resources 資源配置
allocative efficiency 配置效率
all risks 一切險
allowance 折讓
Anticipatory Breach in Common Law 普通法（英美法）上預期違約
Antidumping Authority 反傾銷機構
appreciation 升值
Applicability of the CISG CISG 的適用範圍
Arbitration Agreement and Arbitration Clauses 仲裁協議和合同中的仲裁條款
arbitration tribunals 仲裁機構
assignment 合同權利轉讓
avoidance 解除

arbitrage 套利
assumption 假設
asymmetric information 非對稱性信息
available market 可獲得市場
average cost 平均成本
average fixed cost 平均固定成本
average product of capital 資本平均產量
average product of labour 勞動平均產量
average revenue 平均收益
average total cost 平均總成本
average variable cost 平均可變成本

B

balance-of-payments system 國際收支體系
Balassa-Samuelson Effect 巴拉薩-薩繆爾森效應
Bank For International Settlements（BIS）國際清算銀行
barriers to entry 進入壁壘
barter transaction 易貨交易
base year 基準年
basis 基差，基價
behavioral segmentation 行為細分
below-the-line 線下廣告
benchmark 評估基準，評估標準
benefit 收益
bid-ask spread 買賣差價
bilateral monopoly 雙邊壟斷
black market 黑市
Bligations of the Parties 當事人各方的義務
Bligations of the Seller and the Buyer 買賣雙方的合同義務
blue chip 藍籌股
bond 債券
brand 品牌
brand extension 品牌延伸
brand image 品牌形象
break even point 收支相抵點
budget constraint 預算約束

budget line 預算線
budget set 預算集
business risk 經營風險，商業風險
bulk cargo 散裝貨
by-products 副產品

C

call option 看漲期權
callable bond 可贖回債券
capital 資本
capital formation 資本構成
capital market 資本市場
capital stock 資本存量
capital output ratio 資本產出比率
cardinal utility theory 基數效用論
cartel 卡特爾
cash discount 現金折扣
ceteris paribus assumption「其他條件不變」的假設
change in demand 需求變化
change in quantity demanded 需求量變化
change in quantity supplied 供給量變化
change in supply 供給變化
charter transport 租船運輸
choice 選擇
civil law、code law 大陸法系
Coase theorem 科斯定理
Cobb-Douglas production function 柯布-道格拉斯生產函數
cobweb model 蛛網模型
collectivism 集體主義
collusion 合謀
commodity 商品
commodity combination 商品組合
common law 英美法系
common stock 普通股
Commercial Arbitration 國際商事仲裁
Commodity Arrangements 初級產品/農產品安排

Common Enterprise Liability 企業的一般責任

Comparison of Municipal Legal Systems 國內法系的比較研究

Compulsory Licenses 強制許可

comparative advantage 比較優勢

comparative static analysis 比較靜態分析

compensatory financing facility 補償貸款辦法

competition 競爭

competitive advantage 競爭優勢

competitive market 競爭性市場

competitive strategies 競爭戰略

competitor analysis 競爭者分析

complement goods 互補品

complete information 完全信息

condition for efficiency in exchange 交換的最優條件

condition for efficiency in production 生產的最優條件

concave 凹

concave function 凹函數

concave preference 凹偏好

conceptual skills 概念性技能

conclusion of sales contract 簽約

Consent to the Jurisdiction of the Host State 給予東道國管轄權的許可/同意

constant returns to scale 規模報酬不變

constructive total loss 推定全損

Contemporary International Trade Law 當代國際貿易法

Contract Law for the International Sale of Goods 國際貨物銷售合同法

Contractual Issues Excluded from the Coverage of CISG 排除在 CISG 適用範圍之外的合同問題

contractual joint venture 合作企業

consumer 消費者

consumer behavior 消費者行為

consumer equilibrium 消費者均衡

consumer market 消費市場

consumer preference 消費者偏好

consumer price index (CPI) 消費者價格指數

consumer surplus 消費者剩餘

consumer theory 消費者理論

consumption bundle 消費束

convertible bond 可轉換債券

convex 凸
copyrights 著作權/版權
core capital 核心資本
core product 核心產品
cost 成本
cost minimization 成本最小化
cost-plus pricing 成本加成定價
coupon 息票
counter trade 對等貿易
counter-offer 還盤
country risk 國家風險
Cournot model 古諾模型
covered exposure 保值敞口
covered interest parity 拋補利率平價
crawling peg 爬行式盯住
credit risk 信用風險
crowding out 擠出
cross-border mergers and acquisitions 跨境併購
cross-price elasticity 交叉價格彈性
cross rate 套算匯率
cultural environment 文化環境
currency future 貨幣期貨
currency option 貨幣期權
currency risk 貨幣風險
currency swap 貨幣互換
current yield 本期收益率
customs duties 海關關稅
customer satisfaction 顧客滿意
customer value 顧客價值

D

dead-weights loss 無謂損失
decline stage 衰退階段
default risk 違約風險
deficit 赤字
deflation 通貨緊縮

demands 需求
demand curve 需求曲線
demand elasticity 需求彈性
demand function 需求函數
demography 人口統計
depreciation 折舊
depression 蕭條
derivative 金融衍生工具
derivative credit risks 衍生信用風險
derivative security 衍生證券
desirable products 合意產品
Decision Making within the WTO WTO 內部決定作出機制
Deficiencies in the GATT 1947 Dispute Process 關稅及貿易總協定 1947 爭端解決程序的不足
Delayed Bills of Lading 提單遲延
Denial of Justice 司法不公
developed market 發達市場
diminishing marginal substitution 邊際替代率遞減
diminishing marginal utility 邊際效用遞減
direct effect 直接效力
discounting 貼水、折扣
discount rate 貼現率
diseconomies of scale 規模不經濟
disposable income 可支配收入
Dispute Settlement 爭端的解決
Dissolution by Agreement 協議解散
Dissolution by Court Order 依法院令解散
division of labor 勞動分工
dollar-standard exchange-rate system 美元本位匯率制度
dollarization 美元化
double-layered acculturation 雙重文化適應
Double Taxation 雙重徵稅
duopoly 雙頭壟斷
durable goods 耐用品
duration 持續期
duress 脅迫行為
dynamic analysis 動態分析

E

economic agents 經濟行為者
economic cost 經濟成本
economic efficiency 經濟效率
economic exposure 經濟風險
economic globalization 經濟全球化
economic man 經濟人
economic profit 經濟利潤
economy of scale 規模經濟
Edgeworth box diagram 埃奇沃思圖
Efficiency Markets Hypothesis (EMH) 有效市場假說
elasticities approach 彈性論
emerging market 新興市場
enquiry 詢盤
equities 股票
equity joint venture 合資企業
exchange 交換
exchange rate 匯率
export subsidies 出口補貼
efficiency 效率
elasticity 彈性
elasticity of substitution 替代彈性
Employment Laws in the European Union 歐洲聯盟雇傭/勞工法
endogenous variable 內生變量
endowment 稟賦
Enforcement of Foreign Judgment 外國法院判決的執行
Engel curve 恩格爾曲線
entrepreneurship 企業家精神
entry barriers 進入壁壘
entry/exit decision 進出決策
envelope curve 包絡線
equilibrium 均衡
equilibrium price 均衡價格
equilibrium quantity 均衡產量
equity 公平

Eurobond 歐洲債券
Eurodollar 歐洲美元
exchange-rate band 匯率區間
exchange-rate instability 匯率的不穩定性
exchange rate risk 匯率風險
Excuses for Non-performance 不履行的免責
exercise price 行權價格
explicit cost 顯性成本
external economy 外部經濟
external diseconomy 外部不經濟
externalities 外部性

F

factor 要素
Factor-endowment theory 要素稟賦理論
Failure to Exhaust Remedies 沒有用盡法律救濟
Favorable trade balance 貿易順差
Fed 美國聯邦儲備局
Fed Wire 聯邦結算系統
final goods 最終產品
financial advisor/planner 理財顧問，財務規劃師
financial crisis 金融危機
financial intermediation 金融仲介
firm 企業
fixed costs 固定成本
flexible and sticky prices 有伸縮性和黏性的價格
floating-exchange-rate system 浮動匯率制度
foreign direct investment (FDI) 外國直接投資
foreign exchange risk 外匯風險
Foreign Portfolio Investment (FPI) 外國組合性證券投資，組合證券的國外投資
Force Majeure Clauses 不可抗力條款
Formal and Informal Application Process 正式和非正式申請程序
forward exchange market 遠期外匯市場
forward premium or discount 遠期升水或貼水
Free From Particular Average (FPA) 平安險
Fraudulent Misrepresentation 受詐欺的誤解

free entry 自由進入
free rider 搭便車
freight 運費
frontier market 前沿市場，邊境市場
Fundamental Breach 根本違約
future 期貨

G

game theory 博弈論
GATT（General Agreement on Tariffs and Trade）《關稅與貿易總協定》
GATS Schedules of Specific Commitments 服務貿易總協定減讓表中的特別承諾
GDP：gross domestic product 國內生產總值
GDP deflator GDP 平減指數
General additional risks 一般附加險
General Agreement on Trade in Services 服務貿易總協定
general administrative theorists 行政管理理論
general average 共同海損
general equilibrium 總體均衡、一般均衡
generalization 普遍化
geographic limitations 地區限制
Giffen goods 吉芬商品
globalization 全球化
globalization of market 市場全球化
globalization of production 生產全球化
Government Controls over Trade 政府對貿易的管制
government failure 政府失敗
government purchases 政府購買
GPA 政府採購協議
greenfield investment 綠地投資，新建投資

H

hedge 套期保值
heterogeneous product 異質產品
homogeneous product 同質產品

host country 東道國
human capital 人力資本
human skills 人際關係技能
hypothesis 假說

I

immaterial condition 非物質條件
imperfect competition 不完全競爭
implicit cost 隱性成本
Import 進口
Import surcharge 進口附加稅
Import variable duties 差價稅
income effect 收入效應
income elasticity of demand 需求收入彈性
increasing returns to scale 規模報酬遞增
indifference curve 無差異曲線
industry 產業
Industrial Property Agreements 保護工業產權的協定
Industrial Revolution 產業革命
inelastic 缺乏彈性的
inferior goods 劣品
inflation rate 通貨膨脹率
inflation risk 通貨膨脹風險
informal organization 非正式組織
information asymmetry 信息不對稱
information cost 信息成本
Innocent Misrepresentation 因無知而致的誤解
input-output 投入-產出
insider information 內幕消息
Insider Trading Regulations 內幕交易規則
institution 制度
institutional economics 制度經濟學
insurance premium 保險費
Intellectual Property Right Law 知識產權法
Interest-rate swap 利率互換
intermediate goods 中間產品

international capital markets 國際資本市場
Internationalization 國際化
International Center for the Settlement of Investment Disputes 解決投資爭端國際中心
International Chamber of Commerce 國際商會
International Commercial Dispute Settlement 國際商事爭端的解決
International Court of Justice 海牙聯合國國際法院
international investment environment 國際投資環境
internalization-specific advantage (I-advantage) 內部化優勢
International Labor Standards 國際勞工標準
International Licensing Agreement 國際許可證協議
International Licensing Agreements 國際許可證協定
International Model Law 國際示範法
International Monetary Fund (IMF) 國際貨幣基金組織
International Multimode Transport 國際多式聯運
InternationalPayment 國際支付
International Persons 國際法主體
International Rules for the Interpretation of Trade Terms 國際貿易術語解釋通則
International Trade Customs and Usages 國際貿易慣例和習慣
International Trade Practice：國際貿易實務
International Treaties and Conventions 國際條約和公約
International Tribunals 國際法庭
investment banker 投資銀行，投資銀行家
invisible hand 看不見的手
inward foreign direct investment 外來直接投資
IS—LM model　IS—LM 模型
isocost line 等成本線
isoprofit curve 等利潤曲線
isoquant curve 等產量曲線

L

law of diminishing marginal utility 邊際效用遞減法則
law of diminishing marginal rate of substitution 邊際替代率遞減法則
Law of Foreign Investment Enterprises of China 中國的外商投資企業法
leading 領導
L/C 信用證
Legal System of International Business 國際商事的法律體系

Lender of Last Resort（LLR）最後貸款人
liability 負債
liberalization　自由化
liner transport 班輪運輸
Liquidated Damages 約定的損害賠償金
liquidity risk 流動性風險
liquidity preference 流動性偏好
location-specific advantage（L-advantage）區位優勢
long position 多頭
long run 長期
luxury 奢侈品

M

macroeconomics 總體經濟學
managed or dirty float 管理浮動，或稱骯髒浮動
management 管理學
market risk 市場風險
Markowitz Efficient Frontier 馬科維茨（證券組合）有效前沿
marginal benefit 邊際收益
marginal cost 邊際成本
marginal product 邊際產量
marginal productivity 邊際生產率
marginal propensity to consume（MPC）邊際消費傾向
marginal propensity to save（MPS）邊際儲蓄傾向
marginal rate of substitution 邊際替代率
marginal rate of transformation 邊際轉換率
marginal revenue 邊際收益
marginal utility 邊際效用
market equilibrium 市場均衡
market failure 市場失靈
Marshall-Lerner Conditions 馬歇爾-勒納條件
material condition 物質條件
maximization 最大化
mean-variance analysis 均值方差分析
megabanks 國際性銀行
mergers and acquisitions（M&A）兼併與收購（簡稱「併購」「購並」）

microeconomics 個體經濟學
middle managers 中層管理者
minimum wage 最低工資
Modification of Foreign Investment Agreements 外國投資協議的修改
monetary policy 貨幣政策
monetary policy autonomy 貨幣政策自主性
money market 貨幣市場
money multiplier 貨幣乘數
money supply 貨幣供給
monopolistic competition 壟斷競爭
monopoly 壟斷，賣方壟斷
moral hazard 道德風險
More or Less Clause 溢短裝條款
mortgage backed securities（MBS）以房地產抵押做擔保的證券，按揭證券
Multilateral Trade Agreements 多邊貿易協定
Multilateral Trade Negotiations 多邊貿易談判
multinational corporation（MNC）跨國公司，多國公司
multinational enterprise 跨國企業

N

national income accounting 國民收入核算
Nash equilibrium 納什均衡
natural monopoly 自然壟斷
necessities 必需品
net creditor 淨債權國
net debtor 淨債務國
net exports 淨出口
nominal exchange rate 名義匯率
nominal GDP 名義GDP
nominal return 名義收益
nominal yield 名義收益率
nondiscrimination 非歧視原則
nonimputable Acts 免責行為
nonlinear pricing 非線性定價
Non-tariff barriers 非關稅壁壘
Nonwrongful Dissolution 非不法原因散伙

normal goods 正常品
normal profit 正常利潤
normative economics 規範經濟學
nude cargo 裸裝貨

O

Organization for Economic Cooperation and Development（OECD）經濟合作與發展組織（通常簡稱「經合組織」）
offer 發盤
offshore banking 境外金融
oligopoly 寡頭壟斷
open-market operations 公開市場活動
opportunity cost 機會成本
optimal choice 最佳選擇
optimal currency area 最優貨幣區
optimization 優化
option 期權
ordinal utility 序數效用
organization chart 組織結構圖
organizing 組織
output 產出
outward foreigndirect investment 對外直接投資
Overseas Private Investment Corporation（OPIC）（美國政府）海外私人投資公司
overvalued currency 估值過高的貨幣
ownership-specific advantage（O-advantage）所有權優勢

P

packed cargo 包裝貨
Pareto efficiency 帕累托效率
Pareto improvement 帕累托改進
parity principle 平價原則
partial equilibrium 局部均衡
partial loss 部分損失
Particular Average 單獨海損

par value (face value) 票面價值，面值，平價
pegged-exchange-rate system 釘住匯率制度
perfect substitution 完全替代品
perfect inelasticity 完全無彈性
perfect elasticity 完全有彈性
performance attribution 投資績效歸因分析
perils of the sea 海上風險
Policy Framework for Investment (PFI) 投資政策框架
Phillips curve 菲利普斯曲線
planning 計劃
point elasticity 點彈性
policy instruments 政策工具
political economy 政治經濟
political risk 政治風險
portfolio 投資組合
portfolio approach 投資組合分析法
portfolio manager 投資組合經理
positive economics 實證經濟學
preference 偏好
preferred stock 優先股
price ceiling 最高限價
price control 價格管制
price discrimination 價格歧視
price elasticity of demand 需求價格彈性
price elasticity of supply 供給價格彈性
price floor 最低限價
price maker 價格制定者
price taker 價格接受者
primary market 一級市場，發行市場
principal 本金
principal-agent issues 委託—代理問題
Principles of Scientific Management 科學管理原理
private cost 私人成本
producer equilibrium 生產者均衡
product life cycle 產品生命週期
product line 產品線
production function 生產函數
production possibility curve 生產可能曲線

productivity 生產率
profit function 利潤函數
profit maximization 利潤最大化
property rights 產權
public choice 公共選擇
public goods 公共商品
public relations 公關
publicity 公共宣傳
Purchasing Power Parity 購買力平價
put bond 可賣回債券，可回售債券
put option 看跌期權

Q

quality latitude 品質機動幅度
quality tolerance 品質公差
quasi-rent 準租金

R

rationality 理性
real exchange rate 實際匯率
real GDP 實際GDP
real interest parity 實際利率平價
real interest rate 實際利率
real return 實際收益
recession 衰退
Recognition and Enforcement of Awards 仲裁裁決的承認和執行
Recognition of Foreign Judgments 外國裁決的承認
red chip 紅籌股
Refusal to Exercise Jurisdiction 拒絕執行管轄權
regional economic integration：區域經濟一體化
regulation 調節，調控
relationship marketing 關係行銷
Rremedies for Breach of Contract 違反合約的救濟
rent 租金

resellers 轉賣商
reserves 準備金
reserve requirements 法定準備金
resource allocation 資源配置
returns to scale 規模報酬
revenue 收益
risk premium 風險溢價
rules of Private International Law 國際私法規則

S

safeguards 保障措施協定
sales promotion 促銷
salvage charges 援助費用
Sanitary and Phytosanitary Measures 衛生與植物衛生措施協定
saving 儲蓄
scalar principle 等級原則
scale economy 規模經濟
scientific management 科學管理理論
scarcity 稀缺性
secondary market 二級市場，流通市場
security 證券，有價證券
security analyst 證券分析員
Seller's Obligations 賣方的義務
Seller's Remedies 賣方可以採取的救濟措施
shipping mark 嘜頭
short-run 短期
shut down 倒閉
shortage 短缺
short position 空頭
slope 斜率
social benefit 社會收益
social cost 社會成本
social marketing 社會行銷
spatial arbitrage 空間套利
special additional risk 特殊附加險
Special Drawing Rights 特別提款權

speculation 投機
spot exchange rates 即期匯率
stagflation 滯脹
static analysis 靜態分析
strategy 策略
sterilization 沖銷
stockbroker 股票經紀人
Structure of the WTO　WTO 的組織結構
Subordinate Business Structures 商業分支機構
Subsidies and Countervailing Measures 補貼與反補貼措施協定
subsidy 補貼
substitutes 替代品
substitution effect 替代效應
supply schedule 供給表
Sue and labor charges 施救費用
surplus 盈餘
swap 互換，掉期
SWIFT 環球銀行金融電信協會
symmetry of information 信息對稱
systemicrisk 系統風險
Systems for Relief from Double Taxation 避免雙重徵稅的救濟體制

T

takeover regulations 收購規則
tangency 相切
target zone（匯率）目標區
tariff barriers 關稅壁壘
technical efficiency 技術效率
technical skills 技術技能
technological constraints 技術約束
technological environment 技術環境
terms of trade 貿易條件
the Anglo-American Common Law System 普通法系或者英美法系
the Basel Committee 巴塞爾委員會
the Bretton Woods 布雷頓森林體系
the Federal Deposit Insurance Corporation（FDIC）聯邦存款保險公司

the Founding of GATT 關稅及貿易總協定的成立
the Framework Agreement 協定的框架
the General Agreement on Tariffs and Trade 關稅及貿易總協定
the International Transfer of Intellectual Property 工業產權的國際轉讓
the Law of Agency 國際商事代理法
the Making of International Law 國際法的構成
the National Standard 國民待遇/標準
the Roman-Germanic Civil Law System 大陸法系或者羅馬日耳曼法系
the wealth of Nations《國富論》
the World Trade Organization（WTO）世界貿易組織
theocratic law 宗教法系，神權法系
theory of comparative advantage 比較優勢理論
top managers 高層管理者
total cost 總成本
total fixed cost 總固定成本
total loss 全部損失
total product 總產量
total revenue 總收益
total return 總回報
total utility 總效用
total variable cost 總可變成本
trade barrier 貿易壁壘
trading in securities 證券交易
trade pattern 貿易模式
trait theory 特質理論
tragedy of the commons 公地悲劇
transaction 交易
transaction cost 交易費用
transactions Covered in CISG CISG 適用的交易範圍
transaction exposure 交易風險
transfer payments 轉移支付
transfer pricing 轉移定價
translation exposure 換算風險
target market 目標市場
triangular arbitrage 三角套利
trilimma 三難困境
The Agreement on Trade Related Investment Measures（TRIMs）與貿易有關的投資措施協議 treasury bill 短期國庫券

U

uncertainty 不確定性

United Nations Conference on Trade and Development（UNCTAD） 聯合國貿易與發展會議

undervalued currency 定值偏低的貨幣

underwriter 承銷商

unemployment rate 失業率

Unfair Competition Laws 不正當競爭法

United Nations Convention on Contracts for the International Sale of Goods（CISG） 聯合國國際貨物銷售合同公約

unity of command 統一指令

uniqueness 唯一性

unit elasticity 單位彈性

utility maximization 效用最大化

V

variable cost 可變成本

variable input 可變投入

variables 變量

visible hand 看得見的手

W

wage rate 工資率

Welfare criterion 福利標準

Welfare economics 福利經濟學

wholesaler 批發商

wholly owned enterprise 獨資企業

World Bank 世界銀行

Workforce diversity 勞動力多元化

World Intellectual Property Organization 世界知識產權組織

W. P. A. With particular average 水漬險

WTO 世界貿易組織
WTO Antidumping Agreement 世界貿易組織反傾銷協議
WTO Dispute Settlement Procedures 世界貿易組織的爭端解決程序

Y

yield curve 收益曲線
yield to maturity 到期收益

Z

Zero-sum game 零和博弈

附錄 4
國際經貿專業重點期刊和報紙目錄

4.1 期刊

1. *Journal of Finance*《金融》
2. *Journal of Financial Economics*《金融經濟學》
3. *The Review of Financial Studies*《金融研究評論》
4. *Journal of Accounting Research*《會計研究》
5. *Journal of Applied Econometrics*《應用計量經濟學》
6. *Rand Journal of Economics*《蘭德經濟學》
7. *Bell Journal of Economics*《貝爾經濟學》
8. *Journal of Political Economy*《政治經濟學》
9. *Journal of Financial and Quantitative Analysis*《金融和數量分析》
10. *The Review of Economic Studies*《經濟研究評論》
11. *The American Economic Review*《美國經濟評論》
12. *The Economist*《經濟學家》
13. *Econometrica*《計量經濟學》
14. *The Quarterly Journal of Economics*《經濟學季刊》
15. *Journal of Economic Literature*《經濟文獻雜誌》
16. *Academy of Management Journal*《管理學會雜誌》
17. *Harvard Business Review*《哈佛商業評論》
18. *The Academy of Management Review*《管理學會評論》
19. *Journal of Marketing*《行銷學雜誌》
20. *Strategic Management Journal*《戰略管理學雜誌》
21. *The Milbank Quarterly*《米爾班克季刊》
22. *Economic Journal*《經濟學》
23. *The Journal of Economic Perspectives*《經濟展望雜誌》
24. *Journal of Marketing Research*《行銷研究雜誌》
25. *The Review of Economics and Statistics*《經濟學與統計學評論》
26. *The Journal of Consumer Research*《消費研究雜誌》
27. *Journal of Economic Literature*《經濟文獻雜誌》

28. *Journal of Labor Economics*《勞動經濟學》
29. *Journal of Economic Surveys*《經濟研究》
30. *Journal of Industrial Economics*《產業經濟學》
31. *Review of Development Economics*《發展經濟學評論》
32. *Review of Income and Wealth*《收入與財富評論》
33. *Review of International Economics*《國際經濟學評論》
34. *Australian Economic Papers*《澳大利亞經濟論文》
35. *Oxford Economic Papers*《牛津經濟論文》
36. *Papers in Regional Science*《區域科學論文集》
37. *Journal of Population Economics*《人口經濟學》
38. *Review of Economic Design*《經濟設計評論》
39. *Economic Theory*《經濟理論》
40. *Empirical Economics*《經驗經濟學》
41. *Journal of Evolutionary Economics*《進化經濟學》
42. *International Journal of Game Theory*《國際對策論》
43. *Journal of Economics & Management Strategy*《經濟管理戰略》
44. *Journal of Real Estate Research*《房地產研究》
45. *Review of Economic Dynamics*《動態經濟學評論》
46. *Applied Economics*《應用經濟學》
47. *Applied Financial Economics*《應用金融學》
48. *International Journal of Finance & Economics*《國際經濟與金融》
49. *Journal of Business and Economic Statistics*《商業與經濟統計學》
50. *Canadian Journal of Economics*《加拿大經濟學》
51. *Canadian Public Policy*《加拿大公共政策》
52. *Computational Economics*《計算經濟學》
53. *Economica*《經濟學》
54. *History of Economic Thought Articles*《經濟思想史》
55. *IMF Staff Papers*《國際貨幣基金組織職員報告》
56. *International Economic Review*《國際經濟評論》
87. *Journal of Business*《商業》
58. *Forbes*《福布斯》
59. *Businessweek*《商業周刊》
60. *Fortune Magazine*《財富》
61. *Bloomberg Markets*《彭博市場》
62. *The Accounting Review*《會計述評》
63. *Issues in Accounting Education*《會計教育視點》
64. *Accounting Horizons*《會計新視野》
65. *Accounting and the Public Interest*《會計和公共利益》

66. *Auditing*：*A Journal of Practice & Theory*《審計：實踐與理論》
67. *Behavioral Research in Accounting*《會計學行為研究》
68. *Current Issues in Auditing*《當前審計面臨的問題》
69. *Journal of the American Taxation Association*《美國稅收協會期刊》
70. *Journal of Emerging Technologies in Accounting*《會計中的新興技術》
71. *Journal of Information Systems*《信息系統》
72. *Journal of International Accounting Research*《國際會計研究》
73. *Journal of Legal Tax Research*《法定稅收研究》
74. *Journal of Management Accounting Research*《會計管理研究》

4.2　報紙

1. *The Wall Street Journal*《華爾街日報》
2. *Financial Times*《英國金融時報》
3. *Oxford Bulletin of Economics and Statistics*《牛津經濟統計公報》
4. *Federal Reserve Bulletin*《聯邦儲備公報》

附錄 5
國際經貿專業重要網站

1. 世界銀行官網：
http：//www. worldbank. org
2. 世界銀行數據庫：
http//www. worldbank. org/data/
3. 《計量經濟學》雜誌：
http//qed. econ. queensu. ca/jae/
4. 美國經濟協會官網：
http//www. vanderbilt. edu/AEA/
5. 美國金融協會官網：
http//www. afajof. org/
6. 美國會計學會官網：
http//aaahq. org/
7. 金融研究學會官網：
http//www. sfs. org/
8. 國際金融管理協會官網：
http//www. fma. org/index. htm
9. 國民經濟研究局官網：
http//www. nber. org
10. 經濟政策研究中心官網：
http//www. cepr. org
11. 《蘭德經濟學》雜誌：
http//www. rje. org/
12. 《政治經濟學》雜誌：
http//www. journals. uchicago. edu/JPE/home. html
13. 《金融》雜誌：
http//www. afajof. org/jofihome. shtml
14. 《金融經濟學》雜誌：
http//jfe. rochester. edu/
15. 《金融和數量分析》雜誌：

http//depts. washington. edu/jfqa/

16.《商業周刊》雜誌：

http//www. businessweek. com/

17. CBS 市場觀察：

http//cbs. marketwatch. com/news/newsroom. htx

18. CNBC：

http//www. cnbc. com/

19.《經濟學家》雜誌：

http//www. economist. com/

20. YAHOO 金融：

http//quote. yahoo. com/

21. 紐約聯邦儲備銀行官網：

http//www. ny. frb. org/pihome/mktrates/

22. 投資指南：

http//www. investmove. com/

23. 證券專家：

http//www. stockmaster. com/

24. CNN 金融市場：

http//www. cnnfn. com/markets/index. html

25. Giancarlo Corsetti：

http//www. econ. yale. edu/~corsetti/wami/wami. html

26. 經濟學網路（華盛頓大學）：

http//netec. wustl. edu/NetEc. html

27.《金融時報》：

http//www. ft. com/index. htm

28.《福布斯》：

http//www. forbes. com/

29.《財富》：

http//pathfinder. com/fortune/

30.《紐約時報》：

http//www. nytimes. com/

31.《華爾街日報》：

http//www. wsj. com/

32. Journal of Economic Literature：

http//econpapers. hhs. se/article/aeajeclit/

33. Journal of Economic Perspectives：

http//econpapers. hhs. se/article/aeajecper/

34. American Economic Review：

http//econpapers. hhs. se/article/aeaaecrev/
35. Journal of Business：
http//econpapers. hhs. se/article/ucpjnlbus/
36. Journal of Labor Economics：
http//econpapers. hhs. se/article/ucpjlabec/
37. Journal of Political Economy：
http//econpapers. hhs. se/article/ucpjpolec/
38. Journal of Economic Surveys：
http//econpapers. hhs. se/article/blajecsur/
39. Journal of Industrial Economics：
http//econpapers. hhs. se/article/blajindec/
40. Oxford Bulletin of Economics and Statistics：
http//econpapers. hhs. se/article/blaobuest/
41. Review of Development Economics：
http//econpapers. hhs. se/article/blardevec/
42. Review of Economic Studies：
http//econpapers. hhs. se/article/blarestud/
43. Review of Income and Wealth：
http//econpapers. hhs. se/article/blarevinw/
44. Review of International Economics：
http//econpapers. hhs. se/article/blareviec/
45. Australian Economic Papers：
http//econpapers. hhs. se/article/blaausecp/
46. Oxford Economic Papers：
http//econpapers. hhs. se/article/oupoxecpp/
47. Papers in Regional Science：
http//econpapers. hhs. se/article/sprpresci/
48. Journal of Population Economics：
http//econpapers. hhs. se/article/sprjopoec/
49. Review of Economic Design：
http//econpapers. hhs. se/article/sprreecde/
50. Economic Theory：
http//econpapers. hhs. se/article/sprjoecth/
51. Empirical Economics：
http//econpapers. hhs. se/article/sprempeco/
52. Journal of Evolutionary Economics：
http//econpapers. hhs. se/article/sprjoevec/
53. International Journal of Game Theory：

http//econpapers. hhs. se/article/sprjogath/
54. The Quarterly Journal of Economics：
http//econpapers. hhs. se/article/tprqjecon/
55. The Review of Economics & Statistics：
http//econpapers. hhs. se/article/tprrestat/
56. Journal of Economics & Management Strategy：
http//econpapers. hhs. se/article/tprjemstr/
57. RAND Journal of Economics：
http//econpapers. hhs. se/article/rjerandje/
58. Bell Journal of Economics：
http//econpapers. hhs. se/article/rjebellje/
59. Journal of Real Estate Research：
http//econpapers. hhs. se/article/jreissued/
60. Review of Economic Dynamics：
http//econpapers. hhs. se/article/redissued/
61. Review of Financial Studies：
http//econpapers. hhs. se/article/ouprfinst/
62. Journal of Applied Econometrics：
http//econpapers. hhs. se/article/jaejapmet/
63. Applied Economics：
http//econpapers. hhs. se/article/tafapplec/
64. Applied Financial Economics：
http//econpapers. hhs. se/article/tafapfiec/
65. International Journal of Finance & Economics：
http//econpapers. hhs. se/article/ijfijfiec/
66. Federal Reserve Bulletin：
http//econpapers. hhs. se/article/fipfedgrb/
67. Journal of Business and Economic Statistics：
http//econpapers. hhs. se/article/besjnlbes/
68. Canadian Journal of Economics：
http//econpapers. hhs. se/article/cjeissued/
69. Canadian Public Policy：
http//econpapers. hhs. se/article/cppissued/
70. Computational Economics：
http//econpapers. hhs. se/article/kapcompec/
71. Econometrica：
http//econpapers. hhs. se/article/ecmemetrp/
72. Economic Journal：

http//econpapers. hhs. se/article/ecjeconjl/
73. Economica：
http//econpapers. hhs. se/article/blaeconom/
74. History of Economic Thought Articles：
http//econpapers. hhs. se/article/hayhetart/
75. IMF Staff Papers：
http//econpapers. hhs. se/article/imfimfstp/
76. International Economic Review：
http//econpapers. hhs. se/article/ieriecrev/

國家圖書館出版品預行編目（CIP）資料

國際經貿專業英語：理論、實務與方法 / 黃載曦, 姚星 編著. -- 第一版.
-- 臺北市：財經錢線文化, 2020.05
　　面；　公分
POD版

ISBN 978-957-680-406-9(平裝)

1.商業書信 2.商業英文 3.商業應用文

493.6　　　　　　　　　　　　　　　109005513

書　　名：國際經貿專業英語：理論、實務與方法
作　　者：黃載曦,姚星 編著
發 行 人：黃振庭
出 版 者：財經錢線文化事業有限公司
發 行 者：財經錢線文化事業有限公司
E - m a i l：sonbookservice@gmail.com
粉 絲 頁：　　　　　網　址：
地　　址：台北市中正區重慶南路一段六十一號八樓 815 室
8F.-815, No.61, Sec. 1, Chongqing S. Rd., Zhongzheng
Dist., Taipei City 100, Taiwan (R.O.C.)
電　　話：(02)2370-3310　傳　真：(02) 2388-1990
總 經 銷：紅螞蟻圖書有限公司
地　　址：台北市內湖區舊宗路二段 121 巷 19 號
電　　話:02-2795-3656　傳真:02-2795-4100　　網址：
印　　刷：京峯彩色印刷有限公司（京峰數位）

　　本書版權為西南財經大學出版社所有授權崧博出版事業股份有限公司獨家發行電子
　　書及繁體書繁體字版。若有其他相關權利及授權需求請與本公司聯繫。

定　　價：520 元
發行日期：2020 年 05 月第一版
◎ 本書以 POD 印製發行